Remote Sensing of Urban and Suburban Areas

Remote Sensing and Digital Image Processing

VOLUME 10

For other titles published in this series, go to
http://www.springer.com/series/6477

Remote Sensing of Urban and Suburban Areas

Tarek Rashed
Geospatial Applied Research Expert House (GSAREH),
Austin, TX, USA

and

Carsten Jürgens
Geography Department, Ruhr-University, Bochum, Germany

Editors

 Springer

Editors
Dr. Tarek Rashed
Geospatial Applied Research
Expert House (GSAREH)
Austin, TX
USA
rashed@gsareh.com

Dr. Carsten Jürgens
Ruhr-University
Geography Department
Geomatics Group
Universitätsstr. 150
44801 Bochum
Germany
carsten.jürgens@rub.de

Cover illustrations: Landsat satellite image of San Francisco, CA, USA, combined with photograph taken by Maike Reichardt.

Responsible Series Editor: Freek van der Meer

ISBN 978-1-4020-4371-0   e-ISBN 978-1-4020-4385-7
DOI 10.1007/978-1-4020-4385-7
Springer Dordrecht Heidelberg London New York

Library of Congress Control Number: 2010929276

Cover design: deblik, Berlin

Printed on acid-free paper

Springer is part of Springer Science+Business Media (www.springer.com)

Acknowledgments

The preparation of this volume was possible due to the fact that all authors supported the original idea behind this book. We thank all authors for their contributions and their patience. The publication process did not run smoothly in all stages and we apologize for the resulting time delay. We also thank all reviewers whose valuable comments improved the content of the different chapters. Finally we thank the Springer team for their continuous support and discussions from the beginning to the end of this book project and for the publication in their book series.

We are convinced that with the publication of this book we are making an essential contribution to the knowledge about the different aspects of urban and suburban remote sensing.

Contents

1 **Urban and Suburban Areas as a Research Topic
for Remote Sensing** 1
Maik Netzband and Carsten Jürgens

Part I   Theoretical Aspects

2 **The Structure and Form of Urban Settlements** 13
Elena Besussi, Nancy Chin, Michael Batty, and Paul Longley

3 **Defining Urban Areas** 33
John R. Weeks

4 **The Spectral Dimension in Urban Remote Sensing** 47
Martin Herold and Dar A. Roberts

5 **The Spatial and Temporal Nature of Urban Objects** 67
Richard Sliuzas, Monika Kuffer, and Ian Masser

6 **The V-I-S Model: Quantifying the Urban Environment** 85
Renee M. Gluch and Merrill K. Ridd

Part II   Techniques and Applications

7 **A Survey of the Evolution of Remote Sensing Imaging
Systems and Urban Remote Sensing Applications** 119
Debbie Fugate, Elena Tarnavsky, and Douglas Stow

8 **Classification of Urban Areas: Inferring Land Use
from the Interpretation of Land Cover** 141
Victor Mesev

9   **Processing Techniques for Hyperspectral Data**................................... 165
     Patrick Hostert

10   **Segmentation and Object-Based Image Analysis**................................ 181
     Elisabeth Schöpfer, Stefan Lang, and Josef Strobl

11   **Data Fusion in Remote Sensing of Urban
     and Suburban Areas** 193
     Thierry Ranchin and Lucien Wald

12   **Characterization and Monitoring of Urban/Peri-urban
     Ecological Function and Landscape Structure Using
     Satellite Data**... 219
     William L. Stefanov and Maik Netzband

13   **Remote Sensing of Desert Cities in Developing Countries**.................. 245
     Mohamed Ait Belaid

14   **Remote Sensing of Urban Environmental Conditions**........................ 267
     Andy Kwarteng and Christopher Small

15   **Remote Sensing of Urban Land Use Change in Developing
     Countries: An Example from Büyükçekmece,
     Istanbul, Turkey** 289
     Derya Maktav and Filiz Sunar Erbek

16   **Using Satellite Images in Policing Urban Environments**.................... 313
     Meshgan Mohammad Al-Awar and Farouk El-Baz

17   **Using DMSP OLS Imagery to Characterize Urban
     Populations in Developed and Developing Countries**......................... 329
     Paul C. Sutton, Matthew J. Taylor, and Christopher D. Elvidge

Index.. 349

Contributors

Mohamed Ait Belaid
College of Graduate Studies, Arabian Gulf University, P.O. Box 26671,
Manama, Kingdom of Bahrain
belaid@agu.edu.bh

Meshgan Mohammad Al-Awar
Research and Studies Center,
Dubai Police Academy, 53900 Dubai, United Arab Emirates
meshkan@dubaipolice.gov.ae; drmeshkan@yahoo.com

Michael Batty
Centre for Advanced Spatial Analysis, University College London,
1-19 Torrington Place, London WC1E 6BT, UK
m.batty@ucl.ac.uk

Elena Besussi
Development Planning Unit, University College London,
34 Tavistock Square, London WC1H 9EZ, UK
e.besussi@ucl.ac.uk

Nancy Chin
Centre for Advanced Spatial Analysis, University College London,
1-19 Torrington Place, London WC1E 7HB, UK
n.chin@ucl.ac.uk

Farouk El-Baz
Center for Remote Sensing, Boston University, 725 Commonwealth Avenue,
Boston MA, 02215-1401, USA
farouk@bu.edu

Christopher D. Elvidge
Earth Observation Group, NOAA National Geophysical Data Center,
325 Broadway, Boulder CO, 80305, USA
chris.elvidge@noaa.gov

Filiz Sunar Erbek
Civil Engineering Faculty, Geomatics Department, Istanbul Technical University,
Maslak Campus, Maslak 34469, Istanbul, Turkey

Debbie Fugate
203 Ceret Ct SW, Vienna VA, 22180, USA
fugate.debbie@gmail.com

Renee M. Gluch
Department of Geography, Brigham Young University, Provo, UT 84602, USA
renee_gluch@byu.edu

Martin Herold
Institute of Geo-InformationScience and Remote Sensing, Wageningen University,
Droevendaalsesteeg 3, Gaia, buildingnumber 101, P.O. Box 6708, Wageningen,
The Netherlands
martin.herold@wur.nl

Patrick Hostert
Geographical Institute, Humboldt-Universitat zu Berlin, Unter den Linden 6,
Berlin 10099, Germany
patrick.hostert@geo.hu-berlin.de

Carsten Jürgens
Geography Department, Ruhr-University, Bochum, Universitätsstraße 150,
44801 Bochum, Germany
carsten.jürgens@rub.de

Monika Kuffer
International Institute for Geo-Information Science and Earth Observation,
Urban and Regional Planning and Geo-information Management,
Hengelosestraat 99, Enschede 7514 AE, The Netherlands
kuffer@itc.nl

Andy Kwarteng
Remote Sensing and GIS Center, Sultan Qaboos University,
P.O. Box 50, Al-Khod, Muscat, 123, Oman
kwarteng@squ.edu.om

Stefan Lang
Centre for Geo-informatics, University of Salzburg, Schillerstr. 30,
Salzburg 5020, Austria
stefan.lang@sbg.ac.at

Paul Longley
Department of Geography, University College London, Gower Street,
London WC1E 6BT, UK
p.longley@geog.ucl.ac.uk

Derya Maktav
Faculty of Civil Engineering, Department of Geomatics, Istanbul Technical
University (ITU), Maslak, Istanbul, 34469, Turkey
maktavd@itu.edu.tr

Ian Masser
Centre for Advanced Spatial Analysis, University College London, 1-19
Torrington Place, London WC1E 7HB, UK
i.masser@ucl.ac.uk

Victor Mesev
Department of Geography, Florida State University, Tallahassee, FL
32306-2190, USA
vmesev@fsu.edu

Maik Netzband
Geography Department, Ruhr-University, Bochum, Universitätsstraße 150,
44801 Bochum, Germany
maik.netzband@rub.de

Thierry Ranchin
Center for Energy and Processes, MINES Paris Tech, 1, rue Claude Daunesse,
Sophia Antipolis Cedex, 06904, France
thierry.ranchin@mines-paristech.fr

Tarek Rashed
Geospatial Applied Research Expert House (GSAREH),
Austin, TX, USA
rashed@gsareh.com

Merrill K. Ridd
Department of Geography, University of Utah, 260 S. Central Campus Dr,
Salt Lake City, UT 84112-9155, USA
merrillridd@geog.utah.edu

Dar A. Roberts
Geography Department, University of California, 5832 Ellison Hall,
Santa Barbara, CA 93106-4060, USA
dar@geog.ucsb.edu

Elisabeth Schöpfer
ESA-ESRIN, Directorate of Earth Observation Programmes,
Via Galileo Galilei, Frascati 00044, Italy
elisabeth.schoepfer@esa.int

Richard Sliuzas
International Institute for Geo-Information Science and Earth Observation,
Urban and Regional Planning and Geo-information Management,

Hengelosestraat 99, Enschede, 7514 AE, The Netherlands
sliuzas@itc.nl

Christopher Small
Lamont-Doherty Earth Observatory, Columbia University, 108 Oceanography,
61 Route 9W, Palisades NY, 10964-8000, USA
small@ldeo.columbia.edu

William L. Stefanov
Image Science & Analysis Laboratory/ESCG, Code KX,
NASA Johnson Space Center, Houston, TX 77058, USA
william.l.stefanov@nasa.gov

Douglas Stow
Department of Geography, San Diego State University,
5500 Campanile Dr., San Diego CA 92182-4493, USA
stow@mail.sdsu.edu

Josef Strobl
Austrian Academy of Sciences, Geographic Information Science, Schillerstraße
30, Salzburg 5020, Austria
josef.strobl@oeaw.ac.

Paul C. Sutton
Department of Geography, University of Denver, Denver, CO 80208, USA
psutton@du.edu

Elena Tarnavsky
Geography Department, King's College London, Strand, London WC2R 2LS, UK
elena.tarnavsky@kcl.ac.uk

Matthew J. Taylor
Department of Geography, University of Denver, Boettcher Center West,
2050 E. Iliff Avenue, Denver, CO 80208-0183, USA
mtaylor7@du.edu

Lucien Wald
Center for Energy and Processes, Ecole des Mines de Paris, 1, rue Claude
Daunesse,
Sophia Antipolis Cedex, 06904, France
lucien.wald@mines-paristech.fr

John R. Weeks
Department of Geography, San Diego State University, 5500 Campanile Dr.,
San Diego, CA 92182-4493, USA
john.weeks@sdsu.edu

Chapter 1
Urban and Suburban Areas as a Research Topic for Remote Sensing

Maik Netzband and Carsten Jürgens

This chapter provides an introduction into the book's theme, its relevance for the scientific community as well as for instructors and practitioners. It tries to give an umbrella for the topics that have been chosen to bridge the gap between remote sensing and urban studies through a better understanding of the science that underlies both fields. In so doing, in the second half this first chapter introduces the following 16 chapters written by leading international experts in respected fields to provide a balanced coverage of fundamental issues in both remote sensing and urban studies.

> **Learning Objectives**
>
> Upon completion of this chapter, the student should gain an understanding of:
>
> ❶ Overview of urbanization research issues
> ❷ Introduction to recent developments in Urban Remote Sensing

1.1 Introduction

Starting the theme of research on urban and suburban areas, a recently taken aerial photograph in bird's eye perspective is illustrated. Figure 1.1 pictures a recent suburban development in the City of Rio Vista, California.

As a prosperous plan, 750 houses should be developed here once – most strikingly, these plans originate from a time, when still nobody suspected, what the term "largest economic crisis since 80 years" meant. And in such a way on 20 November 2008, thus few weeks after the collapse of the investment bank Lehman Brothers, the

M. Netzband (✉) and C. Jürgens
Geography Department, Ruhr-University, Bochum, Universitätsstraße 150,
44801 Bochum, Germany
e-mails: maik.netzband@rub.de; carsten.jürgens@rub.de

T. Rashed and C. Jürgens (eds.), *Remote Sensing of Urban and Suburban Areas*,
Remote Sensing and Digital Image Processing 10,
DOI 10.1007/978-1-4020-4385-7_1, © Springer Science+Business Media B.V. 2010

Fig. 1.1 Aerial photograph 'Rio Vista, California/USA' (Credits: Justin Sullivan/AFP)

house development project was already being adjusted in California's Rio Vista. Only the roads are still remaining – and a few sample houses. In the meantime the city in the north of the Sunshine state considers even to announce insolvency bankruptcy. Source: AFP

progress in information and communication technologies and the decentralization of economical activities are modifying the traditional patterns of urban agglomerations

It is a general argument, that every period of socioeconomic development is joined by different effects on population and landscape dynamics, e.g. the transition from agricultural-based economies to industrialization forced the urbanization process and the development of cities, predominantly accompanied by a monocentric urban growth pattern due to the concentration of industries, residences, and commerce in metropolitan areas (e.g., Mexico City, Beijing, London, New York) (Parés-Ramos et al. 2008; Anas et al. 1998). Today, one can observe in many countries a major transformation from an industrial-based economy to a knowledge-based economy (OECD 1996). As a result, innovations in information and communication technologies along with the decentralization of commercial, industrial, and financial activities are altering and diversifying the traditional patterns of urban agglomerations and driving new population and landscape dynamics (Munroe et al. 2005).

The consequences of these current socioeconomic trends comprise changes in the spatial structure of urban and rural areas, such as urban core population decline, the appearance of brownfields, suburban growth, and the urbanization of rural areas (Munroe et al. 2005). Decentralization tendencies are forcing urban sprawl and the conversion of agricultural lands and open spaces into urban land uses. Conversely, urban agglomerations with their manifold employment opportunities in manufacturing, trade, tourism, and other service sectors, attract more people, particularly the young and educated, to urban areas and supports the decline in agriculture jobs (Losada et al. 1998).

The cities today are spreading into their surrounding landscapes, sucking food, energy, water and resources from the natural environment, without taking into due account the social, economic and environmental consequences generated at all levels by their 'urban footprints'. The urban environment itself is profoundly changing the entire global ecosystem. Environmental changes are also expressed in land-use changes. Social, economic or political trends are conveyed spatially. In recent decades, the strongest per capita growth shifted to the more rural areas of the urban fringe (Bugliarello 2003). Open spaces are increasingly included between cities, villages, and traffic axes. An urbanizing landscape, accompanying technical infrastructure, and uncontrolled dynamics of urban growth patterns are the results. The conversion from land cover to land being used progresses, i.e., predominantly agricultural surfaces are transformed into settlement and traffic surfaces, resulting in decreased settlement density, increased traffic, and costly infrastructure development. Especially the increase of imperviousness at the expense of the decrease of green and open spaces must be documented from local to global scale, and it is a 'must' that the knowledge is integrated into climate change investigations and further global change issues. Socio-spatial patterns are expressed in different building activities for single family houses of different strata, with different amounts of green spaces, shopping facilities and infrastructure have driven settlement areas to further expand. The settlement density and, correspondingly, the inner urban densification continue to decrease.

cities today are spawling into their surrounding landscapes, without taking into due account the social, economic and environmental consequences generated by their 'urban footprints'

Merely characterizing and monitoring land-cover and land-use change is of limited use in understanding the development pathways of cities and their resilience to outside stressors (Longley 2002). Geological, ecological, climatic, social, and political data are also necessary to describe the developmental history of an urban center and understand its ecological functioning (Grimm et al. 2000). It is the process of urbanization that must be described, monitored, and even simulated on different scales. Dependent on the issue to be investigated upon, the relevant scale must be selected (see Fig 1.2). Local and regional environmental effects must be documented, analysed, evaluated, and, if possible, predicted. Without researchers and stakeholders exchanging and collaborating, the goal cannot be achieved.

In recent years 'Urban Remote Sensing' (URS) has proved to be a useful tool for cross-scale urban planning and urban ecological research. Remote sensing in urban areas is by nature defined as the measurement of surface radiance and properties connected to the land cover and land use in cities. Today, data from earth observation systems are available, geocoded, and present an opportunity to collect information relevant to urban and periurban environments at various spatial, temporal, and spectral scales.

The urban pattern causes deterioration in air quality, the urban ecosystem processes and biodiversity. In this context URS is a necessary prerequisite to examine how urban forms modify the landscape as a complex system. It can help to detect and evaluate the distribution of impervious or, likewise, sealed surfaces, a key parameter of urban ecology (surface and groundwater availability and runoff,

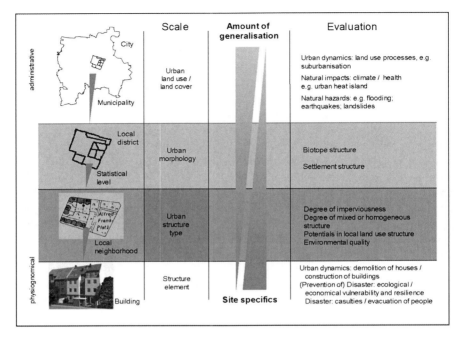

Fig. 1.2 Scale-dependent urban analysis (Banzhaf and Höfer 2008; modified after Wickop et al. 1998)

vegetation dynamics) and planning (storm water runoff, flooding hazards, landslides). Kühn (2003) explains the development of urban landscapes being shaped by the penetration of settlement and open-space structures. Remotely sensed data will be used to detect and evaluate the physical structure and composition of urban areas, such as the structure of residential, commercial or mixed neighborhoods, green spaces or other open spaces.

increase of geographical data availability has not been fully accompanied by an increase of knowledge to support spatial decisions, thus spatial analytical techniques are needed

The growth of 'Spatial Data Infrastructures', Geo-portals and private sector initiatives (e.g. Google Earth, Microsoft Virtual Earth, etc.) produced an increase of geographical data availability at any scale and worldwide. This growth has not been fully coupled by an increase of knowledge to support spatial decisions. Spatial analytical techniques and geographical analysis and modeling methods are therefore required in order to analyse data and to facilitate the decision process at all levels. As cities can be described as a concentration of people it is most striking to find coherence between urban land use and socio-demographic as well as socio-economic parameters. The statistical analysis of census data infers information on the human usage of the land, the human exposure to potential hazards in the city, and the configuration of each neighbourhood indicating the urban quality of life. For example, overlaying choropleth maps of socio-demographic features with land-use maps give information on gender and age distribution connected with proximity to urban green spaces, income and building density, or water consumption and level of provision of infrastructure. In this context URS aids at providing spatial information being linked to social indicators to explain the interrelations between ecological conditions and socio-spatial development (Banzhaf et al. 2009).

In this volume we try to cover most of but not all of the afore-mentioned topics and assembled widely known scholars of urban sciences specializing in the application of geospatial technologies or, vice versa, geo-information specialists with a distinct focus on urban and peri-urban developments to draw a widespread overview of the state-of-the-art knowledge in the growing field of urban remote sensing. "Remote Sensing of Urban and Suburban Areas" has been primarily assembled to introduce scientists and practitioners to this emerging field. Additionally it provides instructors with a text reference that has a logical and easy-to-follow flow of topics around which they can structure the syllabi of their urban remote sensing courses. The following six chapters of this book provide a comprehensive introduction in urban theories adapted to geospatial problems and solutions. In the second part of this book we present techniques and applications of various data sources and methodologies relevant for the analysis of urban status and dynamics.

Remote Sensing for Urban and Suburban Areas

Remote sensing in urban areas is by nature defined as the measurement of surface radiance and properties connected to the land cover and land use in cities. Today, data from earth observation systems are available, geocoded, and present an opportunity to collect information relevant to urban and peri-urban environments at various spatial, temporal, and spectral scales.

1.2   Introduction to the Chapters

Chapter 2 by Elena Besussi, Nancy Chin, Michael Batty and Paul Longley introduce the different *theoretical and methodological approaches* to understand and measure urban growth and urban patterns, their structure and form. The authors emphasize the idea that the contemporary city in both developed and developing worlds needs much more than just one theory or one method of analysis or one typology of data to be fully understood. It clearly appears to be a challenge to traditional analytical methods requiring interaction of social sciences and earth sciences, and urban economics using GIS techniques to understand patterns and trends of urbanization.

In Chapter 3 John R. Weeks reviews the vast literature on *dimensions of urbaneness*, but focuses especially on issues, such as classifying places as urban or rural by adequately capturing changes over time in the characteristics of a place. The urbaneness of a place as a continuum is determined based on a range of elements encompassing population size and density, social and economic organization, and the transformation of the natural and agricultural environments into a built environment. This chapter introduces you to one of such indices, i.e. an urban index that combines census and survey data (to capture aspects of the social environment) with data from remotely sensed imagery (to capture aspects of the built environment).

Martin Herold and Dar A. Roberts describe in Chapter 4 the *spectral properties of urban areas*, how different urban land-cover types are spectrally discriminated, and which sensor configurations are most useful to map urban areas. They also demonstrate potentials of new remote sensing technologies improving capabilities to map urban areas in high spatial and thematic detail. The authors stress the fact that urban areas with roofing materials, pavement types, soil and water surfaces, and vegetated areas represent a large variety of surface compositions. It is emphasized that most suitable wavelengths are characterized by specific spectral features to separate urban land cover.

The purpose of Chapter 5 authored by Richard Sliuzas, Monika Kuffer and Ian Masser is to examine the utility of remote sensing data on urban and suburban areas for *Urban Planning and Management* (UPM) from an *application perspective*. This chapter especially discusses the use of remote sensing at two different spatial scales, city-wide and neighborhood or site specific, the information needed with respect to monitoring planned and unplanned development, and the optimal spatial and temporal requirements for images used in this regard.

Rene M. Gluch and Meryll K. Ridd emphasize in Chapter 6 the *ecological nature of urban places* and introduce the *V-I-S (Vegetation-Impervious surface-Soil) model* to be used for remotely sensed data to characterize, map, and quantify the ecological composition of urban/peri-urban environments. The model serves not only as a basis for biophysical and human system analysis, it also serves as a basis to detect and measure morphological/environmental changes of urban places over time.

In Chapter 7 Debbie Fugate, Elena Tarnavsky, Douglas Stow review the *development of remote sensing systems and their contribution to the emergence of urban remote sensing*, especially how they promoted the pursuit of novel approaches to the study of urban environments. The chapter also covers data availability and requirements for a number of the most common earlier remote sensing applications such as land use and land cover classification, building and cadastral infrastructure mapping and planning, and utility and transportation system analysis. Additionally, the chapter highlights first attempts that have already been made to link the physical and social attributes of urban environments.

In Chapter 8, Victor Mesev explores *the role of ancillary data* (information from beyond remote sensing) *for improving the contextual interpretation of satellite sensor imagery* during spectral-based and spatial-based classification. Supplementary, explanations are given to the distinctions between urban land cover and urban land use, and how the inherent heterogeneous structure of urban morphologies is statistically represented between hard and soft classifications.

Basic knowledge about the differences between multispectral and hyperspectral data is provided by Patrick Hostert in Chapter 9, where the *potential of hyperspectral image analysis* is distinguished. He presents relevant pre-processing steps and different ways to analyze hyperspectral data. Moreover, relevant analysis approaches are explained including material detection techniques, spectral angle mapping, or spectral mixture analysis, to name some. The chapter closes with a short outlook on expected developments with relevance for urban applications.

Chapter 10, written by Elisabeth Schöpfer, Stefan Lang and Josef Strobl, focuses on *segmentation* of remotely sensed image data and *object-based image analysis* of urban areas. It also discusses the differences between the two different approaches 'pixel-based' and 'object-based' image analysis. They explain the main concepts of object-based image analysis: to work on homogeneous image objects rather than on single pixels and to use spectral and spatial information while merging pixels into homogeneous groups (image objects, segments). The chapter depicts very briefly urban applications by means of two case studies.

Thierry Ranchin's and Lucien Wald's Chapter 11 concentrates on techniques related to image and data from different sources with varying spatial and spectral resolutions. It presents and discusses some of the technical issues that influence *data fusion in the urban context*. Several fusion cases studies are discussed here to illustrate the potential of data fusion techniques. The authors emphasize on the importance of the diversity of data fusion. The few examples provided cannot fully describe its complexity and this field is still a strong and active research in urban remote sensing and the other civilian domains.

A case study from Phoenix, Arizona is depicted by William L. Stefanov and Maik Netzband in Chapter 12. They examine the *relationships between ecological variables and landscape structure* in cities. These relationships are assessed using ASTER and MODIS data; and through the techniques of expert system land cover classification and grid-based landscape metric analysis. The authors argue that this multi-scale approach is of great use to urban ecologists and spatial planners, as landscape structural analysis and measures of ecosystem function provide monitoring tools for regional habitat and climatic alteration associated with urbanization. Furthermore, the applied uniform spatial reference systems provided by remotely sensed data permit quantitative evaluation in comparative studies regarding the spatial configuration of existing developed and open spaces.

Chapter 13 by Mohamed Ait Belaid focuses on *remote sensing (RS) of desert cities*, within the context of developing countries. The characteristics of urban areas in the desert environment are described, and the potential of satellite imageries is discussed, how they are used to map and monitor changes in these areas over space and time. Urban and sub-urban landscapes of desert cities are shaped by various factors such as desertification, economic development, and wars and conflicts. In their chapter the authors include photo-interpretation techniques assisted by computer techniques to produce the classified imagery maps of land use categories and the comparison of the classified land use changes in urban areas.

In Chapter 14 Andy Kwarteng and Christopher Small give an overview over urbanization and the urban environment connected to urban vegetation, surface temperature and public health issues. They explain techniques for urban vegetation mapping, urban thermal mapping, and show the results of a comparison of urban vegetation and surface temperature and their impact on environmental conditions in New York City and Kuwait City. The authors advance the opinion that most successful applications of remote sensing to the urban environment generally involve *measurement of physical quantities related to environmental conditions* such as vegetation abundance and surface temperature.

In alignment with other *application oriented* chapters in the book discussing the *context of developing countries*, Derya Maktav and Filiz Sunar Erbek discuss in Chapter 15 the impact of rapid urban growth on land use changes, especially on the agricultural land in Turkey. The way in which remote sensing is used to monitor and assess these changes is pointed using a case study from a suburban area in the greater Istanbul region. The results of the analysis show how it is possible to utilize urban remote sensing in generating reliable measures and new layers of information that are otherwise not readily available in developing countries with relatively simple techniques such as image differencing and vegetation indices.

Demonstrating a case study of Dubai in the United Arab Emirates, Meshgan Al-Awar and Farouk El-Baz discuss in Chapter 16 the role of remote sensing technology in the *monitoring and management of security in cities* and in assuring the timely policing of urban environments. This chapter presents application examples from the Dubai's Police to show how the utilization of geo-referenced satellite images on top of GIS platforms can allow the immediate allocation of the needed response. It is a good example on how these imagery and geospatial technologies can be used for a better command level decision-making and, furthermore, how they are most useful in the reconstruction and enhancement of crime scenes.

Nighttime Satellite imagery examined by Paul C. Sutton, Matthew J. Taylor and Christopher D. Elvidge in Chapter 17 shows great potential for mapping and monitoring many human activities including: (1) population size, distribution, and growth, (2) urban extent and rates of urbanization, (3) Impervious Surface, (4) Energy Consumption, and (5) CO_2 emissions. They argue that the relatively coarse spectral, spatial, and temporal resolution of the imagery proves to be an advantage rather than a disadvantage for these applications.

Within two case studies (first of exurbia in the United States and second in Guatemala) they explain that while nighttime satellite imagery is no substitute for an 'on the ground census' of the population, it can be used in innovative and interesting ways to supplement mapping human presence and activity on earth.

References

Anas A, Arnott R, Small KA (1998) Urban spatial structure. J Econ Lit 36:1426–1464

Banzhaf E, Höfer R (2008) Monitoring urban structure types as spatial indicators with CIR aerial photographs for a more effective urban environmental management. In: Journal of Selected Topics in Applied Earth Observations and Remote Sensing (JSTARS), IEEE. 1(2):129–138. ISSN: 1939-1404. Digital Object Identifier: 10.1109/JSTARS.2008.2003310

Banzhaf E, Grescho V, Kindler A (2009) Monitoring urban to peri-urban development with integrated remote sensing and GIS information: a Leipzig, Germany case study. Int J Remote Sens 30(7):1675–1696

Bugliarello G (2003) Large urban concentrations: a new phenomenon. Heiken G, Fakundiny R, Sutter J (eds) Earth science in the city: a reader. American Geophysical Union, Washington, DC, pp 7–19

Grimm NB, Grove JM, Redman CL, Pickett SA (2000) Integrated approaches to long-term studies of urban ecological systems. Bioscience 70:571–584

Kühn M (2003) Greenbelt and Green Heart: separating and integrating landscapes in European city regions. Landscape Urban Plann 64(1–2):19–27

Longley PA (2002) Geographic information systems: will developments in urban remote sensing and GIS lead to 'better' urban geography? Prog Hum Geogr 26(2):213–239

Losada H, Martinez H, Vieyra J, Pealing R, Cortés J (1998) Urban agriculture in the metropolitan zone of Mexico: changes over time in urban, sub-urban and peri-urban areas. Environ Urbanization 10(2):37–54

Munroe D, Clark J, Irwin E (2005) Regional determinants of exurban land use in the U.S. Midwest. Prepared for the 52nd Annual North American Meetings of the Regional Science Association, Las Vegas, NV, USA

Organisation for Economic Cooperation and Development (OECD) (1996) The knowledge-based economy, Paris, France (online), http://www.oecd.org/dataoecd/51/8/1913021.pdf

Parés-Ramos IK, Gould WA, Mitchell Aide T (2008) Agricultural abandonment, suburban growth, and forest expansion in Puerto Rico between 1991 and 2000. Ecol Soc 13(2):1 (online), http://www.ecologyandsociety.org/vol13/iss2/art1/

Wickop E, Böhm P, Eitner K, Breuste J (1998) Qualitätszielkonzept für Stadtstrukturtypen am Beispiel der Stadt Leipzig. UFZ Bericht 14:156

Part I
Theoretical Aspects

Chapter 2
The Structure and Form of Urban Settlements

Elena Besussi, Nancy Chin, Michael Batty, and Paul Longley

This chapter introduces you to the different theoretical and methodological approaches to the understanding and measuring of urban growth and urban patterns. Particular attention is given to urban sprawl as one of the forms of suburbanization. Urban sprawl today represents a challenge for both scientists and decision makers, due to the complexity of its generative processes and impacts. In this chapter, we introduce ways of measuring the spatial pattern of sprawl noting how remotely sensed imagery need to be integrated with spatial socioeconomic data, and how this integration is essential in making accurate interpretations of very different urban morphologies.

Learning Objectives

Upon completion of this chapter, you should be able to:

❶ Speculate on the range of processes which generate urban growth and its different structures
❷ Differentiate between approaches used to define and measure urban and suburban patterns
❸ Describe some of the zone-based spatial statistical methods available to measure urban growth dynamics and patterns

E. Besussi (✉)
Development Planning Unit, University College London, 34 Tavistock Square, London, WC1H 9EZ, UK
e-mail: e.besussi@ucl.ac.uk

N. Chin and M. Batty
Centre for Advanced Spatial Analysis, University College London, 1-19 Torrington Place, London WC1E 7HB, UK
e-mails: n.chin@ucl.ac.uk; m.batty@ucl.ac.uk

P. Longley
Department of Geography, University College London, Gower Street, London WC1E 6BT, UK
e-mails: p.longley@geog.ucl.ac.uk

T. Rashed and C. Jürgens (eds.), *Remote Sensing of Urban and Suburban Areas*,
Remote Sensing and Digital Image Processing 10,
DOI 10.1007/978-1-4020-4385-7_2, © Springer Science+Business Media B.V. 2010

2.1   Urban Structure and Urban Growth: An Overview of Theories and Methodologies

Cities emerge and evolve from the coalescence and symbiotic interaction of infrastructures, people and economic activities. These interactions are systematic, generally in that they are related to development in the global economy, and more specifically in that they manifest building and transport technologies. But these interactions are also sensitive to local context, in that settlements are individually resilient to constraints in their evolutionary path. Given advances in technology, and the sheer scale and pace of contemporary urban growth, the most rapid changes in urban form, pattern and structure, are taking place where historical roots are weakest – as in the recent suburbs of long established Western cities, or in the new cities of developing countries. A city like London would never have been able to develop its contemporary form, skyline, and density of activity were it not for technological innovations such as its underground transport network and its role in global financial markets. Yet there are local and institutional factors such as the role of "green belt planning policy," peculiar to the UK that has prevented the kind of sprawl characteristic of North American cities taking hold throughout the functional region.

traditional urban theories investigate how cities develop and grow through systematic interactions of infrastructures, people and economic activities

Traditional urban theories investigate how cities develop and grow through these kinds of interactions, and in macro terms are based on advantages that co-location (i.e., the physical location where urban and economic activities are in close spatial proximity to one another) can offer to economies and societies. Agglomeration economies, defined by those economic production systems that benefit from co-location, have been identified as key forces at work in the growth of cities at any time and in every place. However, over the last half century our traditional understanding that the only outcomes of these forces should be an accelerating concentration of population, infrastructures and jobs has been challenged by the evidence of de-concentration, first in the United States and now in Europe. The migration of agricultural populations into the city which has been a centuries old factor in rural depopulation and the dominant force in creating urban agglomerations is now giving way to a reverse migration into the countryside, at least in many western cities, as suburbanization and sprawl become the modus operandi of urban growth.

Of course, the inertia in the skeletal structure of the built form of the city in its buildings and streets are important principally because they accommodate the loci of activities of "urban" populations. There is nearly a century of interest in understanding the socio-spatial differentiation of urban populations, that can be traced back to the 1920s in the work of Park, Burgess and the Chicago School of urban ecologists, if not before in the writings of Max Weber and his nineteenth century contemporaries. Here again, urban research has focused upon the general as well as the specific. The classic ringed socio-economic structure of 1920s Chicago, for example, was deemed by the Chicago School to be a manifestation of general biotic and cultural forces (which lead to the term "urban ecology"), constrained by the particular physical setting of the city.

Underpinning these physical structures and locational patterns is transportation. Cities exist largely because transportation to accessible nodes in space provides the rationale for the agglomeration economies that define them. Sprawl for example is loosely associated with the tradeoff between the desire to live as close to the city as possible against the desire to purchase as much space as possible and still retain the benefits of "urban" or "suburban" living. Sprawl thus comes about through rising wealth and transportation technologies that allow such suburban development and urban morphologies to reflect this tradeoff. The dynamics of the processes defining such spatial interaction and land development are thus central to an understanding of urban form and structure.

underpinning the skeletal structure of the built form of the city is transportation

In both physical and socio-economic terms, the ways in which urban phenomena are conceived very much determines the ways in which they are subsequently measured and then analyzed. Studies concerned principally with urban extent (such as inventory analysis focusing upon the ways in which the countryside might be gobbled up by urban growth) tend to be guided by definitions of the extent of irreversibly urban artificial structures on the surface of the Earth. Such structures support a range of residential, commercial, industrial, public open space and transport land uses.

Remote sensing classification of surface reflectance characteristics allows the creation of simple, robust and directly comparable measures of the dichotomy between natural and artificial land cover (read relative discussions in Chapters 3–5). Of course, such urban development is not necessarily entirely contiguous and, as shown in Chapter 8, techniques of GIS can be used to devise appropriate contiguity and spatial structure rules. In this straightforward sense, it is possible to formulate fairly robust and objective indicators of class and extent through the statistical classification of land cover characteristics and "spatial patterning" of the size, shape and dimension of adjacent land use parcels. These indicators can provide a useful and direct measure of the physical form and morphology of urban land cover that is very useful in delineating the extent of individual urban settlements and in generating magnitude of size estimates for settlement systems (Batty and Longley 1994).

remote sensing can provide a useful and direct indication of the physical form and morphology of urban land cover in cities

Chapter 7 of this book describes how developments in urban remote sensing have led to the deployment of instruments that are capable of identifying the reflectance characteristics of urban land cover to sub-meter precision (also see Donnay et al. 2001; Mesev 2003). In addition to direct uses, remotely sensed measures are also of use in developing countries where socioeconomic framework data such as censuses may not be available. For reasons that lie beyond the scope of this

remote sensing represents a complementary data source to traditional socioeconomic surveys

chapter, improvements in the resolution of satellite images have not been matched by commensurate improvement in the detail of socioeconomic data on urban distributions. This creates something of an asymmetry between our increasingly detailed understanding of built form and our ability to measure the detail of intra urban socioeconomic distributions (and we should not forget that built form is also measurable through national mapping agency framework data (Smith et al. 2005). However, remote sensing and socioeconomic sources increasingly present complementary approaches, in that today's high-resolution urban remote sensing data may also be used to constrain GIS-based representations of socioeconomic distributions (Harris and Longley 2000).

There is considerable research in the patterning of cities but much of this has been focused on explaining urban structure and form at a single point in time, as if cities were in some sort of perpetual equilibrium. Clearly the absence of rigorous data through time has been a major constraint on our ability to manufacture an appropriate science of urban dynamics and thus most of the thinking about urban change has been speculative and non rigorous. This is changing. New data sets, a concern for intrinsically dynamic issues such as how to control and manage urban sprawl rather then simply worrying about the spatial arrangement of growth, and new techniques such as urban remote sensing which are being fast developed to process routine information from satellite and aerial photographic data, are becoming important. This book will deal with these techniques in considerable detail but in this chapter we will set the context in illustrating the kinds of issues that are involved in understanding the most significant aspects of contemporary urban growth: suburban development and sprawl. In the next section we will examine the physical manifestation of suburbanization and this will set the context to a discussion of urban sprawl in Europe where we will focus on how it might be measured and understood.

2.2   Physical Manifestations of Urban Growth: Suburbanization and Sprawl

Whether we envision vast swathes of single-family detached houses, each surrounded by a garden and equipped with a swimming pool as in many parts of North America, the much more fragmented and diversified low density fringes of European cities, or the seemingly uncontrollable slums sprawling around the capital cities in developing countries, it is clear that suburbanization is the distinctive outcome of contemporary urban growth. Urban sprawl is by no means restricted to any particular social or economic group or any culture or indeed any place. It is largely the results of a growing population whose location is uncoordinated and unmanaged, driven from the bottom-up and subject to aggregate forces involving control over the means of production whose impact we find hard to explain in generic terms.

suburbanization is the distinctive outcome of contemporary urban growth

In the following discussion, we will focus upon urban sprawl as a defining characteristic of urban development and growth. Given the difficulties inherent in measuring and monitoring physically-manifest socioeconomic structures, set out above, we will adopt what is essentially a physicalist definition of sprawl as the rapid and uncoordinated growth of urban settlements at their urban fringes, associated with modest population growth and sustained economic growth. What is particularly interesting about urban sprawl is less the quest for an all-encompassing definition of its causes and manifestations, than the challenge it represents for the theoretical and scientific debates. In this respect the fields of science interested in collecting and structuring empirical evidence of urban growth through remote sensing are becoming increasingly important. When it comes to defining and analyzing urban sprawl, urban theories, whether traditional or emergent, descriptive or normative, conflict with each other on almost everything, from their conception and rationale, through to the measurement of sprawl and the recommended policy assessment and analysis which such theories imply in its control.

urban sprawl is generally defined as the rapid and uncoordinated growth of urban settlements at their fringes

While we have defined urban sprawl in general terms, its exact local connotations will always likely be debatable. From this standpoint, as Ewing (1994) implies, it is often easier to define sprawl by what it is not. It is sometimes implicitly defined by comparison to the ideal of the compact city, and for the most part, emerges as its poor cousin. The consequences of urban sprawl remain a hot topic of policy concern, most often because of its perception as a force eroding the countryside, which marks the final passing of an urban–rural world into an entirely urbanized one (see Chapter 3 in this volume) – with all the negative connotations that this implies for the visual environment, as well as a growing concern for the impacts posed to long-term urban sustainability. Though these concerns are not new, the last 20 years of economic growth has fuelled not only rapid urban expansion but associated problems such as crime, unemployment, and local government budget deficits which are all connected to the contrast between the sprawling periphery of the city and its inner decline.

Urban sprawl has thus become a major contemporary public policy issue. During much of the twentieth century, the control of urban growth has been of major concern to planning agencies who have sought to control peripheral development through a variety of rather blunt instruments such as "green belts" and strict development controls which were designed to "stop" growth. But as contemporary accounts of urban sprawl illustrate (Hayden 2004), these instruments have been largely ineffective and now the focus is on much more informed and intelligent strategies for dealing with such growth. Contemporary urban strategies focus more on sustainability of development under different economic scenarios and have come to be called strategies for "smart growth." We have come to the understanding that growth can never be "stopped" per se and thus peripheralization of cities is likely to continue for it is unlikely that even the most draconian strategies to

"smart growth" denotes a range of urban strategies that focuses on sustainability of development under different economic scenarios

control sprawl will lead to high density, compact and more constrained cities, at least in the foreseeable future.

Much of the confusion over the characteristics and impacts of sprawl stems largely from the inadequacies of definition. However it is illusory to believe that more data whether remotely sensed or census based can help in solving the debate over what sprawl is or is not, and whether it has only negative or also some positive impacts. Definitions of sprawl are highly dependent on the cultural, geographic and political context where sprawl is taking place to the point where what is perceived as suburban sprawl in Europe might be described as dense and urban in the US. Differences also exist between different European countries due to their different histories of land use planning. This is to say the solution to the problem of defining urban sprawl does not rest on more data and better methods to treat them, but in the meaning that is assigned to it in different contexts and times. To this purpose the importance of urban sprawl in the public policy agenda has generated an area of misunderstanding between descriptive and explanatory approaches on one side and normative ones on the other. This is a much broader issue than can be addressed within the limits of this chapter, but it should be kept in mind when exploring the literature that has been developing around urban sprawl in the last 20 years. Often, sprawl has been defined in terms of its negative effects and impacts, even though these are sometimes taken as underlying assumptions rather than empirically demonstrated facts.

local connotations of urban sprawl are highly dependent on the cultural, geographic and political context where sprawl is taking place

Here we will present some possible definitions of urban sprawl based on form, density and land use patterns. As a caveat, it must be noted that none of these approaches alone can identify urban sprawl, rather sprawl is comprised of a combination of multiple aspects. Causes of sprawl (e.g., changing location preferences and decreasing costs of private individual transport, for example) and its impacts (e.g., land consumption, traffic congestion, social segregation based on income or ethnic origins) should also be taken into account, especially if the purpose of a definition is to support the design of policy measures to tackle urban sprawl. We will subsequently illustrate these issues at the end of the chapter with reference to the EU SCATTER project.

2.2.1 Defining Sprawl Through Form

The term "urban sprawl" has been used to describe a variety of urban forms, including contiguous suburban growth, linear patterns of strip development, and leapfrog or scattered development. These forms are typically associated with patterns of clustered, non-traditional centers based on out of town malls, edge cities, and new towns and communities (Ewing 1994; Pendall 1999; Razin and Rosentraub 2000; Peiser 2001). These various urban forms are often presented in the literature as poorer, less sustainable or less economically efficient

sprawling forms can be considered to lie along a continuum from fairly compact to completely dispersed developments

Table 2.1 Types of sprawl

Type	High density	Low density
Compact contiguous	Circular or radial using mass transit	Possible but rare?
Linear strip corridor	Corridor development around mass transit	Ribbon development along radial routes
Polynucleated nodal	Urban nodes divided by green belts	Metro regions with new towns
Scattered/discontiguous	Possible but rare?	Metro regions with edge cities

alternatives to the compact ideal of urban development. In practice sprawling forms can be considered to lie along a continuum from fairly compact to completely dispersed developments.

A variety of urban forms can be described using a typology based on two continuous dimensions, which here are made discrete for explanatory purposes: settlement density (high and low) and physical configuration (ranging from contiguous and compact to scattered and discontiguous). This classification system suggests the eight idealized types of sprawl which are presented in Table 2.1.

Galster et al. (2001) have also classified the physical forms associated with urban sprawl into types (Fig. 2.1) and which need to be viewed in the context of the typology presented in Table 2.1. This classification also accommodates considerations of physical configuration and density. This method classifies patterns of urban sprawl according to eight components: *density, continuity, concentration, clustering, centrality, nuclearity, land use mix* and *proximity*. These measures are demonstrably useful to identify the major dimensions of sprawl. At the more compact end of the scale, the traditional pattern of suburban growth has been identified as sprawl. Suburban growth is defined as the contiguous expansion of existing development from a central core. Scattered or leapfrog development lies at the other end of the spectrum (Harvey and Clark 1965). The leapfrog form characteristically exhibits discontinuous development some way from a historic central core, with the intervening areas interspersed with vacant land. This is generally described as sprawl in the literature, although less extreme forms are also included under the term. Other forms that are classified as sprawl include compact growth around a number of smaller centers (polynucleated growth), and linear urban forms, such as strip developments, along major transport routes.

Indeed a vocabulary of different varieties of sprawl is fast emerging due to the fact that growth everywhere seems to be somewhat uncoordinated particularly on the periphery of the city (Hayden 2004). Sprawl in fact exists in very different forms which range from highly clustered centers – edge cities – in low density landscapes to the kinds of edgeless cities that exist where cities grow together into mega-poles of the kind that are characteristic of western Europe and even eastern

the various forms for urban sprawl pose a challenge for urban remote sensing

China. The morphology of these structures ranges from rather distinct edges and peripheries to somewhat more blurred or fuzzy perimeters and these various differences

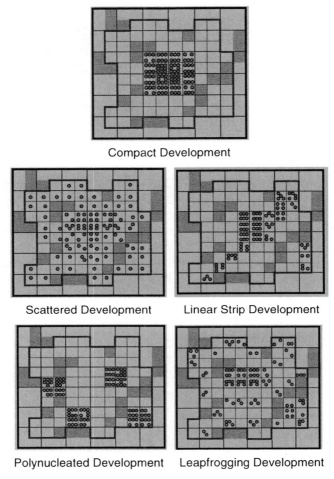

Fig. 2.1 Physical patterns defining sprawl (Galster et al. 2001)

pose a major planning problem for urban remote sensing which can only be resolved by fusing socioeconomic data into their interpretations.

Another classification is that of Camagni (Camagni et al. 2002), who has identified five types of suburban development patterns on the basis of the level of land consumption that each type requires. This classification seeks to gauge impacts, and also makes use of the same criteria (e.g., density and physical configuration) used in the previous two classifications (see Table 2.2). The Camagni classification provides an idealized taxonomy, and real world instances of urban sprawl development may be positioned on a continuum passing through these idealized types. We will present some of these real cases below in our outline of the SCATTER model.

Table 2.2   Types of suburban development (Camagni et al. 2002)

(T1) in-filling, characterized by situations where the building growth occurs through the
in-filling of free space remaining within the existing urban area
(T2) extension which occurs in the immediately adjacent urban fringe
(T3) linear development that follows the main axes of the metropolitan transport infrastructure
(T4) sprawl that characterizes the new scattered development lots
(T5) large-scale projects, concerning the development of new lots of considerable size that are
independent of the existing built-up urban area

2.2.2   Defining Sprawl Through Land Use

Land use patterns provide a second means of describing urban sprawl. A report from
the US Transportation Research Board (1998) lists the characteristics of sprawl
pertinent in the United States setting as: low-density residential development;
unconstrained and non-contiguous development; homogenous single-family residential development with scattered units; non-residential uses such as shopping
centers, strip retail, freestanding industry, office buildings,
schools and other community uses; and land uses which are   **urban sprawl**
spatially segregated from one another. Additionally the report   **is sometimes**
characterizes sprawl as entailing heavy consumption of ex-   **characterized**
urban agricultural and environmentally sensitive land, reliance   **in terms of**
on the automobile for transport, construction by small develop-   **land use**
ers, and lack of integrated land use planning. These character-   **patterns**
istics are very broad-based and typify almost all post-World
War II development in the United States. Thus "sprawl is almost impossible to separate from all conventional development" (Transportation Research Board 1998, pp.
7). Unfortunately, while this ensures that no aspect of sprawl is omitted, it does little
to differentiate sprawl from other urban forms. Sprawl is most commonly identified
as low-density development with a segregation (measured at an appropriate scale) of
uses; however, it is not clear which other land use characteristics must be present for
an area to be classified as sprawl (Batty et al. 2004).

2.3   The SCATTER Project

A recent EU-funded project has developed a definition of sprawl that is based on
the environmental, social and economic impacts of sprawl processes. The literature
generally assumes that these are negative, a perception that is becoming common
in Europe where urban sprawl is a much more recent and rather differently differentiated phenomenon than in the United States, and where its emergence has been
accompanied by an increased public and private sensitivity towards urban sustainability. The SCATTER Project (**S**prawling **C**ities **A**nd **T**ranspor**T** from **E**valuation
to **R**ecommendations) belongs to the sustainability-oriented research and policy
actions sponsored by the European Commission. Its main starting point is once

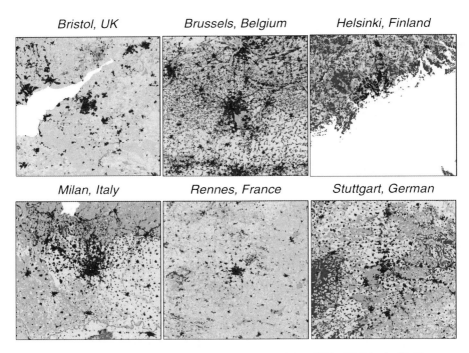

Fig. 2.2 Urban land use (*dark gray*) (from Remotely sensed data (EEA, 1990) in the Six European city regions)

again rooted in the notion that infrastructure, people and economy interact and that transport infrastructures in particular play a key role in reinforcing or constraining sprawl processes. The main goal of the project is to evaluate the impact of new transport infrastructures on sprawl processes and to provide policy recommendations to local authorities that are willing to reduce sprawl and its impacts.

The SCATTER project analyzes sprawl using both qualitative and quantitative methods, and considers a sample of six European cities (Bristol, Brussels, Helsinki, Milan, Rennes and Stuttgart). Figure 2.2 shows the CORINE-based land use maps of these cities, based on the visual interpretation of Landsat and SPOT satellite images. In Fig. 2.3 we show the cities as we have partitioned them into administrative units where we record population and related economic change associating this with land cover change in Fig. 2.2. A number of models have been developed for these cities where it is clear that although all size cities have been characterized by physical sprawl for the last 40 years, population and employment have not been continuously increasing. In Europe we are encountering a phenomenon which has long dominated North American cities, that is, despite continued sprawl, economics and population might actually be declining in such sprawling cities.

At this point, it is worth digressing a little to note how urban remote sensing might be able to provide data that can be complemented by traditional socioeconomic data.

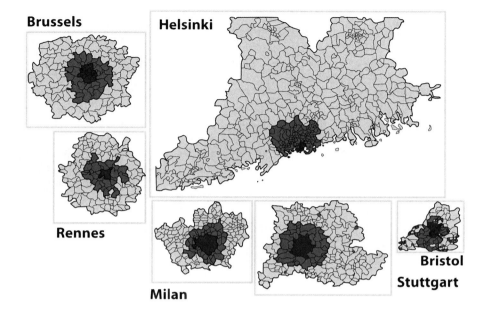

Fig. 2.3 The SCATTER case study cities (shown at the same scale)

In a sense this is what this entire book is about, but such remote sensing is in its infancy and, as discussed in Chapters 3 and 6, as satellite technologies generate higher and higher resolution images, the possibility of getting much more authoritative definitions of urban boundaries, and different urban land uses, enables a step change in our understanding of the patterns and dynamics of suburban growth. The various chapters in this book illustrate the state of the art but a good overview is provided by Mesev (2003) who shows that increasing resolution through ever more elaborate satellite imagery in fact is usually accompanied by an increasing level of noise in the data which tends to confuse interpretation.

higher spatial resolution in remotely sensed images is usually accompanied by an increasing level of noise in the data which tends to confuse interpretation

Fusing of Remote Sensing Images and Socioeconomic Data

Cities are artifacts that exist physically and socially in terms of our own definitions and these exist at different scales. As we get ever fine scale data, the nature of the heterogeneity in spatial patterning changes and far from increasing our ability to detect land use more accurately, it often confounds this.

This is why is it so important to fuse socioeconomic data which is much more scale dependent in terms of the way it is structured and delivered to us than is remotely sensed data. Ways of enabling such fusion depends on new techniques for ingeniously aggregating and disaggregating data, for overlaying data in diverse ways and for calculating multiple indices of scale and correlation which thence need to be interpreted in robust frameworks. In fact one of the most difficult problems with new imagery at finer resolutions from the new generation of airborne scanners and satellites is that the error structures in such data are largely unknown and thus new statistical theories are required before effective post processing of such data sources becomes resilient (Smith 2004). This quest is only just beginning and in terms of urban morphology, socioeconomic patterning is still more distinct than physical patterning from remote sensing imagery.

2.3.1   Qualitative Analysis of Urban Sprawl in Europe

As discussed in our introduction, generalized quantitative measures of urban form, obtained through urban remote sensing, can provide only a partial contribution to our understanding of the efficiency and effectiveness of different urban forms. The SCATTER project has thus encompassed qualitative as well as quantitative analysis. The purpose of the former was to detect and understand the local events and planning processes that led to the emergence of urban sprawl. The relevance of these events and processes in the decision agenda of local authorities and experts was assessed, as was the overall level of awareness of this particular urban phenomenon. This information is necessary if we want to complement quantitative measures with an embedded understanding of sprawl that is relevant to planners and decision makers.

The objectives were therefore achieved by analyzing interviews conducted with local authorities' representatives and experts in our six case cities. The results of the qualitative investigations have revealed that policy makers and local experts provide descriptions of urban sprawl, which are quite different from those available through a literature review. For this reason we have found them valuable in our research and have grouped them to build new typologies of sprawl. Although not centrally relevant to a book concerned principally with remote sensing, it is appropriate to discuss them briefly here, in the interests of balance and completeness of coverage (for a full description of the methodology and of the typology, see Besussi and Chin 2003). Policy makers and implementers essentially see sprawl as:

quantitative measures of urban phenomena from remote sensing and different censuses need to be complemented with input from planners and decision makers

- Emergent polycentric region, characterized by the emergence or development of secondary urban centers
- A scattered suburb, characterized by infill processes through which scattered and low density housing developments locate between centers or around existing transport infrastructures
- Peripheral fringes, characterized by higher densities than suburban developments and inhabited by populations that have relocated because of the increasing costs of life in the urban centers and/or
- Commercial strips and business centers, located following a rationale based on accessibility, low cost of land and agglomeration economies

2.3.2   *Statistical Indicators to Identify and Quantify Urban Sprawl*

The objective of the statistical analysis within SCATTER has been to quantitatively identify and measure urban sprawl in the case cities. The methodology adopted uses statistical techniques based upon *shift-share* analysis (see below), which are applied to time-series of zonal data. The data used in the analysis are mainly population, employment and average commuting distance. The method divides each urban region into two types of sub-regional zoning systems. The first one consists of concentric areas based on commuting patterns, as illustrated in Fig. 2.4 for the

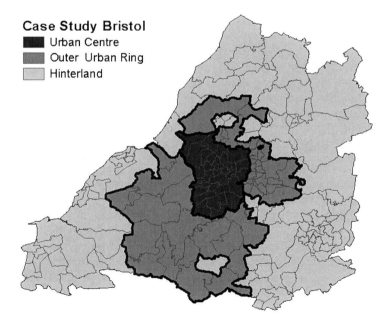

Fig. 2.4 Concentric zoning system for Bristol urban region

Bristol region; this distinction was based on percentage of commuters traveling daily towards the core urban area. The core urban area is identified differently for each of the case cities, on the basis of national classification methods while the first and second rings (suburb and hinterland) consist of zones where more or less than 40% of commuters' trips are directed towards the core area.

The second zoning systems, illustrated for all six cities in Fig. 2.3, consist of sub-zones representing the smallest statistical unit for which consistent and comparable data are available. In the UK context, these sub-zones are based on wards and parishes and aggregations thereof.

The generalized shift-share method computes for each small sub-zone the growth rate of each variable (population, employment and commuting distances). In a second step the deviation of each small sub-zone's growth rate from the regional growth rate is also computed. In the SCATTER project the shift-share method is used to identify the role played by the two growth components, the overall growth rate, $\lambda^a(t)$ and a time depending factor $\gamma_i^a(t)$ representing zonal deviations from the average growth path, in the actual growth of each small zone.

The analysis is carried out in three steps:

1. Estimation of the average growth rate as

$$\lambda^a(t) = \frac{1}{\Delta t} \ln\left(\frac{X^a(t+\Delta t)}{X^a(t)}\right) \tag{2.1}$$

where $X^a(t), X^a(t+\Delta t)$ represent the total volume of the variable over the entire urban region at times t and $t+\Delta t$ respectively.

2. Estimation of the zonal deviations of the average growth path as:

$$\gamma_i^a(t) = \frac{1}{\Delta t} \ln\left(\frac{X_i^a(t+\Delta t)}{X_i^a(t)}\right) - \lambda^a(t) \tag{2.2}$$

3. The estimated parameters $\lambda^a(t)$ and $\gamma_i^a(t)$ may exhibit some noisy structure, due to possible data uncertainties. Therefore appropriate data filters are applied to the mean growth rates and the deviations of the growth rates in order to smooth out such disturbances.

Tables 2.3 and 2.4 show the values of the parameters for population and employment growth rate in the six case cities. The values are smoothed using a Gaussian moving average procedure.

The quantitative analysis has also applied more traditional spatial statistical measures, such as the indicators of local and global spatial autocorrelation. For a value of a particular variable (e.g., population density), indicators of spatial autocorrelation make it possible to estimate whether a zone i is surrounded by zones exhibiting very similar or very dissimilar values, or is surrounded by a heterogeneous, patchy pattern of similar and dissimilar values. To identify local spatiotemporal pattern of variables the correlations between nearby values of the statistics are derived and verified by simulations. There are many possibilities to test for the existence of such pattern. One of the most popular is Moran's I statistic, which is used to test the

Table 2.3 Temporal mean value of $\lambda^a(t)$ and $\gamma_i^a(t)$ for population

| Cities | Years | Smoothed $\lambda^a(t)$ | Smoothed $\gamma_i^a(t)$ | | |
		Whole study area (%)	Urban centre (%)	Outer urban ring (%)	Hinter-land (%)
Milan	1971–2001	−0.1	−1.2	0.6	0.9
Brussels	1981–2001	0.2	−0.4	0.3	0.2
Stuttgart	1976–2000	0.5	−0.5	−0.1	0.4
Bristol	1971–1991	0.1	−0.8	0.8	0.4
Helsinki	1990–1999	1.2	−0.5	0.5	−0.4
Rennes	1962–1999	1.5	−0.7	1.8	−0.2

Table 2.4 Temporal mean value of $\lambda^a(t)$ and $\gamma_i^a(t)$ for employment

| Cities | Years | Smoothed $\lambda^a(t)$ | Smoothed $\gamma_i^a(t)$ | | |
		Whole study area (%)	Urban centre (%)	Outer urban ring (%)	Hinterland (%)
Milan	1961–2001	0.7	−1.0	1.3	1.0
Brussels	1984–1999	1.2	−0.9	1.7	0.6
Stuttgart	1976–1999	0.4	−0.7	0.4	0.3
Bristol	1971–1991	0.4	−1.1	1.2	0.6
Helsinki	1990–1999	0.3	−1.1	1.5	−0.6
Rennes	1982–1999	1.3	−0.7	1.6	−0.6

null hypothesis that the spatial autocorrelation of a variable is zero. If the null hypothesis is rejected, the variable is said to be spatially autocorrelated (see Anselin 1995; Getis and Ord 1996 for a theoretical and formal description of the indicators). As an example, when applied to population density, local indices of spatial autocorrelation might be used to define urban centers (high autocorrelation of density between adjacent units – similar high densities), the rural hinterland (high autocorrelation – similar low densities), urban poles (low autocorrelation – urban poles surrounded by rural zones, with much lower densities), and finally intermediate zones characterized by very low spatial autocorrelation, corresponding to suburban areas, which are a mix of more or less recently urbanized communes and other still rural communes. In Fig. 2.5 we provide a map of the local indicator of spatial autocorrelation for the population densities in the SCATTER case study cities.

2.4   Conclusions

This chapter has provided an overview of some of the issues that are salient to the measurement of urban form and function. In many respects, urban remote sensing provides an important spur to improving our understanding of the way that urban areas grow and change. Certainly there is a sense in which our abilities to routinely monitor incremental accretions and changes to urban shapes are not matched by socio-economic data of similar spatial or temporal granularity. Although increasingly

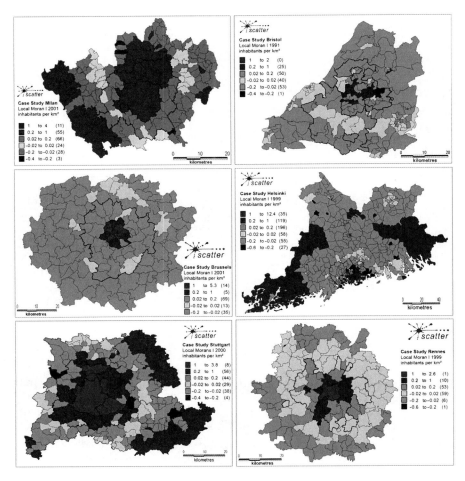

Fig. 2.5 Spatial distribution of Local Moran I for inhabitants per square kilometers

detailed and precise in spatial terms, very high resolution remote sensing images of urban areas tell us rather little about urban lifestyles, unless supplemented by socio-economic data. This chapter has set out some of the ways in which definitions of sprawl may be based upon quantitative measures of urban infrastructure and qualitative impressions of the way that urban policy evolves. An important challenge is to augment such quantitative and qualitative measures with generalized indices of urban lifestyle (e.g., sprawling low density settlements suggest suburban lifestyles). Today there is no single urban "way of life" (if ever there was) and there is a need for a better and more generalized understanding of lifestyles, since they may hold the key to understanding how individual cities evolve and change within systems of cities.

Several challenges arise from the use of remote sensing in the analysis of urban sprawl. More ways of fusing remotely sensed data (see Chapter 11) with socioeconomic data are required so that the definition of different types of urban morphology might be readily identified. The current state of the art is such that the edges of urban land uses are always fuzzy and this makes ground truthing almost impossible. Urban planning and a whole host of urban model applications require much more accurate data than remote sensing has so far been able to deliver. Moreover, although there are now some quite good examples of urban remote sensing interpretation, and although we have quite long time series in many places going back to the 1970s, for example, the quality of this data has continually improved and this makes good time series analysis tricky. Further, such imagery is still more appropriate in situations where fast analysis of rapid urban growth is needed, for example, the exploding cities in developing countries. In developed countries, emerging developments in new remote sensing technologies such as LIDAR that are fused with conventional technologies are providing exciting developments at the local scale (see Chapter 9). At the same time, adding prior geometric information to such interpretations is providing impressive means for advancement in the field. These challenges set a context for applications of these new technologies presented in the rest of this book.

Chapter Summary

In this chapter you have been introduced to key concepts and theories on urban growth and how these have approached the analysis and measurement of suburbanization and sprawl. The main idea is that the contemporary city in both developed and developing worlds needs much more than just one theory or one method of analysis or one typology of data to be fully understood. The contemporary city, of which urban sprawl is one of the most evident aspects, is a challenge to traditional analytical methods and requires that social sciences interact with earth sciences, and urban economics with GIS in order to build a coherent picture of patterns and trends of urbanization. The approach developed by the SCATTER research project and presented in this chapter provides an example of an interdisciplinary method that mixes qualitative and quantitative methods to understand sprawling settlements surrounding European cities and to evaluate the impact of transport on future development.

Learning Activities

Learn to Identify Sprawl

- Using the Internet, search for maps of different cities showing their urban form and structure and learn the differences between sprawl in North America, Europe, developing countries, and cities in other parts of the world. Below are some links you can start with:

- ○ SCATTER Project: http://www.casa.ucl.ac.uk/scatter/
- ○ Modelling Land Use Dynamics: http://moland.jrc.it/
- ○ Earth Science Data Interface: http://glcfapp.umiacs.umd.edu:8080/esdi/index.jsp

- Using remotely sensed images of different cities, reflect on and identify significant morphological differences that tell you something about the social and economic structure of each city. Discuss with your instructor how the size of the area and scale of analysis make a difference.

Study Questions

- What difference does the level of resolution of a remotely sensed image of an urban area make to your interpretation of its form and structure?
- How can socioeconomic data such as that from a Population Census help you in making good interpretations from a remotely sensed image which is overlaid with such data?
- To what extent can state-of-the-art remote sensing imagery enable you to detect different varieties of transportation systems in cities?
- To what extent is city development constrained by physical constraints? How can land cover analysis provide good representations of such constraints?
- Can remote sensing imagery enable you to make coherent estimations of urban density? How?
- How can information on the connectivity of an urban area through the layout of its physical buildings and street patterns be fused into remote sensing data so that interpretations of urban morphology may be enhanced?
- How can zonal based data be merged with pixilated data from urban remote sensing images in GIS?

References

Anselin L (1995) Local indicators of spatial association – LISA. Geogr Anal 27:93–115

Batty M, Longley PA (1994) Fractal cities: a geometry of form and function. Academic, San Diego, CA

Batty M, Besussi E, Maat K, Harts JJ (2004) Representing multifunctional cities: density and diversity in space and time. Built Environ 30:324–337

Besussi E, Chin N (2003) Identifying and measuring urban sprawl. In: Longley PA, Batty M (eds) Advanced spatial analysis. ESRI, Redlands, pp 109–128

Camagni R, Gibelli MC, Rigamonti P (2002) Urban mobility and urban form: The social and environmental costs of different patterns of urban expansion. Ecol Econ 40:199–216

Donnay JP, Barnsley MJ, Longley PA (eds) (2001) Remote sensing and urban analysis. Taylor & Francis, London

European Environment Agency (1990) CORINE-Land Cover data

Ewing R (1994) Characteristics, causes and effects of sprawl: a literature review. Environ Urban Issues 21:1–15

Galster G, Hanson R, Ratcliffe MR, Wolman H, Coleman S, Freihage J (2001) Wrestling sprawl to the ground: defining and measuring an elusive concept. Hous Policy Debate 12:681–717

Getis A, Ord JK (1996) Local spatial statistics: an overview. In: Longley PA, Batty M (eds) Spatial analysis: modeling in a GIS environment. Geoinformation International, Cambridge, pp 261–277

Harris RJ, Longley PA (2000) New data and approaches for urban analysis: models of residential densities. Trans GIS 4:217–234

Harvey EO, Clark W (1965) The nature and economics of urban sprawl. Land Econ 41:1–9

Hayden D (2004) A field guide to urban sprawl. Norton, New York

Mesev V (ed) (2003) Remote sensed cities. Taylor & Francis, London

Peiser R (2001) Decomposing urban sprawl. Town Plann Rev 72:275–298

Pendall R (1999) Do land use controls cause sprawl? Environ Plann B 26:555–571

Razin E, Rosentraub M (2000) Are fragmentation and sprawl interlinked? North American evidence. Urban Affairs Rev 35:821–836

Smith S (2004) Post processing of airborne scanning data (LIDAR). Unpublished Ph.D. thesis, CASA, University College London, London

Smith S, Holland D, Longley PA (2005) Quantifying interpolation errors in urban airborne laser scanning models. Geogr Anal 37:200–224

Transportation Research Board (1998) The costs of sprawl – revisited. National Research Council, National Academy Press, Washington, DC

Chapter 3
Defining Urban Areas

John R. Weeks

What is an urban area? How do we know it when we see it? And how do we measure the concept of urban, so that we can study it? This chapter reviews the many dimensions of urbanness in an attempt to synthesize the vast literature that exists on the topic, but focuses especially on issues of classifying places as urban or rural in such a way that changes over time in the characteristics of a place can be adequately captured by the researcher.

Learning Objectives

Upon completion of this chapter, you should be able to:

❶ Articulate how places are defined as urban
❷ Describe how the urbanness of a place could be measured
❸ Explain how urbanness is used as a predictor of human behavior

---

3.1 What Is Urban?

We all know an urban place when we see it, but defining it is not as easy as it might seem. In other writings, I have defined urban as being a characteristic of place, rather than of people (Weeks 2008). Places are typically defined as "urban," and on the basis of that definition the people living there are thought of as being part of the urban population. But, we do not usually apply the term "urban" to a person. The personal adjective "urbane," still occasionally used to describe a person, is defined by the Oxford English Dictionary as "having the qualities or characteristics associated with town or city life; esp. elegant and refined in manners, courteous, suave, sophisticated"

J.R. Weeks (✉)
Department of Geography, San Diego State University, 5500 Campanile Dr., San Diego, CA 92182-4493, USA
e-mail: john.weeks@sdsu.edu

T. Rashed and C. Jürgens (eds.), *Remote Sensing of Urban and Suburban Areas*,
Remote Sensing and Digital Image Processing 10,
DOI 10.1007/978-1-4020-4385-7_3, © Springer Science+Business Media B.V. 2010

(Brown 1993: pp 3527). Of course, you might well question how well that describes the average urban dweller in the modern world.

If we agree that urban is a place-based characteristic, then we can proceed to define an urban place as a spatial concentration of people whose lives are organized around nonagricultural activities. The essential characteristic here is that urban means nonagricultural; whereas rural means any place that is not urban. A farming village of 5,000 people should not be called urban, whereas a tourist spa or an artist colony of 2,500 people may well be correctly designated as an urban place. You can appreciate, then, that "urban" is a fairly complex concept. It is a function of (1) sheer population size, (2) space (land area), (3) the ratio of population to space (density or concentration), and (4) economic and social organization. As I will discuss below, the changes occurring throughout the world might well call into question this definition that relies on non-agricultural activity as a major criterion, because urban characteristics of place – especially those related to infrastructure – are increasingly (and deliberately) showing up in places that used to be strictly agricultural in nature. In other words, the urban–rural divide is becoming less obvious as the world population grows, as the fraction of humans living in cities increases, and as technology continues to transform human society.

"urban" is a place-based characteristic that incorporates elements of population density, social and economic organization, and the transformation of the natural environment into a built environment

Urban places are now home to virtually one of every two human beings and, by the middle of the twenty-first century, nearly two out of every three people will be urban dwellers (United Nations Population Division 2008). This is a truly remarkable transformation when you consider that as recently as 1850 only 2% percent of the entire population of the world lived in cities of 100,000 or more people. By 1900 that figure had edged up to 6%, and it had risen to 16% by 1950 (Davis 1972). Today the world is dotted by places with 100,000 or more people, and it is so commonplace that a city of that size is considered to be very small. "The present historical epoch, then, is marked by population redistribution as well as by population increase. The consequences of this redistribution – this "urban transition" from a predominantly rural, agricultural world to a predominantly urban, nonagricultural world – are likely to be of the same order of magnitude as those of the more widely-heralded increase in world population" (Firebaugh 1979: pp 199).

So pervasive is the lure of urban places that governments of many developing countries have promoted schemes to bring urban infrastructure to traditionally agricultural villages, in an attempt to keep migrants from overwhelming cities that are already crowded beyond the limits of the infrastructure. "In Vietnam the government is attempting to promote rural industrialization through the encouragement of sideline productions with the slogan 'leaving the land without leaving the village'" (Rigg 1998: pp 502). As non-agricultural work gradually soaks up a larger fraction of a village's labor force, the social and economic life of the village changes and of course the place becomes essentially more urban. "As the relationship between city and countryside becomes ever more entwined, it is becoming ever harder to talk of discrete 'rural' and 'urban' worlds" (Rigg 1998: pp 515).

It is not a coincidence that the urban transition has occurred in concert with the worldwide increase in population over the past 200 years. The urban transition is an inextricable part of the demographic transition because they both have roots in the same sets of technological advances that have rocked the world. The root cause of modern population growth is the massive drop in death rates that has been brought about by scientific control of disease, and by the provision of more and better food, shelter, and clothing. These are part and parcel of the industrial changes occasioned by technological advance. Modern technology allowed an increase in agricultural output per worker, which permitted more people to be freed from agricultural activity and were thus available to move to jobs being created in cities. Technology also helped improve the health of the population, which led eventually to cities being demographically self-sustaining (i.e., having a positive rate of natural increase and thus not being completely dependent upon migration for population increase). At the same time, technology was expanding the possibilities for city size and structure because premodern technology did not permit buildings to be very high or very deep – they were physically restricted to being close to the surface which clearly limits the potential population density and thus city size because: (1) cities had to be compact enough to be traversed easily on foot; (2) roads did not have to be very wide or regular in shape because they did not have to accommodate fast-moving motorized traffic; (3) population size was limited by the ability to supply the city with water and with some way of getting rid of human waste; (4) population size was limited by the ability to supply the city with food, which in the absence of refrigeration limited locations of cities to those places near a ready agricultural supply; and (5) economic activity was labor-intensive and so there were no special spatial advantages to having manufacturing done in the city; rather it could be "farmed" out to people living outside the city, meaning that cities were largely service (including government and finance) and commercial centers, which limited the variability in land use.

> **urban transition is an inextricable part of the demographic transition and both are related to advances in modern technology**

Technology has led to a larger population worldwide through its impact on controlling mortality, but has also led to the need for that population to be increasingly urban – to get out of the way of the mechanization of agriculture which is required to feed the larger population. Thus, only in modern times has it been not only possible but also necessary for any but a small fraction of the population to live in cities. Technology has led to the ability of food to be preserved and shipped farther distances, thus expanding the geographic scope of where cities can be located – thereby creating greater possibilities for the creation of city systems. Technology first led to the ability to house a larger number of people in the same urban space as before and therefore permitted an increase in city size through densification. Technology then, as discussed in Chapter 2, permitted the population of cities to spread out spatially

> **the limits of technology in the preindustrial era represented constraints on the location and size of cities, and led to a greater variety in the organization of social life**

as transportation improved, leading to spatial extensification of urban areas, associated often with a decline in the population density in central areas. The spread of cities beyond traditional core areas is also encouraged by the reluctance to tear down old buildings and widen narrow streets, thus pushing economic activity into newer places where a built environment can be created that is more accommodating of modern technology.

Overall, the limits of technology in the preindustrial era represented constraints on the location and size of cities, and thus on the complexity of the city systems that could develop in any given region. Modern technology, on the other hand, has virtually demanded the growth of the urban population both numerically and as a fraction of the total population, and has given rise to numerous new forms of urban structure in the process. Therefore, we should expect that, all other things being equal, modern technology will be associated with considerably more complex forms of city systems than was true in the preindustrial era. However, once again technology potentially alters the expectation because improved communication means that the advantages and disadvantages of particular city systems are evaluated quickly and that information can be disseminated in a way that can influence policy makers and market forces in the same way in disparate places. In preindustrial societies the slowness of communication led to a greater variety of cultural responses to the organization of social life and to a much slower diffusion of innovations.

Technology and Self-Sufficiency

The importance of modern technology is that it provides us with a set of clues about how to redefine, or at least improve our definition of, an urban place. Urban places are increasingly characterized by the kinds of infrastructure they provide to their residents. To a degree, this is a function of self-sufficiency. A truly non-urban place is one in which its residents are completely self-sufficient, in that they grow their own food, have their own water supply, create their own energy (largely from wood fires) and deal with their own waste products. This mode of living represents the life that most humans who have ever lived were born into. Yet, it is also a life that is precarious, because it is associated with high death rates and low levels of innovation. At the other end of the continuum is a place in which residents are completely dependent upon strangers for virtually all of their needs – piped water, piped sewage, landfills beyond the city limits, food brought to them from elsewhere in the world, and energy sources that are generated from outside the region.

3.2   Urban–Rural Is Not Really a Dichotomy

The idea of a continuum suggests that urban and rural are, in fact, ends of a continuum, rather than representing a dichotomy. Nevertheless, most countries employ a dichotomy in the definition of urban. "Of the 228 countries for which the United Nations (UN)

compiles data, roughly half use administrative considerations – such as residing in the capital of the country or of
a province – to designate people as urban dwellers. Among
the other countries, 51 distinguish urban and rural populations based on the size or density of locales, 39 rely on

**urban and rural
are ends of a
continuum**

functional characteristics such as the main economic activity of an area, 22 have no
definition of 'urban,' and 8 countries define all (Singapore, for example) or none
(several countries in Polynesia) of their populations as living in urban areas"
(Brockerhoff 2000: Box 1).

In the United States in the nineteenth and early twentieth centuries, rural turned
into urban when you reached streets laid out in a grid.
Today, such clearly defined transitions are rare. Besides,
even living in a rural area in most industrialized societies
does not preclude your participation in urban life. The
flexibility of the automobile combined with the power of
telecommunications put most people in touch with as
much of urban life (and rural life, what is left of it) as they
might want. In the most remote areas of developing coun-

**people create an
urban place,
and then are
influenced by
the place that
has been created**

tries, radio and satellite-relayed television broadcasts can make rural villagers
knowledgeable about urban life, even if they have never seen it in person (Critchfield
1994). There is probably more variability among urban places, and within the populations in urban places, than ever before in human history. This variability has
important consequences for the relationship between human populations and the
environment, because populations become urban through the transformation of the
natural environment into a built environment, and as urban places evolve, the subsequent changes in the built environment may well have forward-linking influences
on human behavior: Humans transform the environment; and are then transformed
by the new environment.

As long ago as 1950, when less than 30% of the world lived in urban places, the
United Nations Population Division was already making the case that a rural–urban
continuum would be preferable to a rural/urban dichotomy (Smailes 1966). "We
recognize, of course, that there will undoubtedly always be political and administrative uses to which dichotomies such as urban/rural and metropolitan/non-metro-
politan will be put, but we argue that such dichotomies are increasingly less useful
in social science research. Instead, we must move more intensively to the construction of a variable – a continuum or gradient – that more adequately and accurately
captures the vast differences that exist in where humans live and thus how we organize
our lives" (Weeks et al. 2005: pp 267).

In order to build an ecological model of the rural–urban continuum, we must
recognize that most social science literature that describes the nature and character
of urban populations focuses almost exclusively on the measurement of the social
environment, often drawing upon census data to describe this milieu. But variations
in the social environment are dependent, at least in part, upon variability in the built
environment. For example, high population density – an index that is often used as
a measure of urbanness – can be achieved with some kinds of physical structures,
but not others. The idea that people create an urban place, and then are influenced

by the place that has been created, leads to the hypothesis that some variability in human behavior may be captured in surrogate form by knowledge of the variability of the built environment, along with data from the census that provide surrogate measures of the social environment. In this conceptualization, the built and social environments are intimately entwined, but not completely dependent upon one another. The same built environment can host variation in the social environment, and the same social environment can exist within a range of built environments, but I would suggest that a relatively narrow range of combined values of the built and social environments will describe a unique set of urban populations.

3.3   Remotely-Sensed Data as Proxies for the Built Environment

Census and survey data provide most of the knowledge that we have of the social environment of places. Yet, one of the difficulties of using only census or survey data is that people are enumerated or surveyed at their place of residence. Since urban residents typically work in a different location than where they live, this spatial mismatch has the potential to produce a bias in the classification of the urbanness of place. An example might be a central business district which has only a small residential population, characterized largely by lower-income persons in single-room occupancy hotels. Census data might yield an index that indicates a relatively low degree of urbanness, based on a fairly small population and/or low density. Yet, the daytime population might represent a large number of commuting workers, and if they were to be counted the place would score much higher on an urban index. However, to accommodate that daytime population there must be a substantial built environment that includes a range of structures, infrastructures and other features indicative of urban lifestyle.

The built environment could be described by databases that document the type of structures and infrastructures comprising each parcel of land in every place. The cost of generating and maintaining such a database is enormous, however, and we do not really expect that any but the wealthiest of cities will be in a position to do that. In the meantime, it turns out that remotely-sensed data offer a way of generating reasonable proxy variables of the built environment, and thus of an important part of the way that places differ from one another with respect to urbanness. The modification of the physical environment that is characteristic of urban places can be inferred from the classification of multispectral and panchromatic satellite images. A place that is distinctly urban can be determined from the imagery regardless of the characteristics of the residents and we then have an indirect way to capture the characteristics both at the place of residence and at the presumptive place of work.

remotely-sensed data offer indirect ways to measure the urbanness of a place regardless of who resides in that place

The creation of an urban–rural dichotomy requires that the researcher decide upon the criteria that will go into an algorithm for assigning each place to either the urban or rural category. The creation of an urban–rural gradient requires that we adapt such an algorithm to tell us how urban or how rural a place is (a "soft" classification), rather than simply assigning it to one category or the other (a "hard" classification). There are several issues that must

> **the creation of an urban–rural gradient requires a knowledge of how urban or how rural a place is**

be dealt with in the creation of an index, including: (1) the spatial unit of analysis to be used; (2) the variables to be combined in the index; and (3) how the variables will be combined to create an index.

3.3.1 What Spatial Unit of Analysis Should Be Used?

If we are able only to circumscribe some large geographic zone (e.g., the contiguously built-up area in a region) then the ends of the rural–urban spectrum will be relatively close to one another. On the other hand, if we are able to define the attributes for relatively small and regular zones, such as a half-kilometer grid of land, then we could better understand variability both between and within human settlements. Furthermore, if we had a clearly defined spatial grid, then we could more accurately measure change over time – to understand the process of urban change and evolution that almost certainly has an important impact on human attitudes and behavior. However, the preliminary set of calculations that helps to establish the utility of this approach must of necessity be based on geographically irregular administrative boundaries because the census data that we are using in the creation of the index are readily available only at the level of those administrative boundaries.

3.3.2 What Variables Should Be Used to Define Urbanness?

I have suggested elsewhere (Weeks et al. 2005) that the urban index should combine census and survey data (to capture aspects of the social environment) with data from remotely-sensed imagery (to capture aspects of the built environment). Let me focus here on the latter part of the equation. The classification of an image is done at the level of the individual picture element (pixel), but in the creation of an index of urbanness we are less interested in each pixel than we are in the *composition* and *configuration* of all of the pixels within a defined geographic region (read further discussions in Chapters 5 and 12). This is the realm of landscape metrics, which are quantitative indices that describe the structure of a landscape by measuring the way in which pixels of a particular land cover type are spatially related to one another (Herold et al. 2002; Lam and DeCola 1993; McGarigal et al. 2002). The structure of a scene is inferred by calculating indices that measure composition and configuration of the pixels within an area.

3.3.3   How Will the Variables Be Measured?

Composition refers to the proportional abundance in a region of particular land cover classes that are of interest to the researcher. We employed Ridd's (1995) V-I-S (vegetation, impervious surface, soil) model to guide the spectral mixture analysis (SMA) of medium-resolution multi-spectral images for Cairo for 1986 and 1996, in a manner similar to methods used by Phinn and his colleagues for Brisbane, Australia (Phinn et al. 2002), and by Wu and Murray (2003) for Columbus, Ohio. The classification methods are described elsewhere (Rashed and Weeks 2003; Rashed et al. 2001, 2003, 2005; Roberts et al. 1998) and so will not be discussed here in any detail. The

the built environment is quantified by measures of composition and configuration of land cover within an area

V-I-S model (see Chapter 6) views the urban scene as being composed of combinations of three distinct land cover classes. An area that is composed entirely of bare soil would be characteristic of desert wilderness, whereas an area composed entirely of vegetation would be dense forest, lawn, or intensive fields of crops. At the top of the pyramid is impervious surface, an abundance of which is characteristic of central business districts, which are conceptualized as the most urban of the built environments.

   We added another component to Ridd's physical model – shade/water – following the work of Ward et al. (2000) suggesting that the fourth physical component improves the model in settings outside of the United States. When combined with impervious surfaces in urban areas it becomes a measure of the height of buildings (based on the shadows cast by buildings). When combined with vegetation it provides a measure of the amount of water in the soil and the shade cast by tall vegetation (largely trees that may serve as windbreaks in agricultural areas). In combination with bare soil it is largely a measure of any shadows cast by trees, although there could be some component of shade from large buildings in heavy industrial areas. Spectral mixture analysis permits a "soft" classification of a pixel into the likely fraction of the pixel that is composed of each of the four physical elements of vegetation, impervious surface, soil, and shade. By summing up these fractions over all pixels contained within each area of interest, we have a composite measure of the fraction (the "proportional abundance") of the area that is covered by each of the four land cover types.

   These compositional metrics build on the qualitative sense that each of us has about what an urban place "looks like." Even today in highly urbanized countries in Europe and North America it is visually very evident when you move from a largely rural to a predominantly urban place and, of course, the change in the built environment is the principal index of that. Even within non-urban areas it is usually quite evident when you have passed from a wilderness area into a largely agriculture area. Once again, it is the configuration of the environment that provides the clue. Figure 3.1 shows this in a schematic way. Wilderness areas can, at the extreme, be expected to be composed especially of bare soil, since deserts tend to

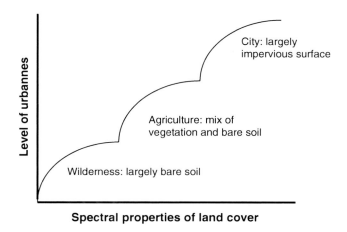

Fig. 3.1 The urban gradient may be discontinuous

be the places least habitable by humans. As the fraction of vegetation increases, there is an implicit increase in the availability of water and where there is sufficient water the possibility of agricultural increases and agriculture creates a signature on the ground that is typically distinct from areas that have not been modified by humans. However, the nature of urban places is that the built environment is dominant, and so cities are distinctly noticeable from the air because vegetation gives way immediately, discontinuously, to impervious surfaces.

The proportional abundance of impervious surface is the baseline measure of urbanness, as suggested by the Ridd V-I-S model, but shade is also a factor, especially in areas dominated by tall buildings. Thus, in areas that are generally urban, the simple addition of the impervious surface and shade fractions should provide an appropriate measure of the proportional abundance of land cover most associated with an urban place. In agricultural areas, where shade may indicate canopy cover or water-saturated ground, it would be less appropriate to combine the impervious surface with the shade fraction.

The other aspect of landscape metrics is the quantification of the spatial configuration of the patches comprising each land cover class. We may know that 60% of a given area is covered by impervious surface (the measure of composition), but we would also like to know how those patches are arranged within the area under observation. McGarigal et al. (2002) notes that configuration is much more difficult to assess than composition and over the years a large number of measures have been developed in an attempt to capture the essence of landscape configuration. However, it is important to keep in mind that most measures of landscape configuration were developed for the purpose of describing landscape ecology and have only recently been shown to have an adaptation to the measurement of the urban environment (Herold et al. 2002). One of the more interpretable measures in the context of urban places is the contiguity index, which is a measure of the "clumpiness" of land cover classes. In particular, we are interested in the clumpiness of impervious surfaces because we hypothesize that high levels of clumpiness (where pixels of the impervious surface land cover class are all in close proximity to one another) represent one or only a few buildings, characteristic of central cities and other dense areas. On the other hand, low levels of clumpiness of impervious surface should represent a disaggregation of pixels of the same land cover class, representing a greater number of buildings, associated with lower density, more suburban areas.

Two areas might have identical fractions of impervious surface, but the one with a high contiguity index would probably represent a "more urban" area than the one with the lower contiguity index. In general, we would expect that city centers would have the highest abundance of impervious surface and also the highest level of contiguity of that impervious surface. At the other extreme, a place that is not very urban will have a low proportion of impervious surface, but that surface might be highly contiguous (one small building) or only moderately so (three small buildings), but the degree of contiguity would matter less than it would when the proportion of impervious surface is high. This suggests that the configuration of the pixels increases in importance as the proportional abundance of impervious surface increases, implying the existence of an exponential relationship.

The way in which these several measures of composition and configuration can be most satisfactorily put together is still under investigation (see Weeks 2004; Weeks et al. 2005). However, the research conducted thus far suggests the utility of this approach to the creation of an urban index that can be combined with census data to characterize the nature of urbanness of a place.

3.4   Using the Urban Index as a Predictor Variable

An urban index of the type that I have suggested may be of inherent interest on its own, but its greatest value in social science research is almost certainly that it provides a way of contextualizing the environments in which people live. Places that are different in terms of urbanness are likely to be different in other ways that will affect the lives of the people there. Similarly, changes over time in urbanness can be expected to be related, both causally and consequentially, to the lives of the people who comprise the residents and/or workers in those changing environments.

As long as the researcher is careful to use the same measurements from the satellite imagery and census data over space and time, then differences in the urban index can be proxies for differences between places and changes over time in the social and economic aspects of the people being studied. This characteristic of a place can than be introduced into a regression analysis as a predictor variable, or even into multi-level analyses as a community-level factor that may be related to individual behavior taking place in different places and/or at the same place at different times.

Chapter Summary

Urban is a place-based characteristic that describes the degree to which the lives of a spatial concentration of people are organized around nonagricultural activities. The urbanness of a place is determined based on a range of elements encompassing population size and density, social and economic organization, and the transformation of the natural and agriculture environments into a built environment. Because of the spatial and temporal variability of such elements, the degree of urbanness varies across space (and through time), suggesting that urban and rural are, in fact, ends of a continuum, rather than representing a dichotomy. The idea of an urban–rural continuum or gradient lends itself to the development of indices to help describe how urban (or how rural) a place is at a given point of time. This chapter has introduced you to one of such indices, an urban index that combines census and survey data (to capture aspects of the social environment) with data from remotely sensed imagery (to capture aspects of the built environment). Focusing mainly on the latter part of the equation, this chapter has discussed several issues to be considered in using remote sensing to define the urbanness of a place, including: (1) the spatial unit of analysis to be used (pixel versus zonal units); (2) the variables to be combined in the index (composition and configuration of the built environment); and (3) how the variables will be combined to create an index (spectral mixture analysis and landscape metrics).

LEARNING ACTIVITIES

Internet Resources

- Explore the changing nature of urbanness

  - The Timeline of New Urbanism http://www.nutimeline.net/. Features several way to search key events in the history of new urban since the nineteenth century.
  - USGS Urban Dynamics Program.
  - http://landcover.usgs.gov/urban/intro.asp. Features temporal maps and data resources, animations, articles, and timelines for selected metropolitan regions in the United States.

- Explore the different ways used to define the urbanness of places:

    ○ The World's Bank urban environmental indicators http://www.worldbank. org/urban.
    ○ The Human Settlements page on the website of International Institute for Environment and Development http://www.iied.org/HS/index.html. Features several discussions of and resources for the rural–urban divide and free access to the international journal of Environment and Urbanization.
    ○ The Global Urban Indicator Program at the UN-HABITAT http://www.unchs. org/programmes/guo/guo_indicators.asp.

- Links to some regional and country-specific urban indicators programs:

    ○ Central and Eastern Europe http://greenpack.rec.org/urbanisation/seeing_a_ city/05-01-03.shtml
    ○ Canada
        - Montreal http://www.ecoplan.mcgill.ca/?q=node/view/102
        - Toronto http://tui.evcco.com/
    ○ India http://www.cmag-india.org/programs_urban_indi_prog.htm

- FRAGSTAT for landscape metrics http://www.umass.edu/landeco/research/ fragstats/fragstats.html

Study Questions

- What do you understand about the terms "urban" and "self-sufficiency? How do they connect to each other?
- Use census data to plot the population of your own city or a city of your choice over time. Identify and explain significant trends. Using two or more of remotely sensed images of the same city, identify urbanization trends in the city and whether they correspond to population trends. Use the procedures described in Weeks et al (2005) and Rashed et al (2005) to develop an index of urbaness at one or more points of time for your study city.

References

Brockerhoff MP (2000) An urbanizing world. Popul Bull 55:3–44
Brown L (1993) The New Shorter Oxford English Dictionary on Historical Principles. Oxford: Clarendon Press
Critchfield R (1994) The villagers: changed values, altered lives. Anchor Books, New York
Davis K (1972) World urbanization 1950–1970 vol(2): Analysis of trends, relationship, and development. Institute of International Studies, University of California, Berkeley, CA
Firebaugh G (1979) Structural determinants of urbanization in Asia and Latin America, 1950–1970. Am Sociol Rev 44:199–215
Herold M, Scepan J, Clarke KC (2002) The use of remote sensing and landscape metrics to describe structures and changes in urban land uses. Environ Plann A 34:1443–1458

Lam N, DeCola L (1993) Fractals in geography. Prentice-Hall, Englewood Cliffs, NJ

McGarigal K, Cushman SA, Neel MC, Ene E (2002). FRAGSTATS: spatial pattern analysis program for categorical maps, computer software program produced by the authors at the University of Massachusetts, Amherst. Available at www.umass.edu/landeco/research/frag-stats/fragstats.html

Phinn SR, Stanford M, Scarth P, Murry AT, Shyy PT (2002) Monitoring the composition of urban environments based on the Vegetation-Impervious Surface-Soil (VIS) model by subpixel techniques. Int J Remote Sens 23:4131–4153

Rashed T, Weeks JR (2003) Assessing vulnerability to earthquake hazards through spatial multi-criteria analysis of urban areas. Int J Geogr Inf Sci 17:547–576

Rashed T, Weeks JR, Gadalla M, Hill A (2001) Revealing the anatomy of cities through spectral mixture analysis of multispectral satellite imagery: a case study of the greater Cairo region, Egypt. Geocarto Int 16(4):5–16

Rashed T, JR, Weeks DA, Roberts J, Rogan, and Powell R (2003) Measuring the Physical Composition of Urban Morphology using Multiple Endmember Spectral Mixture Analysis. Photogrammetric Engineering and Remote Sensing 69 (9):1111–1120

Rashed T, Weeks JR, Stow DA, Fugate D (2005) Measuring temporal compositions of urban morphology through spectral mixture analysis: toward a soft approach to change analysis in crowded cities. Int J Remote Sens 26:699–718

Ridd M (1995) Exploring a V-I-S (Vegetation-Impervious Surface-soil) model for urban ecosystem analysis through remote sensing: comparative anatomy of cities. Int J Remote Sens 16:2165–2185

Rigg J (1998) Rural–urban interactions, agriculture and wealth: a Southeast Asian perspective. Progr Hum Geogr 22:497–522

Roberts DA, Batista GT, Pereira JLG, Waller EK, Nelson BW (1998) Change identification using multitemporal spectral mixture analysis: applications in Eastern Amazonia. In: Lunetta RS, Elvidge CD (eds) Remote sensing change detection: environmental monitoring applications and methods. Ann Arbor Press, Ann Arbor, MI, pp 137–161

Smalies AF (1966) The geography of towns. Aldine, Chicago, IL

United Nations Population Division (2008) World urbanization prospects: the 2007 revision. United Nations, New York

Ward D, Phinn SR, Murray AT (2000) Monitoring growth in rapidly urbanization areas using remotely sensed data. Prof Geogr 52:371–385

Weeks JR (2004) Using remote sensing and geographic information systems to identify he underlying properties of urban environments. In: Ag C, Hugo G (eds) New forms of urbanization: conceptualizing and measuring human settlement in the twenty-first century. Ashgate, London

Weeks JR (2008) Population: an introduction to concepts and issues, 10th edn. Wadsworth Thomson Learning, Belmont, CA

Weeks JR, Larson D, Fugate D (2005) Patterns of urban land use as assessed by satellite imagery: an application to Cairo, Egypt. In: Entwisle B, Rindfuss R, Stern P (eds) Population, land use, and environment: research directions. National Academies, Washington, DC, pp 265–286

Wu C, Murray AT (2003) Estimating impervious surface distribution by spectral mixture analysis. Remote Sens Environ 84:493–505

Chapter 4
The Spectral Dimension in Urban Remote Sensing

Martin Herold and Dar A. Roberts

Urban environments are characterized by different types of materials and land cover surfaces than found in natural landscapes. The analysis of remote sensing data has to consider these unique spectral characteristics. This chapter describes the spectral properties of urban areas, how different urban land cover types are spectrally discriminated, and which sensor configurations are most useful to map urban areas. We also show how new remote sensing technologies improve our capabilities to map urban areas in high spatial and thematic detail.

> **Learning Objectives**
>
> Upon completion of this chapter, you should be able to:
>
> ❶ Distinguish the unique spectral characteristics of urban areas
> ❷ Explain the separability and most suitable spectral bands in discriminating urban land cover type
> ❸ Speculate on the potential of hyperspectral, multispectral and LIDAR remote sensing data in urban mapping

4.1 Introduction

The spectral signal is one of the most important properties of land surfaces measured with remote sensing (Fig. 4.1). The amount and spectral qualities of energy acquired by the remote sensing system (at sensor radiance) is dependent

M. Herold (✉)
Institute of Geo-Information Science and Remote Sensing, Wageningen University, Droevendaalsesteeg 3, Gaia, building number 101, P.O. Box 6708, Wageningen, The Netherlands
e-mail: martin.herold@wur.nl

D.A. Roberts
Geography Department, University of California, 5832 Ellison Hall, Santa Barbara, CA 93106-4060, USA
e-mail: dar@geog.ucsb.edu

T. Rashed and C. Jürgens (eds.), *Remote Sensing of Urban and Suburban Areas*, Remote Sensing and Digital Image Processing 10, DOI 10.1007/978-1-4020-4385-7_4, © Springer Science+Business Media B.V. 2010

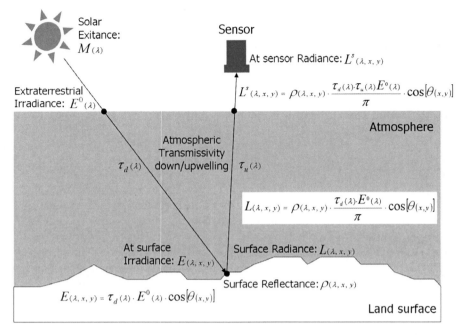

Fig. 4.1 Spectral signal acquired by a remote sensor (λ = wavelength, θ = local incidence angle, x/y = location on earth surface) (Schowengerdt 1997)

spectral resolution of a sensor is determined by the number of spectral bands, their bandwidths and locations along the electromagnetic spectrum

upon the source function (sun), the extent to which the radiation is modified by the atmosphere (downwelling and upwelling atmospheric transmission) and the physical and geometric structure and chemical constituents present at the surface (reflectance). The sensor measures the radiation in spectral bands, i.e., at a specific wavelength or over a defined wavelength range (bandwidth). The number of spectral bands, their bandwidths and locations along the electromagnetic spectrum determine the spectral capabilities or spectral resolution of sensors.

Most satellite sensors are multispectral systems like LANDSAT Thematic Mapper (TM) or IKONOS. They sense the earth surface with a few broad spectral bands. Sensors that are able to acquire a large number of spectral bands with narrow bandwidths are called hyperspectral systems. They have high spectral sampling but so far are limited to airborne or ground based systems except the only spaceborne hyperspectral sensor HYPERION on the EO-1 satellite. Such detailed spectral measurements, however, potentially allow for precise identification of the chemical and physical material properties as well as surface geometry of surfaces (Clark 1999). Related analyses are usually referred to as spectroscopy or imaging spectrometry.

hyperspectral systems sense the earth surface in a large number of narrow spectral bands for precise identification of the chemical and physical material properties

The optical spectral range is separated into the visible (VIS: 400–700 nm) and near infrared (NIR: 700–2,500 nm). The NIR is commonly divided into two broad regions based on detectors used to measure radiation, including the photographic NIR (700–1,000 nm), a region that can be sampled by photographic film or silicon-based detectors and the short-wave-infrared (SWIR: 1,000–2,500 nm), which requires other detector materials. In general, imaging spectrometry supports our spectral understanding of different land surfaces and thereby improves our ability to discriminate materials using remote sensing.

Remote sensing of urban environments is particularly challenging. The land surface objects (e.g., buildings/roofs, roads) have a small spatial extent. Given this large amount of spatial heterogeneity most analyses in urban areas have relied upon aerial photography as a data source. Recent advances in spaceborne systems, such as IKONOS and QUICKBIRD provide cost effective alternatives to aerial photography. For example, IKONOS provides 4 m multispectral data, thereby meeting the minimum spatial resolution of 5 m considered necessary for accurate spatial representation of urban materials such as buildings and roads (Jensen and Cowen 1999). But urban environments also possess a high spectral heterogeneity (Ben-Dor et al. 2001; Roberts and Herold 2004). They are characterized by a large diversity of materials such as human-made features, vegetation, soils, and others. Since spectral understanding is essential for remote sensing applications, the following sections will focus on this issue and give an introduction to the spectral dimension in urban environments. Based on principles of spectrometry, we discuss the spectral characteristics and the spectral discrimination between urban land cover types. We show examples of remote sensing mapping applications to highlight the effects of different sensor configurations on the urban mapping accuracies.

4.2   Spectral Characteristics of Urban Surfaces

The generic study of reflectance characteristics or spectral signatures is usually based on spectral libraries. These libraries contain pure spectral samples of surfaces, including a wide range of materials over a continuous wavelength range with higher spectral detail, and additional information and documentation about surface characteristics and the quality of the spectra (i.e., metadata). They can be derived from laboratory and ground spectral measurements, and from hyperspectral remote sensing observations. Figure 4.2 highlights this with example spectra of urban surfaces. In field spectrometry the sample measurement is taken along with a calibration signal of a 100% reflectant material. The ratio of both

spectral libraries include spectral samples of surface materials derived from laboratory, ground measurements or hyperspectral imagery

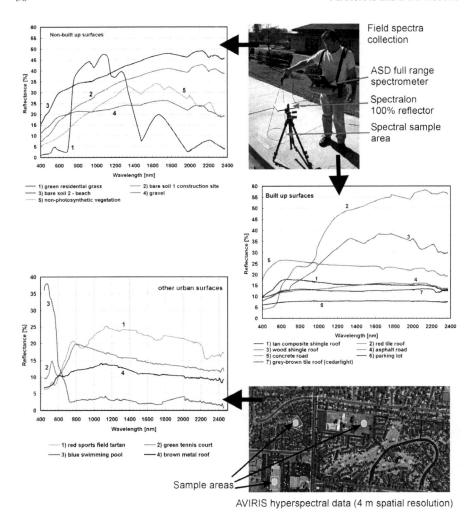

Fig. 4.2 Examples of urban spectra derived from two different sources: ground spectrometry (*diagrams in the upper left and middle right*) and hyperspectral image data (*diagram in the lower left*). The major water absorption bands (1,340–1,480 and 1,770–1,970 nm) are linearly interpolated

measurements describes the bi-directional reflectance factor, also referred to as the reflectance of the sample surface expressed in percent. Hyperspectral remote sensing measurements (in radiance) have to be converted to reflectance values by correcting for atmospheric influences and wavelength-dependent changes in solar irradiance. This process can be quite challenging and is one of the main reasons why image spectra are in general noisier than ground measurements.

The spectral signatures shown in Fig. 4.2 are best explained and understood with absorption processes that represent the specific material properties of the surfaces. One of the non-built spectra indicates the classic spectral characteristics for green vegetation. Characteristic features are the reflectance peak in visible green (550 nm; chlorophyll absorptions at 450 and 680 nm) and the red edge (~750 nm), and the near infrared with absorption features at 980, 1,200, 1,400 and 1,900 nm caused by water content, and ligno-cellulose absorptions at 2,100 and 2,300 nm in the SWIR. The spectra of bare soil show general similarities with non-photosynthetic vegetation (NPV) in the visible (VIS) and NIR region. NPV spectra have significant ligno-cellulose absorption bands that clearly identify them as vegetation.

The bare soil spectra show clay mineral absorption features at 2,200 nm. Gravel surfaces also reflect mineral absorption features similar to the soil surfaces. The roof spectra indicate distinct spectral signatures for red tile roofs and wood shingle roofs. Both roof types show a significant reflectance increase in the NIR and SWIR region. The wood shingle signature contains the ligno-cellulose absorption feature in the SWIR that is common for all non-photosynthetic vegetation. Liquid water and hydroxyl absorptions, typically found in clays are lacking in fired red tile bricks due to the loss of water in the production firing process. Red tile roofs and gravel roads show significant iron oxides absorptions in the visible and near infrared (near 520, 670 and 870 nm). Concrete road materials are comprised of cement, gravel, water, and various other ingredients. Significant absorptions appear in the SWIR due to calcium carbonate with a feature at 2,300 nm for calcite and at 2,370 from dolomite. The feature near 2,170 nm can be related to content of clays in the concrete. The diagram showing built up spectra further contains several materials with constant low reflectance and no or only minor unique small-scale absorption features like asphalt roads, parking lots, composite shingle roofs and dark tile roofs. Swimming pools (AVIRIS spectra of other urban surfaces) have a large reflectance in the visible blue near 450 nm. In this wavelength the radiation penetrates the water body and reflects off the light blue/cyan painted bottom. Towards longer wavelengths the water absorbs increasing amounts of radiation. The reflectance drops and basically no radiation is reflected in the NIR and SWIR region. The sport surfaces red sport tartan and tennis court, nicely show the absorptions in the visible region that cause their significant color. In the case of red tartan this color is caused by the iron oxide content with absorptions near 520, 670 and 870 nm.

spectral absorption features are important indicators of material characteristics

4.3   Urban Materials Versus Land Cover Types

From a remote sensing mapping perspective it is important to view the spectral signatures (Fig. 4.2) in terms of the spectral contrast. If a surface type has a unique spectral signal it is easily distinguished in remote sensing images, i.e., the expected

spectral similarities among some of urban materials pose challenge for urban remote sensing application

mapping accuracy is high. Some surfaces indicate such unique characteristics like green vegetation, red tile roofs, or swimming pools. Other land cover types seem spectrally alike. In particular, darker built up surfaces like asphalt roads and dark tile and composite shingle roofs have quite similar signatures and no distinct absorption characteristics. Considerable confusion between these surface types is anticipated in remote sensing applications.

One remote sensing mapping approach to test this is shown in Fig. 4.3. It presents the results of a matched filter analysis (ENVI image processing software) applied to hyperspectral AVIRIS data (4 m spatial resolution) acquired over Goleta, California.

Matched filter analysis compares the spectrum in each image pixel to a known reference spectrum from a spectral library. If the spectral signal of a specific pixel and the reference spectra perfectly match, hence the reflectance properties of the surfaces are the same, the matched filter score becomes 100; if they are absolutely dissimilar the score is 0. The first frame in Fig. 4.3 shows the matched filter scores for red tile roofs. It indicates that only a few areas have similar material characteristics than the sample red tile roof spectra. The areas with higher matched filter scores actually are red tile roofs. These roofs can be mapped with high accuracies from the AVIRIS data. Wood shingle roofs also have fairly unique spectral signals and the matched filter approach identifies these roofs with high accuracy. There are, however, a few areas in the left part of the image that show intermediate to higher matched filter scores. The most significant false positives appear to be open fields consisting of senesced grass in more rural areas. Both wood shingle roofs and senesced grasslands contain non-photosynthetic vegetation and therefore their spectral signals are somewhat similar (see also Fig. 4.2). In contrast, the matched filter for asphalt road shows considerable spectral confusion. In this map (lower frame), the matched filter correctly identifies most road surfaces, but also maps large areas of other surface types like composite shingle roofs, parking lots, and tar roofs. The spectral similarity was already indicated in the spectral signatures of Fig. 4.2. In fact, these land cover types are composed of similar materials like dark asphalt, rocky components, and other tar/oil products. This fact shows that we have to consider whether we are interested in mapping urban materials or urban land cover types.

urban surfaces are generally made of one or more of four material components: minerals, vegetation, oil products, and human-made/ artificial materials

From a material perspective urban areas are composed of four main components: (1) rocks and soils (i.e., minerals), (2) vegetation, (3) oil products such as tar and asphalt, and (4) other human-made materials such as refined oil products like paints and plastics. Most surfaces represent an aggregate of these components, i.e., concrete includes rock aggregates, limestone, and portlandite. The matched filter approach is best suited to identify and map urban materials using the full spectral range. Other methods focus more on small-scale spectral absorption features to derive detailed material characteristics (e.g., mineral compositions,

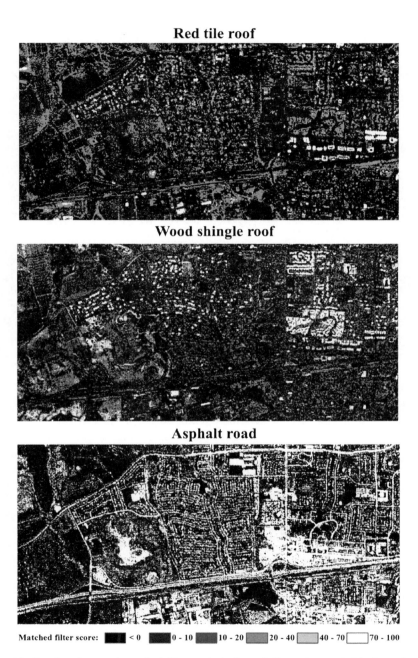

Fig. 4.3 Results of spectral matched filter analysis for red tile roofs, wood shingle roofs and asphalt roads from high resolution AVIRIS data

grain size effects). Methods to extract, compare and analyze such spectral features are described in Swayze et al. (2003). A successful application of hyperspectral remote sensing for material mapping in urban areas is presented by Clark et al. (2001). This research utilized high-resolution AVIRIS data with continuum-removed spectral feature analysis to derive a detailed map of material/dust accumulations in connection with the World Trade Center attack on September 11, 2001.

Land Cover versus Materials

Many urban remote sensing applications focus on mapping urban land cover rather than materials. Although both are related, i.e., red tile roofs have unique material characteristics, mapping land cover types requires a different perspective in urban area remote sensing. Land cover considers characteristics that not only come from the material itself. The surface structure (roughness) affects the spectral signal as much as usual variations within the land cover type (i.e., cracks in roads, buildings with different roof angles, age or cover). Two different land cover types (e.g., asphalt roads and composite shingle roofs) can be composed of very similar materials. From a material perspective these surfaces would map accurately (Fig. 4.3). The discrimination of the land cover types roads and roofs, however, is limited by this similarity.

The land cover perspective is the most common remote sensing based approach in urban area mapping and will be the focus of the rest of this chapter. Table 4.1 presents a hierarchical scheme to structure the diversity of land cover types within urban areas. Broad land-cover classes are considered Level 1. Level 2 further subdivides the Level 1 classes based on their use, function or other generic surface characteristics. Level 3 further separates the functional land cover classes based on their material properties for built up classes. This classification schemes does not claim to be complete and even more detailed levels of categories might be differentiated, e.g., based on surface color or other characteristics. However, remote sensing data with higher spectral and spatial resolution, and more specialized image analysis techniques, are usually required when a more detailed level of land cover class discrimination is desired in the mapping process.

4.4   Spectral Separability of Urban Land Cover Types

Measures of spectral separability are used to quantify the degree of discrimination between land cover types. Several of these statistical measures have been developed (see Schowengerdt 1997) with the Bhattacharyya distance (B-distance) being very useful in the analysis of hyperspectral data (Landgrebe 2000). The spectral comparison is based on several spectral samples for each land cover class from spectral libraries

Table 4.1   A land cover classification scheme for urban environments

Level 1	Level 2	Level 3
Built up	Buildings/roofs	Composite shingle roof
		Gravel roof
		Metal roof
		Asphalt roof
		Tile roof
		Tar roof
		Wood shingle roof
	Transportation areas	Asphalt roads
		Concrete roads
		Gravel roads
		Parking lots
		Railroad
	Other built surfaces	Tennis court
		Basketball field
Vegetation	Green vegetation	
	Non-photosynthetic vegetation (NPV)	
Non-urban bare surface	Bare soil	
	Bare rock	
Water bodies	Natural/quasi-natural water bodies	
	Swimming pools	

	1: Com_sh	2: Grav_rf	3: Tar_rf	4: Gr_tile	5: Rd_tile	6: Wd_sh	7: Asp_rd	8: Concr	9: Grav_rd	10: P_lot	11: Gr_veg	12: NPV	13: Soil_dk	14: Soil_be
1: Composite shingle		56	19	14	75	61	8	18	106	13	80	70	133	285
2: Gravel roof	405		36	46	109	189	51	17	88	84	97	52	184	480
3: Tar roof	190	599		30	69	127	17	20	135	26	66	58	145	285
4: Gray tile roof	92	178	67		34	32	35	16	61	31	59	31	99	237
5: Red tile roof	549	581	559	375		84	90	52	147	130	92	59	248	748
6: Wood shingle roof	315	359	171	172	197		218	31	152	249	119	10	378	899
7: Asphalt road	244	693	119	99	1331	351		28	68	7	97	64	48	91
8: Concrete road	687	735	1325	423	1247	977	1151		29	11	59	42	27	20
9: Gravel road	2533	2514	1733	2460	927	4370	3047	1799		117	79	105	485	632
10: Parking lot	194	700	98	81	1499	436	194	897	3832		53	171	104	278
11: Green veg.	992	1066	1023	779	609	426	1614	1589	1106	588		88	64	144
12: Non-photos. veg.	585	646	439	366	511	156	880	887	2288	953	1266		72	84
13: Bare soil (dark)	438	627	330	230	652	542	542	840	2196	801	638	731		218
14: Bare soil (beach)	1152	780	1145	477	1568	1073	1413	1035	1249	1614	889	881	354	
Coding of values:	Bold: Average separability (lower left part of matrix) / *Italic*: Minimum separability (upper right part of matrix)													
Coding of background:	Average value ≤ 150 / Minimum value ≤ 20					151 ≤ Average value ≤ 300 / 21 ≤ Minimum value ≤ 40								

Fig. 4.4   Matrix of B-distance values for minimum and average spectral separability between different urban land cover types derived from the Santa Barbara urban spectral library (Herold et al. 2004)

or derived from sample areas in remote sensing image data. This spectral separability technique calculates the spectral distance for all samples of a class to the ones of another category. The result is an average and minimum separability score between these two classes with higher B-distance values reflecting larger spectral separation. The resulting matrix in Fig. 4.4 containing the separability scores for several urban land cover types are presented in Table 4.1.

The separability scores emphasize the generally higher degree of separability for green vegetation, gravel roofs and roads, and red tile roofs. Lower average and lower minimum separability are obvious between asphalt roads, parking lots, and specific types of roof materials such as composite shingle, tar and gray tiles. These classes are spectrally very similar on the material scale. The spectral similarity of these classes has already been observed in their spectral signatures (Fig. 4.2). Concrete roads have fairly high average separability scores for all classes. However, their low minimum separability with specific classes such as bare soil, parking lots and several roof types emphasizes some degree of spectral similarity caused by the large heterogeneity of concrete road surfaces. Figure 4.4 also emphasizes that the spectral confusion of road materials is mostly with non-road surface types, e.g., specific roofs, bare soil and parking lots. Wood shingle roofs and NPV represent another pair of classes with good average but low minimum separability. These materials are similar and some confusion between these classes may be observed in remote sensing data (Fig. 4.5, see also Fig. 4.3). Basically, the analysis of spectral separability quantifies low spectral separation between some urban land cover types. Particularly, large confusion exists between specific roof types, and between some road materials, specific roof materials, bare

some urban land cover types can hardly be separated based on their spectral characteristics

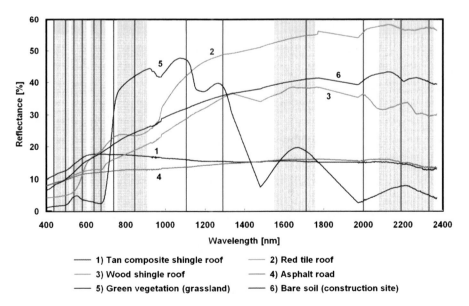

Fig. 4.5 Most suitable spectral bands for urban mapping derived from the ground spectral library and hyperspectral AVIRIS data compared to spectral signatures of several urban land cover types and the spectral coverage of LANDSAT ETM satellite sensor (*gray in the background*). The major water absorption bands (1,340–1,480 and 1,770–1,970 nm) are linearly interpolated

soil and parking lots. These land cover types might only be derived with insufficient accuracy from spectral remote sensing mapping.

4.5   Most Suitable Wavelengths in Mapping Urban Land Cover

One of the purposes of spectrometry is to identify the most important wavelengths or spectral bands for mapping urban areas. This helps to evaluate limitations of current multispectral systems or to design appropriate sensor systems, and is required for specific image analysis algorithms. In land cover classification, for example, having 224 bands from a sensor like AVIRIS provides "too much" spectral information. Map accuracy can actually decrease if too many spectral bands are applied for such purposes (Landgrebe 2000). The derivation of the most suitable bands also uses the B-distance spectral separability measurements. The B-distance values identify the wavelengths that contribute the most spectral contrast and rank the band combinations that are optimal for separating between the land cover classes. Related investigations of a ground spectral library and AVIRIS data resulted

most suitable spectral bands can be selected from hyperspectral datasets based on measures of spectral separability

in a set of most suitable bands that allow the greatest spectral separability of urban land cover classes (Herold et al. 2003). In this case, the number of most suitable bands was 14. This number was sufficient for this study, but does not imply that exactly 14 bands should always be selected for hyperspectral urban mapping. In this context, it is more important to look at the location of these bands. They appear in nearly all parts of the spectrum with a fair number in the visible region (Fig. 4.5). Narrow spectral bands are important in resolving small-scale spectral contrast in the visible spectrum, e.g., from color, iron absorption features. Additional bands appear in the near and short wave infrared. They represent the larger dynamic range of reflectance values related to the increase in object brightness towards longer wavelengths for several land cover types (e.g., tile roofs, wood shingle roofs, vegetation, soils, gravel surfaces). There also are specific absorption features that correspond to some of the most suitable bands (Herold et al. 2003).

4.6   Effects of Spectral Resolution on Urban Land Cover Mapping

The distribution of most suitable bands indicates that some of them are located outside or near the edges of the Landsat TM spectral configuration. This indicates possible limitations of this and similar sensor systems in mapping urban areas. To evaluate the effects of spectral sensor resolution on urban mapping accuracies,

Table 4.2 Overall classification accuracies for mapping 26 urban land cover classes in three different spectral sensor resolutions in 4 m spatial resolution (simulated IKONOS and Landsat TM, and AVIRIS data represented by 14 most suitable bands shown in Fig. 5, Herold et al. 2003)

	Mean accuracy (%)	Kappa coefficient (%)	Area weighted accuracy (%)	Built classes accuracy (%)
IKONOS (4 bands)	61.8	60.2	66.6	37.7
Landsat TM (6 bands)	68.9	67.7	75.8	53.9
AVIRIS (14 bands)	73.5	72.5	82.0	66.6

multispectral sensor configurations (e.g., IKONOS and Landsat TM) can be simulated from hyperspectral data. The results for simulated IKONOS and Landsat TM data, and AVIRIS data using 14 most suitable bands classifying 26 urban land cover classes (Maximum Likelihood classification) are shown in Table 4.2.

The classification results for simulated IKONOS and Landsat TM versus AVIRIS (selected 14 bands) show the improvement in map accuracy with increasing spectral resolution. The area-weighted accuracies are higher since major parts of the study area are covered with vegetation; these areas

**landcover map-
pings accuracy
increases with
increasing spectral
resolution**

map with little error. The differences between IKONOS and AVIRIS accuracy ranges from ~12% for the mean overall accuracy and Kappa to ~15% for the area weighted accuracy and nearly 30% for only the built categories. The improvements for built-up class mapping clearly confirm the limitations of IKONOS in detailed separation of urban land cover types. Landsat TM data provided intermediate accuracies with much better performance than IKONOS in particular for the built land cover classes; the improvements in classifying these cover types was more than 16% compared to IKONOS. This especially highlights the importance of the SWIR region. The four bands in the VIS/NIR region of IKONOS and Landsat TM have similar spectral coverage so the two additional SWIR bands in Landsat TM are the main reason for the difference in classification accuracy (Table 4.2).

As expected, AVIRIS with the 14 most suitable bands shows the best mapping performance. While IKONOS and Landsat TM are broadband multispectral systems, the narrow AVIRIS bands can resolve small-scale absorption features and the increasing number of bands better separate more cover types. In general, the spectral sensor characteristics of IKONOS and Landsat TM were designed for mapping a variety of land surfaces, especially for acquisition of natural and quasi-natural environments. Different spectral sensor configurations are anticipated to resolve the unique spectral properties and complexity of urban environments. However, the AVIRIS land cover classification of 26 different urban land cover classes illustrated general limitations in mapping the urban environment even using hyperspectral optical remote sensing data. These limitations reflect the similar spectral characteristics of certain land cover types indicated in interpretation of spectral signatures and separability analysis. Considering the high degree of within-class variability due to roof geometry, condition, and age, their classification accuracy was low, reaching only 66.6% for the built up categories (Table 4.2, Herold et al. 2003).

4.7   Effects of Spatial Resolution and Three-Dimensional Information

One approach for reducing spectral confusion between some land cover types would be to incorporate a third dimension into the analysis. Urban areas have a distinct three-dimensional structure that can be mapped with LIDAR systems (Jensen 2000). LIDAR (LIght Detection and Ranging) uses a laser pulse (usually in the NIR) to measure the time distance from the sensor source to the reflecting object. Based on the position of the sensor and the pointing direction, the LIDAR signal can be used to accurately calculate the three-dimensional position and reflectance characteristics of the object. The LIDAR pulse is first reflected on the top of the surface object (*first return*) representing the object elevation (tree top or top of buildings). The *last return* LIDAR signal is similar to the first one if the surface is flat (e.g., parking lot). Differences between first and last return appear if the sensed surface is rough or the LIDAR beam partly penetrates through the surface material, e.g., vegetation is partly transparent for infrared radiation. In this case the *last return* elevation signal represents the ground elevation in contrast to the first return that provides the surface signal (Fig. 4.6). For

topographic variations and surfaces structure can be incorporated in the classification to reduce the spectral confusion among urban land cover types

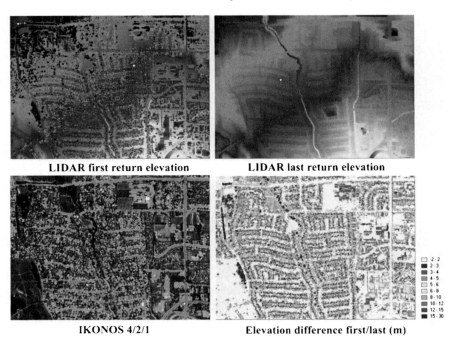

LIDAR first return elevation   LIDAR last return elevation

IKONOS 4/2/1   Elevation difference first/last (m)

Fig. 4.6 Examples of the LIDAR data (first and last response elevation and difference) compared to IKONOS false color composite

urban mapping, the building elevations can be removed in the last return signal using different acquisition and processing techniques, i.e., minimum filters, large footprint LIDAR data or existing ground elevation models. Then the difference between the first and the last response elevations normalizes large-scale topographic variations and emphasizes the three-dimensional surface structure of the urban environment from buildings and vegetation. The LIDAR elevation difference can be used as a pseudo spectral band in image classifications.

Another important consideration for reducing spectral confusion among urban land cover types is spatial resolution. In fact, spatial and spectral resolutions are strongly related since a distinct spectral signal can only be acquired if the spatial resolution is sufficiently fine enough to represent the land cover object in "pure" pixels. Figure 4.7 presents classification results to highlight spatial-spectral resolution effects, and the contribution of LIDAR information on the urban land cover mapping process. In Fig. 4.7, the producers accuracy measures the percentage of test areas of a specific land cover type sampled in the real world that were classified right. The users accuracy describes the percentage of all areas classified as a specific class, and actually belong to this class in the sampled test data (Jensen 2000).

AVIRIS (14 most suitable bands) data shows significant spectral improvements as indicated in the previous sections. However, including the LIDAR elevation difference in the image classification provides an additional important level of information. Especially for mapping buildings/roofs, the combination IKONOS/LIDAR resulted in better classifications than using AVIRIS data. For most classes, the accuracy variations for different sensor configurations are larger than the changes in spatial resolutions. The producers accuracy of green vegetation is high for all sensor configurations and spatial resolutions. The unique spectral signal of vegetation is well represented in all sensor signals and, in terms of the producers accuracy, allows for very accurate classification results. The green vegetation users accuracy on the other hand shows a tremendous decrease in accuracy for all sensor types especially from 4 to 10 m spatial resolution, i.e., vegetation gets increasingly over-mapped at coarser spatial resolutions. Pixels adjacent to green vegetation areas increasingly merge with non-vegetation land cover types and form mixed pixels. The strong spectral vegetation signal leads to increasing amounts of vegetation being classified in the image. This trend is evident for all sensor configurations and reflects the general limitations of lower spatial resolution data in mapping urban land cover. For coarser spatial resolution, the use of spectral mixture analysis helps to map urban land cover on the sub-pixel level (Rashed et al. 2001).

the quality of urban land cover mapping strongly depends on the spatial and spectral characteristics of the remote sensing data

In contrast to the other classes, bare soil classification accuracies show improvements for lower spatial resolutions. Bare soil usually represents areas with larger spatial extents that do not require high spatial resolutions for their accurate mapping. The users accuracy also indicates the importance of the detailed spectral information for accurate separation of bare soil from other land cover types. The signal from IKONOS, and IKONOS and LIDAR elevation difference, is quite limited in this

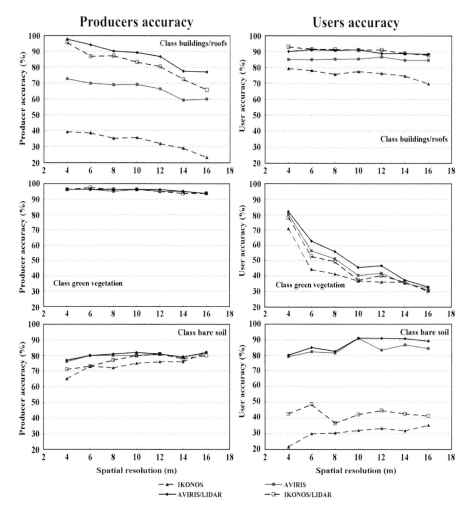

Fig. 4.7 Producer and user classification accuracies for three land cover classes and four different sensor configurations with varying degraded spatial resolutions. All original sensor resolutions are 4 m with IKONOS four multispectral bands, the AVIRIS 14 most suitable bands, and the LIDAR data with the difference between the first and the last response elevations

context and results in significant over-mapping of bare soil. In general, the results of Fig. 4.7 indicate the importance of fine spatial resolution data for detailed mapping of urban land cover. This fact is emphasized in Fig. 4.8, which shows the overall classification accuracies and the KAPPA coefficients. Unquestionably, the combination AVIRIS/LIDAR provides the best classification result with an overall decrease of 11% from 4 to 16 m spatial resolutions. The decrease in accuracy from 4 to 16 m of the AVIRIS data is only 7%. For the combination of IKONOS/LIDAR, on the other hand, this change is nearly 20% overall accuracy. At 16 m spatial

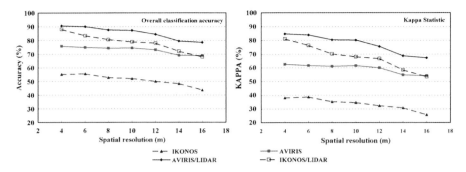

Fig. 4.8 Overall accuracies and KAPPA coefficient for different sensor configurations and varying spatial resolution

resolution the classification performance of IKONOS/LIDAR drops below the AVIRIS accuracy. Hence, AVIRIS data analyses are less sensitive to changes in spatial resolution. Although the trends certainly vary for different land cover classes, IKONOS and LIDAR data strongly depend on the accurate representation of individual urban land cover objects and should only be used in fine spatial resolution. If only coarse spatial resolution data are available fine spectral resolutions should be preferred for urban land cover mapping.

It should be noted that these results reflect a purely pixel-based spectral mapping perspective. Thus, if the remote sensing mapping objective is focused on the spatial and geometric properties of land cover structures (e.g., the shape and size of buildings), fine spatial resolution data on the order of 3–5 m are required for a clear representation of the urban environment (Jensen and Cowen 1999). Also, the use of object-oriented, segmentation image analysis approaches can add an additional level of information to the image classification and help to resolve some of the limitations shown here for spatial-spectral resolution dependent mapping approaches.

Chapter Summary

This chapter has provided an overview of the spectral character of urban attributes. Urban areas with roofing materials, pavement types, soil and water surfaces, and vegetated areas represent a large variety of surface compositions. The spectra of urban spectral libraries reflect these properties in characteristic spectral signatures and related absorption features. Specific urban land cover types show low spectral separability over the whole spectral range from 350 to 2,400 nm (i.e., specific roof and road types). This results in lower remote sensing mapping accuracies. The most suitable wavelengths for separation of urban land cover emphasized specific spectral features that provide the best separation and highlight the importance of hyperspectral urban remote sensing. Spectral limitations exist for current multispectral sensors where the location and broadband character of the spectral bands only marginally resolved the complex spectral characteristics of urban environments, especially for built surface types. The use of three-dimensional information from LIDAR can help to overcome some of these spectral limitations. Important for mapping urban land cover from remote sensing are data with high spatial resolution. However, hyperspectral data provide the most stable spectral mapping accuracies for coarser spatial resolution data.

LEARNING ACTIVITIES

- Review the main terms and laws related to electromagnetic radiation:
  - Plancks law, Boltzmanns equation, Wiens law
  - Exitance, irradiance, radiance, reflectance and their units
  - Atmosphere: transmission and absorption processes, scattering (Rayleigh and Mie), atmospheric windows
  - Surface interactions: reflection and transmission, absorption (electronic and vibrational processes), Snells law, Fresnels equation
- Download and explore some example spectra (vegetation/soil/build surfaces) from existing spectral libraries:
  - Santa Barbara urban spectral library: www.geogr.uni-jena.de/~c5hema/spec/ieee_fig2.zip
  - USGS spectroscopy lab: http://speclab.cr.usgs.gov/
  - JPL/ASTER spectral library including man made materials: http://speclib.jpl.nasa.gov/
- IKONOS and Landsat (E)TM are commonly used for mapping urban areas. Recapitulate their capabilities and limitations for urban remote sensing for specific applications. In this context discuss multispectral versus hyperspectral remote sensing of urban areas.
- Learn more about concepts and applications of image analysis techniques for hyperspectral data. A suitable software system to do so is ENVI:
  - Spectral angular mapper (SAM)
  - Spectral mixture analysis (SMA)
  - Matched filter analysis (MFA)
- Review major fields of application in detailed urban mapping:
  - Impervious surfaces for flood and urban water quality management
  - Infrastructure and land use for urban planning and management

- ○ Roof types for energy use and fire danger prevention
- ○ Urban energy balance and local climate in the urban boundary layer
- ○ Urban ecology and biotope/habitat assessment

Exercises

- Consider you follow a photon emitted by the sun that travels through the atmosphere, reflects at the earth's surface and is then acquired by a satellite remote sensor (see Fig. 4.1). In this context, discuss the main factors and processes influencing the spectral signal acquired by the remote sensing systems.
- Download a set of example spectra from the Santa Barbara urban spectral library that have been published (see Herold et al. 2003, http://www.geogr.uni-jena. de/~c5hema/spec/ieee_fig2.zip)
- The spectra are best viewed with MS EXCEL or similar programs. Interpret the spectral signatures to gain understanding of their characteristics resulting from known absorption/reflection processes. What similarities and differences exist between the different spectral signatures and how does this affect remote sensing applications?
- You are asked to map the roads within an urban area. What spectral challenges would you face and what kind of remote sensing data would you choose to be successful?
- Wood shingle roofs are important in assessing the fire hazard danger in urban areas. Your local fire department is interested in looking at remote sensing to update their maps. What potentials and limitations of this technology would you offer them?
- From your understanding of spectral urban characteristics, how would you go about mapping urban land uses like residential, commercial and industrial, and recreational from remote sensing data. Is a pure spectral signal sufficient enough to separate them? What other sources of image information might be useful for this task?
- Obviously there is no "standard" spectrum for urban impervious surfaces. Most linear spectral unmixing approaches, however, require a representative endmember spectrum for built areas. How would you approach spectral mixture analysis in urban areas? You might want to look at the work of Rashed et al. (2001) and Wu and Murray (2003) to help you with your answer.

References

Ben-Dor E, Levin N, Saaroni H (2001) A spectral based recognition of the urban environment using the visible and near-infrared spectral region (0.4–1.1 m): a case study over Tel-Aviv. Int J Remote Sens 22:2193–2218

Clark RN (1999) Spectroscopy of rocks and minerals and principles of spectroscopy. In: Rencz AN (ed) Manual of remote sensing. Wiley, New York, pp 3–58

Clark RN, Green RO, Swayze GA, Meeker G, Sutley D, Hoefen TM, Livo KE, Plumlee G, Pavri B, Sarture C, Wilson S, Hageman P, Lamothe P, Vance JS, Boardman J, Brownfield I, Gent C,

Morath LC, Taggart J, Theodorakos PM, Adams M (2001) Environmental studies of the World Trade Center area after the September 11, 2001 attack. U. S. Geological Survey, Open File Report OFR-01-0429, http://speclab.cr.usgs.gov/wtc/. Accessed Feb 2009

Herold M, Gardner M, Roberts DA (2003) Spectral resolution requirements for mapping urban areas. IEEE Trans Geosci Remote Sens 41:1907–1919

Herold M, Roberts DA, Gardner M, Dennison P (2004) Spectrometry for urban area remote sensing: development and analysis of a spectral library from 350 to 2400 nm. Remote Sens Environ 91:304–319

Jensen JR (2000) Remote sensing of the environment: an earth resource perspective. Prentice Hall, Upper Saddle River, NJ

Jensen JR, Cowen DC (1999) Remote sensing of urban/suburban infrastructure and socio-economic attributes. Photogramm Eng Remote Sens 65:611–622

Landgrebe, DA (2000) Information extraction principles and methods for multispectral and hyperspectral image data. In: Chen CH (ed) Information processing for remote sensing. World Scientific Publishing Co., http://dynamo.ecn.purdue.edu/~landgreb/whitepaper.pdf. Accessed Feb 2009

Rashed T, Weeks JR, Gadalla MS, Hill A (2001) Revealing the anatomy of cities through spectral mixture analysis of multispectral imagery: a case study of the Greater Cairo region, Egypt. Geocarto Int 16(4):5–16

Roberts DA, Herold M (2004) Imaging spectrometry of urban materials. In: King P, Ramsey MS. Swayze G (eds) Infrared spectroscopy in geochemistry, exploration and remote sensing, vol 33. Mineral Association of Canada, Short Course Series, London, Ontario, pp 155–181, http://www.ncgia.ucsb.edu/ncrst/ research/pavementhealth/urban/imaging_spectrometry_of_urban_materials.pdf

Schowengerdt RA (1997) Remote sensing: models and methods for image processing. Academic, San Diego, CA

Swayze GA, Clark RN, Goetz AFH, Chrien TG, Gorelick NS (2003) Effects of spectrometer band pass, sampling, and signal-to-noise ratio on spectral identification using the Tetracorder algorithm. J Geophys Res 108(E9):5105. doi:10.1029/2002JE001974

Wu C, Murray AT (2003) Estimating impervious surface distribution by spectral mixture analysis. Remote Sens Environ 84:493–505

Chapter 5
The Spatial and Temporal Nature of Urban Objects

Richard Sliuzas, Monika Kuffer, and Ian Masser

The purpose of this chapter is to examine, from an application perspective, the utility of remote sensing to collect data on urban and suburban areas for Urban Planning and Management (UPM). Specifically, the chapter discusses the use of remote sensing at two different spatial levels, the information needs with respect to monitoring planned and unplanned development, and the optimal spatial and temporal requirements for images used in this regard.

Learning Objectives

Upon completion of this chapter, you should be able to:

❶ State the spatial and temporal nature of different types of urban objects and attributes

❷ Describe the relationship between the characteristics of urban objects and attributes

❸ Explain how to select the image source with the most appropriate spatial and temporal properties for the acquisition of specific kinds of urban data

---

5.1 Introduction

Urban Planning and Management (UPM) is a professional domain that seeks to guide, coordinate and regulate urban development processes. It entails activities in which spatial data plays a vital role. UPM processes typically operate at two different

R. Sliuzas (✉) and M. Kuffer
Faculty of Geo-Information Science and Earth Observation of the University of Twente, Hengelosestraat 99, Enschede 7514 AE, The Netherlands
e-mails: sliuzas@itc.nl; kuffer@itc.nl

I. Masser
Centre for Advanced Spatial Analysis, University College London, 1-19 Torrington Place, London, WC1E 7HB, UK
e-mail: i.masser@ucl.ac.uk

T. Rashed and C. Jürgens (eds.), *Remote Sensing of Urban and Suburban Areas*, Remote Sensing and Digital Image Processing 10, DOI 10.1007/978-1-4020-4385-7_5, © Springer Science+Business Media B.V. 2010

spatial levels: that of strategic urban planning; and that of detailed local area planning (also known as neighborhood planning) in which site development and control functions are performed. Furthermore, UPM draws a distinction between planned and unplanned development. The aforementioned distinction is crucial because it provides a basis for exploring and understanding the spatial and temporal characteristics of the typical objects (and hence the kinds of remote sensing image data) that need to be considered at both levels of planning. Broadly speaking, the strategic planning level requires spatial data with lower spatial and temporal resolutions than the local area planning and development control level. Nevertheless, a detailed examination of the specific activities is needed to determine the optimal spatial and temporal requirements for a given situation.

urban planning and management operate at two different levels, strategic and local, that provide a basis for exploring the requirement of RS imagery needed for monitoring urban phenomena

5.2   Planning and Management of Urban Development

UPM is a discipline that seeks to create institutional frameworks and regulatory processes that can lead to the creation and maintenance of high-quality, sustainable urban environments. Providing and maintaining basic urban infrastructure (e.g., roads, and networks for energy, communication, water supply and sanitation), and regulating the construction and use of buildings and land are key UPM process components that are concerned with changes in the physical or built environment. As physical urban development generally entails considerable private and public investments with long economic life spans, formal planning and coordinated development is generally desirable.

5.2.1   A Model of Urban Development and Planning

The urban planning process seeks to allocate available and developable land for urban functions, facilitate its servicing, and regulate access to the land by various public and private actors. These actors then pursue their development goals within the spatial development framework or plan.

The entire urban planning process is generally divided into two phases (Baross 1987):

• Phase 1 – City Development: Public authorities undertake broad zoning of land (land use planning) and provide trunk infrastructure. Public and private developers consolidate land in preparation for actual site development.

• Phase 2 – Site Development: Detailed planning of subdivision layouts and land use allocation (P), provision of minor infrastructure to service individual plots (S), construction of required buildings (B), and occupation by the users (O).

developable land includes greenfield locations at the edge of urban areas, brownfield areas (e.g., derelict industrial sites), and land for other special projects

Figure 5.1 shows how the two phases are incorporated in the PSBO model (Planning-Servicing-Building-Occupation), a formal model that is the basis of most urban planning systems. The PSBO model consists of a sequential series of processes designed for the orderly conversion of rural land to urban land. Phase 1 activities are usually performed at a cycle of approximately 10 years and are used to guide a series of subsequent site development cycles throughout Phase 2. Decisions made during each phase are based on a set of norms and regulations that together define the minimum acceptable standards that new development must comply with before the approved activity can formally commence.

Each phase has different requirements for spatial data. In the city development phase, which is more strategic in nature and concerned with establishing the main spatial framework for future development, the focus is on general descriptions of land use, environmental conditions, and main infrastructure elements. The city development phase thus requires data that describes existing conditions and also projects the demands of future residents and organizations for land, services, and employment. Such data is typically at a relatively small scale (1:10,000–1:25,000), depending on

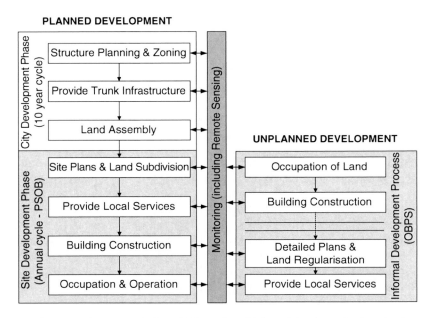

Fig. 5.1 Phases and processes in formal (planned) and informal (unplanned) urban development

the size and complexity of the city, and usually includes the environmental conditions, the location and nature of roads and other trunk infrastructure networks, past and present land use, as well as existing plans and policy areas. In contrast, the site development phase requires more detailed data as this phase relates to the detailed planning and realization of projects that provide serviced land for development. Such data is typically collected at a large scale (1:500–1:1,000), and includes plot boundaries, buildings, encumbrances on land use, detailed zoning or planning ordinances, etc.

In principle, relevant detailed datasets are maintained and updated by the development control authority, both throughout the site development phase and following its realization, so that some generalized and aggregated data can be provided to support the next strategic plan cycle. This does not mean that all data required for the general city development phase can be obtained from such detailed datasets but a considerable amount of data exchange should be feasible. The need for specific monitoring activities, including the use of remote sensing-based approaches, is also shown in Fig. 5.1. The monitoring function should exploit, where possible, opportunities for exchanging data between the city and site development phases (as shown by the arrows connecting the various activities).

A Model for Informal Development

The PSBO model is based on a set of assumptions about the behavior of actors in urban development that may bear little resemblance to their actual behavior. It therefore requires modifications especially when it is adopted in developing countries where a parallel informal development process develops and may even be the dominant form of development. In general, it can be said that informal development lacks the formal approval of the public authorities in one or more of the following terms: land occupancy, orderly layout, building construction, land use, or servicing, and as such can exhibit varying degrees of informality. The informal process is more flexible and enables low-income groups to find affordable shelter in urban areas that is impossible to find via the formal urban development process. The informal approach is, the OBPS model (see Fig. 5.1), in which land occupation and building precede planning and servicing. The first stage, land occupation, may take a variety of forms. In Latin America and South Africa often large, organized over-night land invasions occur while cities in Sub-Saharan Africa tend to develop through a gradual incremental process of land transactions between traditional rural land owners and urban households. As the O-B components are not regulated through development control procedures the linkages to formal data sets are only established if and when a decision is made to upgrade an informal settlement or when city-wide mapping or monitoring is performed. This may be infrequent, and Abbott's (2001) reference to informal settlements as "holes" in the cadastre, is a rather fitting description in many cases. Where informal processes are the main avenue for urban development, the data available for both levels of planning will be partial at best and could be a major barrier for effective decision making and urban management.

5.3   Basic Objects, Attributes, and Planning Level

The implications of planned vs. unplanned urban development are shown in Fig. 5.2 for a piece of vacant land that is urbanized over a period of time. Planned development begins with site planning, land subdivision, and provision of infrastructure. In this process, land is often cleared of much of its vegetation, and earth works give rise to expanses of bare soil. Roads are constructed and surfaced, and are generally easily discernible on remote sensing images as linear or area features depending on the sensor's spatial resolution. Thereafter, building construction commences, which is followed by occupation and completion of landscaping. The process is often carried out in phases (see Time 5 in Fig. 5.2, which shows the extension of the original site

local knowledge of urban development processes is crucial for the successful extraction of urban data from remote sensing images

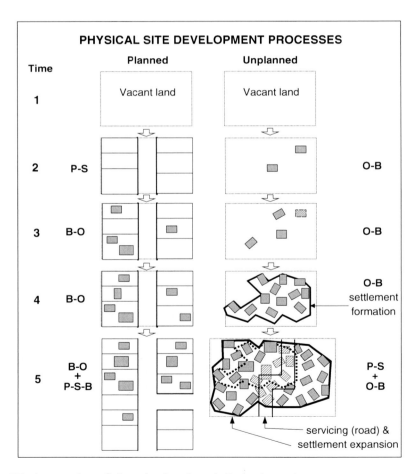

Fig. 5.2 A comparison of planned and unplanned site development processes

development to neighboring land). The distinctly defined spatial structure with the ordered distribution of roads, buildings, etc., clearly distinguishes planned areas from most unplanned areas.

Unplanned development often commences with land occupation and the hasty erection of makeshift buildings, which are gradually improved or replaced by more permanent structures as resources allow and provided there is no expectation of demolition (Fig. 5.2). Over time, settlements are formed, and as the time passes they may be upgraded through the provision of roads and other infrastructure that require the demolition of some buildings.

Different types of informal development are found in different contexts. In Peru and other Latin American countries, well-organized invasion (occupation) of public land by hundred of families has been known to occur overnight (Hardoy and Satterthwaitec 1989). In contrast, most cities in Tanzania experience extensive incremental, unplanned development (Sliuzas and Brussel 2000; Sliuzas 2001, 2004). For example, in Dar es Salaam informal settlements expanded by 5,600 ha (9% p.a.) while planned residential areas expanded by only 780 ha (2% p.a.) between 1992 and 1998, respectively (Sliuzas 2004).

In well-organized land invasions, care is taken to create an informal plan with a clear road pattern and plot demarcation as a means of decreasing the risk of demolition. This strategy has had some success. For urban monitoring purposes using remote sensing, it is essential to be aware of such contextual information as informal settlements in Peru may appear much like formally planned developments on images. Local knowledge therefore remains an important element in the successful extraction of urban data from remote sensing images.

5.4 Area-Based and Object-Based Approaches to Urban Data Extraction

Traditionally, urban remote sensing applications have focused on classifying areas of homogeneous land cover (surface material) or land use (function) (see related discussion in Chapters 4 and 6). However, cities are so complex that large areas of homogeneous land cover often cannot be readily detected, even when using high-resolution images. Most urban land uses are associated with surfaces that are characterized by combinations of various kinds of land cover: buildings, vegetation, roads, water, bare soil, etc. It is therefore not possible **there exists a many-to-many relationship between land cover and land use in cities** to relate one specific form of land use to one specific form of land cover (see Chapter 4 in this volume). The many-to-many relationship between land cover and land use (Gorte 1998; Weber 2001; Ehlers et al. 2002) leads to the poor performance of standard automated pixel-based classifications of urban land use.

The identification and delineation of land uses based on visual interpretation of remote sensing images by trained human interpreters, who supplement the spatial and spectral image information with contextual information, is therefore often preferred. Even so, the identification and delineation of homogeneous land use units is subjective. Further, the position of the unit boundaries and the nature of the mixed uses within each unit are difficult to precisely define or describe. These problems are particularly apparent when a group of interpreters works on the same area or when multi-temporal data extraction is performed.

Mapping the morphology of urban areas, which intuitively appears to be a rather straightforward process, entails complex technical issues related to the state of development at a given location and time. For example, new urban areas where servicing and building are occurring (see Fig. 5.2) tend to have a high spatial and spectral variance due to the great variety of land cover types within relatively small areas. Nevertheless, several authors emphasize the usefulness of remote sensing data for detecting and measuring elements related to urban morphology (Mesev et al. 1995; Webster 1995; Yeh and Li 2001). Yeh and Li (2001) used the concept of entropy to analyze urban sprawl and different growth patterns in the Pearl River Delta, China. Entropy is a measure of disorder within a certain system and has been used by Yeh and Li (2001) to measure the degree of spatial concentration or dispersion of urban sprawl. Low entropy relates to concentrated development while high values indicate more scattered development patterns.

> low entropy value relate to concentrated development while high values indicate more scattered development patterns

Since the availability of high spatial resolution data (<5 m), feature recognition- and object-based approaches for data extraction (application examples of these approaches are given in Chapter 10) are becoming increasingly important in urban applications. A basic consideration in these approaches is the ability to recognize and demarcate discrete/individual objects (Laurini and Thompson 1996). An object with a discrete spatial extent such as a building can be detected and demarcated pending on the spatial resolution of the image (e.g., a large building of 50 × 50 m can be delineated in an image of 5 m spatial resolution or smaller, whereby the accuracy and precision of the delineation improves with increasing spatial resolution). However, the ability to detect and demarcate an object is also affected by other properties of the object. For example, a building with highly reflective roof material such as corrugated galvanized iron sheets, may be difficult to detect if the surrounding environment of the building is bare sand.

The most recent techniques for object extraction and classification typically use spectral information from individual pixels in conjunction with information on the texture, shape, color and/or height properties of the objects of interest (Thurston 2002). A multi-resolution, multi-sensor approach using such characteristics is a feature of some recent work. For example, Ehlers (2002) uses a hierarchical approach that combines existing GIS data with elevation data and multi-spectral imagery, obtained simultaneously from the TopoSys II system, to develop methods for object

> feature recognition and object-based approaches are becoming increasingly important in urban remote sensing applications

extraction from an urban environment. Object-based approaches for identifying and classifying land use objects are also being developed (Zhan et al. 2002; Zhan 2003). In the field of transportation, Tao et al. (1998) used an object-based approach to create a road network database containing information on road surface conditions for inspection and maintenance. Such approaches are also now being explored to map and monitor informal areas and slums (Niebergall et al. 2007; Sliuzas et al. 2008).

5.5 Data Sources for Urban Applications

The usefulness of different approaches is highly dependent upon the data sources that are available. Basically, four generations of sensors for urban studies can be distinguished: first-generation *low-resolution* sensors such as LANDSAT MSS (80m); second and third generation *medium-* to *high-resolution* sensors such as LANDSAT TM (30m), SPOT 4 (10–20 m), SPOT 5 (5–10 m), or IRS (5.8–23 m); and most recent, fourth-generation *very high-resolution* sensors such as IKONOS and QUICKBIRD (1 m and less) (Donnay et al. 2001; Chapter 7 in this volume). Since the availability of high and very high resolution sensors the interest for using remote sensing data for urban application has increased (Ehlers 2002), because these sensors now facilitate the identification of urban objects, such as individual buildings and details of road networks (Brussel et al. 2003).

Considerable interest is also being shown in ultra-high resolution data from airborne platforms, laser scanners, and digital cameras. For example, Small Format Aerial Photography (SFAP) is used for rapid, low cost data capture (Sliuzas 2004). Laser data is also used for obtaining a high-resolution digital terrain model (DTM), including 3D-models of cities (Vosselman et al. 2005). Furthermore, the use of laser data to detect changes on buildings and other urban objects has been explored in the recent study of Steinle and Baehr (2002).

5.5.1 Selection of an Appropriate Resolution

One of the oldest but still useful schemes for considering the relationship between the spatial resolution of remote sensing data and land use/land cover is that developed by Anderson et al. (1976) (Table 5.1). This scheme divides urban land uses into four hierarchical levels and provides an approximate indication of the sensor resolution required for a given land use/land cover classification. Although this scheme continues to be useful for many remote sensing users, it is primarily concerned with general land use and land cover classes. In contrast, the more

> **Anderson et al. (1976)'s scheme provides an approximate indication of the sensor spatial resolution required for a given land use/land cover classification**

Table 5.1 Land use/cover classification levels (Anderson et al. 1976)

Level	Resolution (m)	Example of class
I	≤100	Built-up urban
II	≤20	Residential, industrial, commercial etc.
III	≤5	Single family units, apartments, etc.
IV	≤1	Additional information e.g. condition of the building

recent work of Jensen and Cowen (1999) incorporates other aspects and categories, including hierarchical object classes.

In order to give a visual impression of how the spatial resolution of a sensor influences object and land use identification in urban areas, several examples of different urban land uses and sensors for the city of Enschede, The Netherlands, are shown in Fig. 5.3 and discussed in Table 5.2. The examples shown in Fig. 5.3 follow a similar work done by Radnaabazar et al. (2004) for Ulaanbaatar, Mongolia.

Clearly, for identifying small urban objects or objects in a complex environment, very high resolution data is a prerequisite. Data of 10 or 15 m spatial resolution may provide an overview of urban areas and general land cover/use classes. However, object recognition requires a minimum of 5 m resolution or less, in addition to any case-specific consideration of other characteristics such as culture or morphology.

The diversity of urban morphology becomes apparent when comparing the formal urban development of Enschede (Fig. 5.3) with various types of urban development found in Dhaka, Cairo, and Dar es Salaam (Fig. 5.4). Informal areas in Dar es Salaam typically consist of single-story buildings that were constructed in an incremental and haphazard manner. On the other hand, many informal areas in Cairo follow the regular pattern of former agricultural fields and contain buildings that are densely packed and frequently exceed 5 floors in height, resulting in extensive shadows (see, for example, the prominent shadows cast by buildings in the 100 × 100 m window). In order to distinguish individual buildings in such cities, very high-resolution images are necessary. This is demonstrated by the examples in Fig. 5.4, which shows images ranging from a spatial resolution of 30 m (LANDSAT ETM+) to 20 cm (SFAP).

object recognition requires very high spatial resolution imagery (minimum of 5 m resolution or less)

In practice, the selection of a particular data source is a compromise between costs, required spatial resolution, date of the image, other image characteristics such as the number of bands, and data availability (Harris and Ventura 1995). The accuracy of a classification (e.g., a land use classification) is highly dependant upon the selected spatial resolution (Welch 1982). The desired accuracy and the required information are therefore valid criteria for the selection of sensor data with an appropriate spatial resolution (Atkinson and Curran 1997).

The spatial resolution required for a given study could be determined by the size of the smallest target objects (see, for example, Forster 1985; Cowen and Jensen 1998).

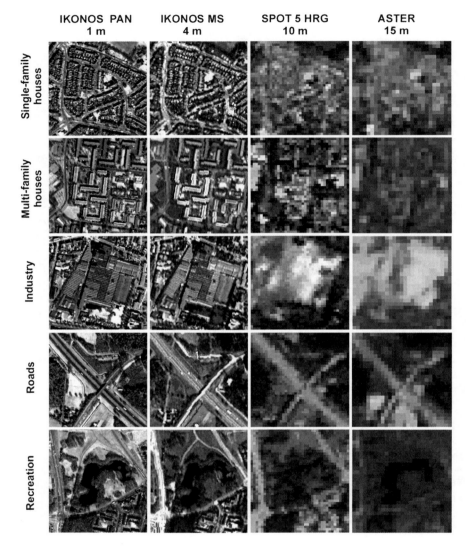

Fig. 5.3 Comparison of urban objects and land uses in Enschede, The Netherlands, by sensor and spatial resolution (each window represents a 400 × 400 m area on the ground)

However, due to several factors, the required spatial resolution is not sufficient to detect urban objects. First, the radiation measured for one pixel is affected by the radiation of its neighboring pixels (scattering), causing a "blurring" effect that complicates land cover classification (Baudot 2001). Second, an object can only be positively identified if it is represented by several pixels. If accurate measurements of an object's spatial properties are required, this must also be considered when selecting an appropriate spatial resolution (Laurini and Thompson 1996).

Table 5.2 Explanation of the usefulness of different sensors for the identification of objects and land use in the city of Enschede, The Netherlands

	IKONOS PAN 1 M	IKONOS MS 4 M	SPOT 5 HRG 10 M	ASTER 15 M
Single family housing	– Shape and size of all buildings can be identified – Land use can be most easily classified	– Small buildings are difficult to identify – Shape of buildings can only be approximated – Land use more easily classified	– Individual buildings are not visible – Only whole neighborhood area can be delimited – Land use is difficult to classify, even with local knowledge	– Less texture information than SPOT – Difficult to delineate residential neighborhoods – Land use is difficult to classify, even with local knowledge
Multi-family housing	– Individual building blocks are visible – Shape and size of buildings is easy to identify – Land use can be most easily classified	– Individual buildings (building blocks?) are visible – Basic shape of buildings can be identified – Land use more easily classified	– Shape of largest buildings can be approximated – Whole neighborhood area can be delimited – Likely land use may be derived from ancillary data	– Multi-family house areas can be identified and whole neighborhood areas delineated – Land use is difficult to classify without local knowledge
Industry	– Individual buildings are visible – Shape of buildings is easy to identify – Land use can be most easily classified	– Individual buildings are visible – Shape of buildings is easy to identify – Land use more easily classified	– General shape and size of large complex is visible – Details are not visible – Land use can be classified	– Shape of big complex can be approximated – No details visible – Land use can be derived if area is part of a larger industrial complex
Roads	– All levels of roads are visible – Lanes and width of roads can be measured – Individual vehicles are visible	– All levels of roads are visible (only minor roads in residential areas may be problematic)	– Major roads are clearly visible – Small and complex road patterns are difficult to identify	– Only major roads can be accurately detected – Small and complex road patterns are mostly not visible
Recreation	– All objects can be clearly identified – Land use can be most easily classified	– Almost all objects can be clearly identified – Land use more easily classified	– Major objects can be identified – Land use can be classified	– Major objects can be identified – Land use can be classified

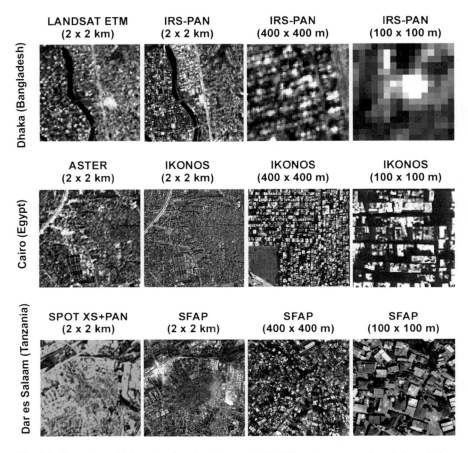

Fig. 5.4 Examples of informal urban development with different sensors and spatial resolution

The ideal spatial resolution of an image for a given application will therefore be several times smaller than the size of the smallest object that needs to be identified.

desired accuracy is a common criterion for the selection of sensor data needed in a given urban analysis application

In this context, it is important to note that average object sizes differ between regions. Welch (1982) suggested that a spatial resolution of 5–10 m is required for performing a reliable urban land use classification in Asian cities, while a resolution of 30 m could be sufficient in the USA. Many studies with medium-resolution data (e.g., 30 m) can be found for cities in the USA (e.g., Gluch 2002; Qiu et al. 2003).

Some authors suggest the use of geo-statistical techniques for selecting an appropriate spatial resolution (Woodcock and Strahler 1987; Atkinson and Curran 1997). The main assumption underlying these techniques is that a scene consists of discrete objects. Consequently, an image resolution that is larger than the object size results in a low local variance, and an image resolution that is similar to the dominant object size results in a maximized local variance. The maximum of the local variance is thus an indication of the object size and can consequently aid in determining a good spatial resolution.

> **spatial resolution required for a given study is determined by the size of the smallest objects that need to be identified**

5.6 The Life Cycle of Planning Processes and Urban Objects

Strategic plans often have a life span of about 10 years and a requirement to be reviewed and updated every 5 years. The review process monitors the implementation of the plans (Masser 1986) and provides an opportunity to use remote sensing data. Site development plans, on the other hand, have very different temporal data requirements, depending on the nature and speed of development. Planned developments are normally facilitated through very detailed planning activities in an initial phase and potentially in intermediate phases depending on the scale of the project. Remotely sensed data can assist in the initial phase of site development processes, and in updating information on the city level after completion. Unplanned developments have higher temporal requirements (varying from days to years) for monitoring. For example, if a planning agency is intervening in an unplanned area (e.g., through an upgrading project), the time span between monitoring will ideally decrease, reflecting an increased level of control of development during such intervention. However, the availability of resources for data acquisition and processing may override considerations.

The temporal resolution of currently operating sensor systems normally used for urban applications ranges from 3 to 24 days. While this resolution is generally sufficient, the availability of usable, cloud-free data may actually be significantly lower. For example, the number of usable images per year may be as low as one for cities in humid climatic zones. In such cases, the use of radar data, which have the ability to penetrate could cover, is an option, either as a single data source (Stabel and Fischer 2001; Grey et al. 2003) or in combination with optical data (Chen et al. 2003). The recent availability of very high resolution radar data from the DLR's TerraSARX system may improve opportunities to reduce the impact of the cloud cover problem.

> **life-cycles of planning processes determine the temporal resolution requirements of remotely sensed data**

With respect to the temporal domain of data, and depending on the remote sensing application, it is also crucial to select imagery acquired during an appropriate season. For example, while winter images with a minimum of vegetation cover are well-suited for topographic mapping in temperate zones, such images may hinder or prevent the classification of land use or the performance of environmental studies. As another example, while the number of useable rainy-season images in tropical areas may be scarce, dry-season images are likely to create problems related to the spectral distinction between highly reflective surfaces (e.g., buildings and bare soil).

5.7   Implications for Urban Applications of Remote Sensing

The spatial and temporal resolutions of remote sensing data have implications for its potential usefulness in urban planning and management. Although aerial photography is still an important remote sensing technology for urban spatial data acquisition, the very high-resolution sensors are now providing interesting alternatives for many spatial data requirements.

Much research is in progress to improve the ability to extract useful data of urbanized areas from these new sensors. The need to deal with the issue of mixed-pixels (e.g., Ridd 1995; Hung and Ridd 2002; Chapters 3, 6 and 8 in this volume) in moderate and even high-resolution images of urban areas remains important. Another problem frequently encountered in urban environments is related to the fact that the accuracy of automated image classifications is still smaller than what could be provided by a human interpreter (Coulter et al., 1999). Barnsley and Barr (1996) also pointed out that, due to the very complex nature of urban areas, even pixel-based classifications of very high-resolution images do not necessarily meet the demands for monitoring urban land use. In fact, the use of higher resolution data can even reduce the accuracy of an automatic urban land use classification. A variety of classification methods such as knowledge based systems, artificial neural networks (Yang 2002) texture and spatial metrics (Herold et al. 2003) and in particular object-oriented feature extraction (Benz et al. 2004) are under development. However, to date, the highest accuracy for urban data extraction and classification is generally still the result of visual interpretation.

In addition to such technical considerations related to urban remote sensing it is worthwhile to also consider the institutional aspects of urban data capture, exchange and use. The MOLAND project seeks to provide a spatial planning tool that can be used for assessing, monitoring and modelling the development of urban and regional environments (http://moland.jrc.it/). Typically, data on urban areas will be collected by several organizations and this creates opportunities for the development of Spatial Data Infrastructures at various spatial levels (Williamson et al. 2003), from local to global. A particularly interesting example of cooperation in Europe is the INfrastructure for SPatial InfoRmation in Europe (INSPIRE) initiative which was launched in December 2001 with a view 'to making available relevant, harmonized and quality geographic information to support formulation, implementation, monitoring and evaluation of Community policies with a territorial dimension or impact.' (http://inspire.jrc.ec.europa.eu/). INSPIRE is seen as the first step toward a broad multi sectoral initiative which focuses initially on the spatial information that is required for environmental policies. A Directive 'establishing an infrastructure for spatial information in the Community' was approved by the European Parliament and the Council Of Ministers (Directive 2007/2/EC) in March 2007. As a result of this legislation all 27 member states are be required to modify existing legislation or introduce new legislation to implement its provisions by May 2009.

Chapter Summary

Urban Planning and Management (UPM) processes are performed at both city-wide, and neighborhood or site levels. Depending on the spatial level, data with different spatial and temporal characteristics are required to support UPM activities. While general land use data is typically required at the city-wide planning level, data on specific objects composing the built environment are required for detailed site planning. The degree of planning and the regional context (e.g., developed vs. developing countries) also influence the data requirements and the usefulness of specific sensors. The very high-resolution remote sensing data that has become available fairly recently has been stimulating and encouraging research into remote sensing methods to obtain/collect data for urban planning and management.

LEARNING ACTIVITIES

Study Questions

Table 5.3 below contains links to internet sites of satellite imagery suppliers. The listed sites provide sample images and valuable information of different sensors, and allow you to search for the availability of images for any location. First examine the sites and then complete the following tasks and answer related questions.

- Which data types and sensors does each site include? Create a list of the image data provided on the sites.
- For each sensor included in the table, find: (a) the swath width (i.e., the area captured by a satellite sensor on a single image); (b) the time span of data availability (this is useful to determine if a sensor can provide data for a time period over which land-use changes occurred, e.g., past few years or decade); (c) the accuracy; and (d) the cost of the data product.
- Select a city of your interest (e.g., your home city). Search the different catalogues provided on the sites and identify available, cloud-free image data. List the potential data for your city of interest, note which part of the city is covered by clouds, and record the exact date of the image acquisition.
- Contact the urban planning agency in your city of interest and request details of its standard land use classification system. Which of the available image data that you identified do you consider most appropriate for the collection of land use data of your city at a 5-year interval? Write down indicators upon which you based your sensor choice and develop arguments to justify your choice.

Table 5.3 Some currently operating space-born sensor systems commonly used for urban applications

System/sensor and Internet link	Spatial and temporal resolution
QuickBird	PAN: 0.6/0.7 m
www.digitalglobe.com	MS: 2.4/2.8 m
	1–3 days
GEOEYE and IKONOS	PAN: 0.5–1 m
OSA (Optical Sensor Assembly)	MS: 1.6–4 m
www.geoeye.com	1–3 days
IRS-1D	LISS: 23.5 m
LISS 3 (linear imaging self-scanning system)	PAN: 5.8 m
www.isro.org	24 (5) days
SPOT 5	VNIR:10 m
HRG (high resolution geometric)	SWIR: 20 m
http://sirius.spotimage.fr	PAN: 2.5–5 m
	5 days
Terra	VNIR: 15 m
ASTER	SWIR: 30 m
http://edcimswww.cr.usgs.gov/pub/imswelcome/	TIR: 90 m
	16 days
Landsat 7; ETM+ (Enhanced Thematic Mapper plus)	Band 1–5.7: 30 m (VNIR and SWIR)
http://glovis.usgs.gov	TIR: 60 m
http://glcfapp.umiacs.umd.edu:8080/esdi/index.jsp	PAN: 15 m
	16 days

References

Abbott J (2001) The use of spatial data to support the integration of informal settlements into the formal city. Int J Appl Earth Obs Geoinformatics 3:267–277

Anderson GL, Hardy EE, Roach JR, Witmer RE (1976) A land use and land cover classification system for use with remote sensor data. USGS Professional Paper, Washington, DC

Atkinson PM, Curran PJ (1997) Choosing an appropriate spatial resolution for remote sensing investigations. Photogramm Eng Remote Sens 63:1345–1351

Barnsley MJ, Barr SL (1996) Inferring urban land use from satellite sensor image using kernel-based spatial reclassification. Photogramm Eng Remote Sens 62:949–958

Baross P (1987) Land supply for low income housing: issues and approaches. Reg Dev Dialogue 8(4):29–45

Baudot Y (2001) Geographical analysis of the population of fast-growing cities in the Third World. In: Donnay JP, Barnsley MJ, Longley PA (eds) Remote sensing and urban analysis. Taylor & Francis, London, pp 225–241

Benz UC, Hofmann P, Willhauck G, Lingenfelder I, Heynen M (2004) Multi-resolution, object-oriented fuzzy analysis of remote sensing data for GIS-ready information. ISPRS J Photogramm Remote Sens 58:239–258

Brussel MA, Belal WE, Rahman MM (2003) Extracting urban road information from IKONOS high resolution imagery. In: Proceedings of the 4th international conference of urban remote sensing, Regensburg, Germany, 27–29 June 2003, pp 29–34

Chen CM, Hepner GF, Forster RR (2003) Fusion of hyperspectral and radar data using IHS transformation to enhance urban surface features. ISPRS J Photogramm Remote Sens 58:19–30

Coulter L, Stow D, Kiracofe B, Langevin C, Chen D, Daeschner S, Service D, Kaiser J (1999) Deriving current land-use information for metropolitan transportation planning through integration of remotely sensed data and GIS. Photogramm Eng Remote Sens 65:1293–1300

Cowen DJ, Jensen JR (1998) Extraction and modeling of urban attributes using remote sensing technology. In: Liverman D, Morna EF, Rindfuss RR, Stern PC (eds) People and pixels: linking remote sensing and social science. National Academy Press, Washington, DC, pp 164–188

Donnay JP, Barnsley MJ, Longley PA (2001) Remote sensing and urban analysis. In: Donnay JP, Barnsley MJ, Longley PA (eds) Remote sensing and urban analysis. Taylor & Francis, London, pp 1–18

Ehlers M, Schiewe J, et al. (2002) Urban remote sensing: new developments and challenges. Remote Sensing of Urban Areas, Istanbul, Turkey. 11–13 June 2002, pp 130–137

Forster BC (1985) An examination of some problems and solutions in monitoring urban areas from satellite platforms. Int J Remote Sens 6:139–151

Gluch R (2002) Urban growth detection using texture analysis on merged Landsat TM and SPOT-P data. Photogramm Eng Remote Sens 68:1283–1288

Gorte BGH (1998) Probabilistic segmentation of remotely sensed images. Published Ph.D. dissertation, Wageningen Agricultural University, Wageningen

Grey WMF, Luckman AJ, Holland D (2003) Mapping urban change in the UK using satellite radar interferometry. Remote Sens Environ 87:16–22

Hardoy JE, Satterthwaitec D (1989) Squatter citizen: life in the urban Third World. Earthscan, London

Harris PM, Ventura SJ (1995) The integration of geographic data with remotely sensed imagery to improve classification in an urban area. Photogramm Eng Remote Sens 61:993–998

Herold M, Liu XH, Clarke KC (2003) Spatial metrics and image texture for mapping urban land use. Photogramm Eng Remote Sens 69:991–1001

Hung MC, Ridd M (2002) A subpixel classifier for urban land-cover mapping based on a maximum-likelihood approach and expert system rules. Photogramm Eng Remote Sens 68: 1173–1180

Jensen JR, Cowen C (1999) Remote sensing of urban/suburban infrastructure and socio-economic attributes. Photogramm Eng Remote Sens 65:611–622

Laurini R, Thompson D (1996) Fundamentals of spatial information systems. Academic, London

Masser I (1986) Strategic monitoring for urban planning in developing countries: some guidelines from British and Dutch experience. Int J Inf Manag 6:17–28

Mesev V, Longley PA, Batty B, Xie Y (1995) Morphology from imagery: detecting and measuring the density of urban land use. Environ Plann A 27:759–780

Niebergall S, Loew A, Mauser W (2007) Object-oriented analysis of very high-resolution QuickBird data for mega city research in Delhi/India. In: Proceedings of urban remote sensing joint event 2007 (URS/URBAN), IEEE, Paris, France

Qiu F, Wolter KL, Briggs R (2003) Modelling urban population growth from remotely sensed imagery and TIGER GIS road data. Photogramm Eng Remote Sens 69:1031–1042

Radnaabazar G, Kuffer M, Hofstee P (2004) Monitoring the development of informal settlements in Ulaanbaatar, Mongolia. In: Proceedings of CORP2004, Vienna University of Technology, Vienna, 25–28 Feb, pp 333–339

Ridd M (1995) Exploring a V-I-S (Vegetation-Imperious Surface-Soil) model for urban ecosystem analysis through remote sensing: comparative anatomy for cities. Int J Remote Sens 16:2165–2185

Sliuzas RV (2001) Understanding the dynamics of informal housing in Dar es Salaam: an exploration of professional knowledge and opinions. TRIALOG 69(2):41–45

Sliuzas RV (2004) Managing informal settlements: a study using geoinformation in Dar es Salaam. Faculty of Geosciences, University of Utrecht, Utrecht, Tanzania

Sliuzas RV, Brussel M (2000) Usability of large scale topographic data for urban planning and engineering applications: examples of housing studies and DEM generation in Tanzania. ISPRS, vol XXXIII, Part B4/3, Commission IV. Amsterdam, 16–23 July

Sliuzas RV, Kerle N, Kuffer M (2008) Object-oriented mapping of urban poverty and deprivation. In: Proceedings Joint Workshop GISDECO 8 and EARSeL special interest group on developing countries, Istanbul, Turkey, 2–6 June

Stabel E, Fischer P (2001) Satellite radar interferometric products for the urban application domain. Adv Environ Res 5:425–433

Steinle E, Baehr HP (2002) Detectability of urban changes from airborne laserscanning data. ISPRS, vol VII. Hyderabad, India, 3–6 Dec 2002

Tao C, Li R, Chapman MA (1998) Automatic reconstruction of road centerlines from mobile mapping image sequences. Photogramm Eng Remote Sens 64:709–715

Thurston J (2002) GIS and remote sensing: new trends in feature and object recognition. http://www.integralgis.com/pdf/GIS%20and%20Remote%20Sensing.pdf. Accessed in Nov 2004

Vosselman G, Kessels P, Gorte BGH (2005) The utilisation of airborne laser scanning for three-dimensional mapping. Int J Appl Earth Obs Geoinf 6:177–186

Weber C (2001) Urban agglomeration delimitation using remote sensing data. In: Donnay JP, Barnsley MJ, Longley PA (eds) Remote sensing and urban analysis. Taylor & Francis, London, pp 145–159

Webster CJ (1995) Urban morphological fingerprint. Environ Plann B 23:279–297

Welch R (1982) Spatial resolution requirement for urban studies. Int J Remote Sens 3:139–146

Williamson I, Rajabifard A, Feeney MEF (eds) (2003) Developing spatial data infrastructures: from concept to reality. Taylor & Francis, London

Woodcock CE, Strahler AH (1987) The factor of scale in remote sensing. Remote Sens Environ 21:311–332

Yang X (2002) Satellite monitoring of urban spatial growth in the Atlanta metropolitan area. Photogramm Eng Remote Sens 68:725–734

Yeh AGO, Li X (2001) Measurement and monitoring of urban sprawl in a rapidly growing region using entropy. Photogramm Eng Remote Sens 67:83–90

Zhan Q (2003) A hierarchical object-based approach for urban land use classification from remote sensing data. Published Ph.D. thesis, Enschede, ITC, University of Wageningen

Zhan Q, Molenaar M, Tempfli K (2002) Hierarchical image-object based structural analysis toward urban land use classification using high-resolution imagery and airborne LIDAR data. In: Proceedings of the 3rd international conference on remote sensing of urban areas, Istanbul, Turkey, 11–13 June 2002, pp 251–258

Chapter 6
The V-I-S Model: Quantifying the Urban Environment

Renee M. Gluch and Merrill K. Ridd

This chapter emphasizes the ecological nature of urban places and introduces the V-I-S (Vegetation-Impervious surface-Soil) model for use by remote sensing to characterize, map, and quantify the ecological composition of urban/peri-urban environments. The model serves not only as a basis for biophysical and human system analysis, but also serves as a basis for detecting and measuring morphological/environmental change of urban places over time.

Learning Objectives

Upon completion of this chapter, you should be able to:

❶ Define an ecosystem and explain urban/peri-urban environments as an ecosystem, with examples from personal experience

❷ Explain the V-I-S Model concept and how it engages remote sensing in quantifying and mapping variations in urban/peri-urban ecosystems over space and time

❸ Explain how spatial and spectral resolutions affect mapping V-I-S patterns in urban/peri-urban environments

❹ Plot and explain V-I-S values on (a) a feature space plot, (b) a linear transect diagram, (c) a linear time-dependent diagram, and (d) a ternary V-I-S diagram, both static and dynamic

❺ Discuss how the V-I-S model may be used to link with environmental/engineering models such as urban heat island and storm runoff

R.M. Gluch (✉)
Department of Geography, Brigham Young University, Provo, UT 84602, USA
e-mail: renee_gluch@byu.edu

M.K. Ridd
Department of Geography, University of Utah, 260 S. Central Campus Dr., Salt Lake City, UT 84112-9155, USA
e-mail: merrillridd@geog.utah.edu

T. Rashed and C. Jürgens (eds.), *Remote Sensing of Urban and Suburban Areas*, Remote Sensing and Digital Image Processing 10, DOI 10.1007/978-1-4020-4385-7_6, © Springer Science+Business Media B.V. 2010

6.1   Introduction

Growing interest in urban systems as ecological entities provides rich opportunities for remote sensing research and application. The wide array of environmental variables within the city and in surrounding landscapes calls for a means of identifying and quantifying land cover types in a systematic and repetitive manner. To the extent this can be done simply and effectively through satellite-borne digital multispectral remote sensing, several positive results may follow: (a) environmental comparisons may be established between cities of the world, (b) dynamic changes of environmental patterns within cities can be mapped and quantified, (c) environmental change consequent to urban expansion into adjacent landscapes can be mapped and quantified, (d) various science and engineering models relating to energy, moisture, and atmospheric quality dependent on land cover can be applied, (e) population, demographic, and other human dimensions of urban places may be estimated, and (f) urban and environmental planning can be enriched by effective, dynamic application of remote sensing.

satellite remote sensing provides a standardized means of identifying and quantifying land cover types in a systematic and repetitive manner

Given that urban land cover features are so complex in terms of size, shape, and physical composition, and so mixed in terms of complex patterns, the challenge to remote sensing is great. In recent years improved spatial and radiometric resolution from satellite-borne sensors has given a boost to mapping in greater detail. However, the challenge of coping with the great variety of land cover types remains (see former discussions in Chapters 4 and 5). In primitive settlements of the developing world, the problem is relatively simple as building materials are less varied. However, in cities of the developed world, an infinite variety of cover materials may be present. Thus the problem of standardizing identification, characterization, quantification, and mapping is a significant matter. If a model for such standardization can be obtained then a global "comparative anatomy for cities" can be accomplished. Further, dynamic changes and environmental impact over time can be measured and mapped.

the complexity of urban morphology in contemporary cities and the great variety of urban materials and land cover types introduce significant challenges to urban remote sensing analysis

In an effort to produce such standardization for urban/peri-urban ecosystems the V-I-S (Vegetation-Impervious surface-Soil) model was introduced (Ridd 1995). This chapter deals with the theory supporting the model, and some applications of the model in different urban settings.

The City as an Ecosystem

To establish a foundation for the V-I-S model we must first establish that the urbanized area and its surroundings, referred to herein as the urban/peri-urban environment, as an ecosystem, in the same sense that any portion of the earth surface, natural and/or human may be considered an ecosystem. The peri-urban area may be thought of as the urban fringe where a dynamic transition is converting pre-urban land to a built environment, while the area beyond is tied to the city through a demand for resources and other activities (Chapters 2 and 3 in this volume, also see Douglas 1983; Dorney and McLellan 1984; Quattrochi 1996; UNFPA 2001).

As in any ecosystem, energy and moisture flux are the primary drivers in environmental dynamics. In the case of urban ecosystems, human-based technologies impose significant influence on the energy/moisture balance through such factors as anthropogenic energy, impervious surfaces, and various structures (Ridd and Hipple 2006). Cities are perhaps the most dramatic expressions of the transformation of natural environments by the hand of humans. Yet, in primitive societies the degree of transformation is minimal and the character of the settlement is quite different from that of highly urbanized places. As settlements grow from early beginnings into villages, towns, cities, and metropoli, the environmental nature of the settlement becomes more complex as technology advances. The reach outward for resources expands as does the problem of disposition of effluent. Remote sensing technology is increasingly capable of monitoring the consequent environmental transformation. Through remote sensing the V-I-S model provides a means of documentation of the imperative and changing nature of human settlements worldwide (Ridd 1995).

6.2   The V-I-S Model

The V-I-S model suggests that the great variety of urban land cover types can be grouped into the three general categories – vegetation, impervious surface, and soil – plus water. These four cover types exhibit highly contrasting influences on the two most important factors in an ecosystem: energy and moisture flux. Variations within each category can be recognized as well by identifying sub-categories of vegetation, impervious surface, soil, and water.

In the context of the model, **V** stands for green vegetation, **I** includes surface pavement as well as impervious roofs or elevated parking structures, and **S** represents exposed soil. Graphically, the three are presented as a ternary diagram, where the sum of all three components is equal to 100% as shown in Fig. 6.1. Water is handled independently. Shadowing cast by vertical

the V-I-S model plays directly in the urban/peri-urban ecosystem

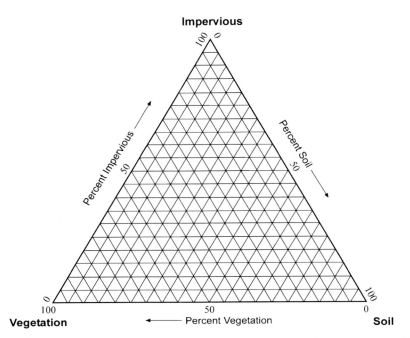

Fig. 6.1 The vegetation-impervious surface-soil (V-I-S) model of urban cover composition (Ridd 1995)

structures can be treated in various ways, for example grouped with impervious surface (especially dark impervious) in areas of multiple stories as commercial and apartment buildings, or with water (especially if water features are rare, etc.).

At this point it is important to emphasize the environmental/ecological basis of the model. Consider the ways in which the three terrestrial cover types and water differ from one another in response to energy and moisture in the ecosystem (see Table 6.1).

It is the variation in energy and moisture response, as they interact with the four surface materials, or sub-categories of them, that gives the model substance as a foundation for environmental or human studies of urban systems. Consider the difference between a central business district (CBD) and a residential area or a city park in terms of temperature. Consider the differing moisture regimes, runoff rates, moisture storage, and the effect of evapotranspiration on cooling. Consider the differences in human habitability and diurnal population dynamics.

The model may be applied at various scales of observation: sub-pixel, pixel, groups of similar pixels, or even the city/settlement as a whole. Depending on the interplay between spatial resolution (pixel size) and spatial complexity of the urban environment (object size), a given pixel may be "pure" (100%) V, I, or S, or, more commonly, mixed – made up of some combination of the three types and/or water. Pure pixels (other than water) occur at the vertices of the diagram. Mixed pixels lie elsewhere in the diagram. A point in the center of the ternary diagram consists of 33% of each of the three cover types. Sampling sites for calibration may be a group

Table 6.1   Comparative energy and moisture exchange rates

	Healthy green vegetation	Impervious surface	Exposed soil	Water
Energy				
Insolation	Drives photosynthesis and evapo-transpiration	Reflects some, but absorbs most[a]; becomes a heat radiator	Generally absorbs and becomes a heat radiator	Absorbs and stores energy; radiates very little
Diurnal heat flux	Minimal	Great	Moderate	Miniscule
Moisture				
Precipitation	Intercepts and stores in plants and soil; retards runoff	Promotes rapid runoff miniscule storage	Intercepts and stores; intermediate runoff rate	Accepts and stores
Evapo-transpiration	Significant transpiration and evaporation	Miniscule evaporation; no transpiration	Moderate evaporation; no transpiration	Considerable evaporation; no transpiration

[a] Depending on brightness

Table 6.2   Sample sites of various urban/peri-urban environments

Sample site	Percent			Description of environment
	V	I	S	
1	8	87	5	Commercial area
2	36	60	4	High density residential
3	65	30	5	Low density residential
4	90	5	5	City park
5	5	65	30	Sparsely vegetated industrial
6	70	5	25	Forest or farm field in crop
7	23	1	76	Desert shrub land
8	10	20	70	Primitive village in arid land

of homogeneous pixels, an area such as a city block, a spatial unit such as a CBD, residential area, golf course, city park, or industrial complex, for example, or an entire settlement. Sampling strategy may be based on random pixels, a grid of pixels, all the pixels in the area, or a linear transect of pixels through an area of interest (AOI).

Table 6.2 shows several hypothetical sites consisting of various combinations of cover composition. Figure 6.2 places each of the sites in the V-I-S model. In each case the combination sums to 100%.

Note that for most built-up urban landscapes (especially in the developed world) there is typically very little exposed soil, usually not more than 5% (aside from some industrial sites or land undergoing conversion). Consequently, the V-I-S difference between sections of the city typically results in a tradeoff between vegetation and impervious surface. Industrial areas may be an exception, especially in arid lands, where exposed soil

the V-I-S model provides a scalable means of revealing the anatomy of a targeted object, whether this object represents a pixel, a neighborhood, or an entire urban region

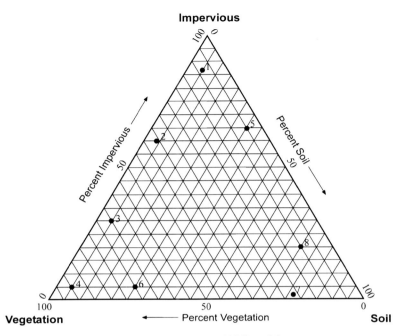

Fig. 6.2 Placement of eight hypothetical sites on the V-I-S model

may be greater than in the rest of the city. Note that sample sites 1–4 represent segments of a typical urban area, site 5 may represent an industrial area in an arid region, site 6 may represent a forest or a farm field with a cover crop, site 7 may represent a desert shrub area, and site 8 may represent a whole village arising in a desert region.

Regarding classification, sites 1 through 4 could be done at the sub-pixel or per pixel (dominant cover type) level. Sites 5 through 8 most likely would be approached at a per-pixel level or groups of pixels in a somewhat homogeneous environment (see Forster 1985).

On a spectral feature space plot the seven sites might appear as in Fig. 6.3, where a visible waveband (e.g., red) is placed on the x-axis and near infrared (NIR) is placed on the y-axis. The commercial core (site 1) with its dominance of impervious surface lies near the non-vegetation line, dark surfaces to the left and lighter surfaces grading to the right. High density residential areas (2) lie above the non-vegetation line, with, in this case, 36% vegetation. Low density residential areas (3) lie closer to the green point on the plot, where a city park may be nearly completely vegetated (site 4). Industrial areas (5) may vary considerably along the I-S axis, not far off the non-vegetation line, with little vegetation, except for so-called industrial parks which may be similar to residential areas. Here again the virtue of the V-I-S model appears, exhibiting the variable environmental character of industrial areas, which land-use maps do not differentiate. A forest or a farm in production (6) rises close to the green point, perhaps exactly on it, if 100% covered. In a densely forested

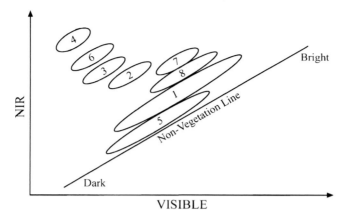

Fig. 6.3 Placement of the eight sites in spectral space, in a visible/near infrared plot

region, soil gives way to vegetation. Desert shrub lands (7) typically are devoid of impervious surface, and are dominated by exposed soil although some desert shrub areas may rise to 60% or more in vegetation cover. Here again variations are documented by the model. A primitive village in arid land (8) usually has little vegetation, and impervious surfaces may be limited to rooftops.

6.2.1   Relationship to Urban Spatial Models

Models of urban form have typically been based on human activities and land use patterns (Burgess 1925; Hoyt 1939; Harris and Ullman 1945). Chapter 2 in this book continues that tradition with more-or-less concentric rings or zones outward from a city core or center to suburbs and beyond to a hinterland. Parameters such as commuting patterns, population density, and employment are used to define the character of the rings outward from the city center. These rings or zones are convenient for discussing V-I-S composition because of our general familiarity with the association of observable land cover with various human factors. However, the V-I-S model is prepared to quantify land cover patterns regardless of variations in rings, zones, radial sectors, or whatever land use or socioeconomic patterns exist.

To simplify and generalize for the present purpose, a typical transition from a commercial core outward might be through a sequence of residential zones of decreasing housing density to the outlying hinterland or peri-urban environment adjacent to the built-up urbanized area (Fig. 6.4). Residences nearest the city center tend to be high density apartment districts with relatively little space for vegetation. In a series of "concentric rings" outward, high density residential areas grade into medium and low

spatial variations in urban land use patterns can be quantified by the V-I-S model in terms of variations in the V-I-S compositions

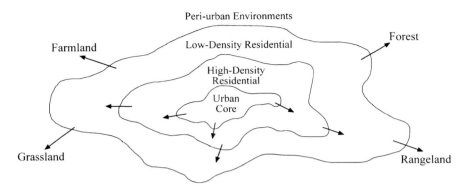

Fig. 6.4 A generalized model of expanding urban spatial patterns

density residential areas. Just as there is typically an inverse relationship between impervious surface and vegetation outward from the core, there tends to be an inverse relationship between housing density and income levels, sizes, and costs of homes. At the urban fringe, scattered patterns of built-up urban features reach into adjacent farmlands or natural landscapes of forest, grassland, wetland, or barren land. Clearly there are numerous exceptions to this generalization, but one of the virtues of the V-I-S model is that whatever the pattern, or changing character of the pattern, the model is well equipped to identify and document the variation or change.

6.2.2   Time and Change

Urban environments are by their nature dynamic places. Clearly, if they are to be chronicled over time, a mechanism by which the change in environmental character can be documented is required. The V-I-S model is well suited to the task. Cities that are growing generally exhibit both internal changes and outward expansion. Internally, impervious surfaces intensify and enlarge around commercial centers, usually replacing vegetation and perhaps some soil as commercial areas invade the adjacent high density residential area in transition. Each concentric ring outward

growing cities exhibit both internal changes and outward expansion

is generally expanding in turn. At the outer fringes of the urban area, the expanding city commonly invades farmland or, in the case of natural environments, the invasion replaces forests in humid areas, grassland in semiarid regions, or desert soils in arid areas, or perhaps wetland. The environmental impact of such invasion is easily and accurately documented by the V-I-S model.

Long term historical change may require reaching back through earlier, low resolution imagery or photography. However, the model still applies. Monitoring contemporary change is likewise implemented with the model. Also, building a time-line as a base for predicting changing patterns is well-served by the model.

Coincidently, the ready integration of these data with other layers of information through GIS may serve many purposes, including forecasting changing patterns of urban environments. That is, factors influencing change and growth, such as terrain, land value, and zoning, etc., may be implemented to predict V-I-S patterns into the future.

How much the (V-I-S) environment changes when cities are built upon pre-urban landscapes depends not just on the character of the prior landscape, but also upon the type of urban form it changes into. Typically the pre-urban environment lies along the V–S axis, from bare soil on the right to 100% vegetation on the left, with little or no impervious surface. This is indicated along the base line in Fig. 6.5. The figure suggests a variety of ways the V-I-S composition may change based on a green landscape prior to urbanization. The most radical change is usually toward an increase in impervious surface. However, that may range **pre-urban environment lies along the V–S axis of the V-I-S model** from 100% impervious in a commercial center (CBD) or highway corridor to a city park at 100% vegetation. Where green fields are cleared for urbanization there is usually a rapid reduction in vegetation and an increase in exposed soil. This is implied by the arrows trending immediately to the right at their origin in the figure. Gradually roads and roofing cover increasing amounts of the landscape. The amount of vegetation that creeps back into the area depends on the type of urban development taking place. For typical industrial and commercial areas, that will be minimal, as they tend toward the I–S axis. Ultimately, residential areas come into full maturity along the V–I axis, with

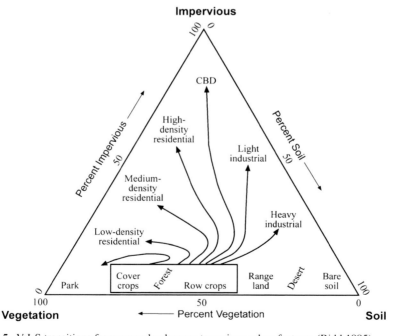

Fig. 6.5 V-I-S transitions from green landscapes to various urban features (Ridd 1995)

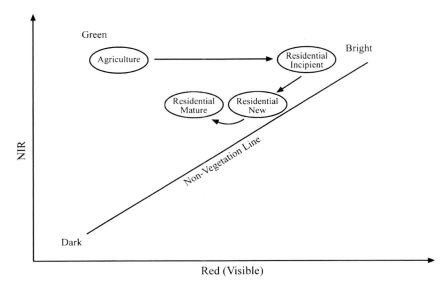

Fig. 6.6 Spectral changes as agricultural land is converted to urban residential land use (Ridd et al. 1983)

increased housing density toward the upper (I) vertex near commercial districts, and grading into low density (typically higher income) residential sites toward the vegetation vertex.

The spectral transition associated with an agricultural to residential development is illustrated in Fig. 6.6 (Ridd et al. 1983). With the clearing of farmland (Agr), the spectral signature moves swiftly to the non-vegetation line, suggesting residential development is about to begin. Roads, driveways, and houses begin to cover bare soil, darkening the signature. As landscaping commences, and eventually matures, the signature migrates back toward the green corner, settling somewhere near the center of the visible/NIR envelope. How near the center depends on, again, the resulting housing density.

Figure 6.7 illustrates the conversion of arid landscapes to urban use assuming water is available to establish sufficient landscaping. The residential sequence is similar but with little or no clearing before development takes place. End results lie along the V-I and I-S axes as well. Cities rising from arid landscapes with insufficient water may end up on a more vertical axis from the I vertex downward toward the center of the ternary diagram.

6.2.3   The Role of Remote Sensing in Urban Systems

As indicated in Fig. 6.8 remote sensing of urban systems provides a basis for investigations in (1) urban morphology, (2) biophysical systems, and (3) human systems.

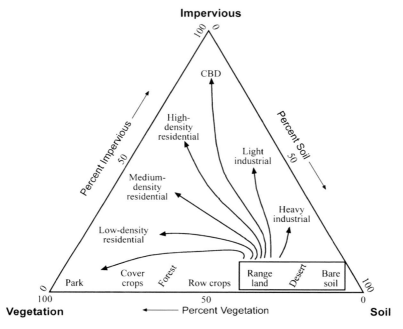

Fig. 6.7   V-I-S transitions from arid landscapes to various urban features

The V-I-S model serves all of these. For example in urban morphology it is recognized that the basic building block is the pixel. However, determination of the composition of a single pixel is usually of little value in and of itself. It is the aggregation of multiple adjacent pixels of similar composition that make ecological analysis possible. The logic by which pixels are grouped into spatial units depends on the objective of investigation. Aggregating similar pixels, or sub-pixel characteristics, into a polygon creates an "eco-unit" (Ridd 1995), an ecologically significant landscape unit emphasizing the biophysical character

the V-I-S "eco-unit" refers to an ecologically significant landscape unit emphasizing the biophysical character as derived from the V-I-S composition

as derived from the V-I-S composition. The eco-unit may represent a "community" as in classical ecology. The central business district (CBD) of a large city is a kind of community or eco-unit dominated by impervious surfaces. Within the CBD there may be a city park or other anomalies quite different from the general character of the unit. Outside the CBD residential eco-units may range from high density to low density and on to surrounding landscapes – each as part of the morphology of the urban/peri-urban region. Based on V-I-S, the composition may be monitored for change and modeled for predictive growth and change over time. Urban morphology becomes the foundation for effective biophysical system and human system analysis.

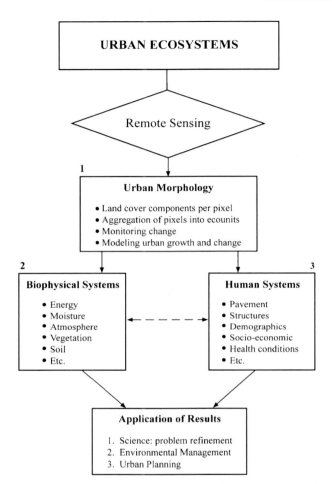

Fig. 6.8 A three-part role for remote sensing of urban ecosystems (Ridd 1995)

On the basis of morphological character – that is, patterns of V-I-S composition, many biophysical analyses may proceed. Patterns of *energy* flux and consequent air temperature are based on varying types of surface composition. The urban heat island (UHI) effect results from the dominance of impervious surfaces, coupled with anthropogenic heat in the city center. Air quality models relate to cover composition, along with traffic patterns on transportation corridors that may be mapped with V-I-S data. The dynamics of *moisture* in the urban ecosystem are driven by variation in morphological composition which can be determined by V-I-S data (see Table 6.1). Models of runoff, evapotranspiration, and moisture uptake from the soil are supported by V-I-S data.

Human system studies also benefit from morphological analysis based on V-I-S composition. Pavement and structure patterns relate to development and transportation analysis. Demographic, population, and socio-economic estimates may be related to morphological composition. Health conditions relate to air quality patterns which are influenced by V-I-S cover composition. Finally, the payoff for remote sensing applied to urban/peri-urban areas leads to a refinement of scientific problems and provides a basis for environmental management and urban planning.

6.3   The Influence of Resolution on V-I-S Analysis

Previous chapters in this book have treated resolution – spectral, spatial, and temporal – in general application to urban remote sensing. This section establishes some principles of resolution that influence the effectiveness of the V-I-S model, and sets a stage for the following sections summarizing various applications of the model. Clearly, adequate spectral and spatial resolutions are vital to detection, mapping, and quantification of vegetation, impervious surface, and exposed soil, as well as water. The need for improvement in resolution increases whenever the objective calls for dividing each of the V-I-S groups into sub-categories. As resolution increases, the opportunity for further subdivision improves.

> **effectiveness of the V-I-S model is directly dependent upon resolution – spatial, spectral, temporal, and radiometric**

Herold et al. (2003) identify 14 spectral bands most suitable for separability of urban land cover classes (also see Chapter 4). Those bands are distributed across the visible (VIS: 400–700 nm) and near infrared (NIR: 700–2,500 nm) spectral range. That is paramount because healthy green photosynthesizing vegetation is distinctly identifiable in the 700–1,000 nm range, and environmental moisture is particularly deflectable in the short wave infrared (SWIR) range from 1,000 to 2,500 nm. To have spectral sensors strategically placed in all three of those zones is vital to detection of vegetation, impervious surface, and soil. Landsat thematic mapper (TM) is designed to take advantage of this spectral range with its six reflective bands: three in the visible range (blue 450–520 nm, green 520–600 nm, and red 630–690 nm); one in the NIR (760–900 nm) region and two in the SWIR region (1,550–1,750 and 2,080–2,350 nm). These are, respectively TM bands 1, 2, 3, 4, 5, and 7 (see Fig. 4.6). TM band 7 was added for its particular sensitivity to certain minerals some of which are significant in the urban environment for detecting various soils and concrete, for example. This band was added after the thermal band (10,400–12,500 nm), band 6. The reflective bands are set at a nominal 30 m spatial resolution. The thermal band at 120 m has limited utility in urban analysis. Furthermore, at the time of Landsat over flight, in mid-morning, temperatures are crossing over from morning cool to mid-day heating.

The earliest earth remote sensing satellite system, Landsat 1, with its multispectral scanner (MSS) data, 1972, had only four spectral bands, in the general range of TM bands 2, 3, and 4. The minimal 80m IFOV (instantaneous field of view) further restricted its utility to urban analysis, but for more than a decade MSS was the only multi-spectral satellite sensor system available. Still, today it is used in historic applications. The French SPOT (Systeme Probatoire d'Observation de la Terre) system added much needed spatial detail for urban use. The three multispectral bands (SPOT-X) roughly equivalent to TM bands 2, 3, and 4, have a nominal 20 m IFOV. A very useful co-registered panchromatic (SPOT-P) band was set at 10 m. These data have frequently been merged with multispectral data such as TM for urban studies to gain spatial detail along with environmental (land cover) information. Landsat 7 was launched in 1999, with an improved sensor ETM+, with a co-registered panchromatic 15 m band and improved thermal band at 60 m spatial resolution.

A new generation of satellite-based sensors has recently brought multispectral remote sensing to 4 m (IKONOS) and panchromatic remote sensing to 1 m or less (IKONOS, Quickbird, GeoEye, and others). These advances have added a robust capability to V-I-S performance for large scale objectives. Meanwhile airborne sensors have provided hyperspectral data: for example, AVIRIS (Advanced Visible InfRared Imaging Spectrometer) yielding 224 spectral bands from the VIS and NIR/SWIR range at a nominal 10 nm band width. Most of the 224 bands are of little value for urban analysis. However, for certain regions of the spectrum, the narrow bands are very useful, for example, in determining vegetation types and conditions. The narrow bands also provide subtle details of absorption features specific to certain minerals in soil and help to distinguish concrete from asphalt.

refined spatial and spectral resolution of new remote sensing systems add robust capability to the V-I-S model performance

Clearly, refined spectral resolution is vital to detection of the many cover types in urban areas. Refined spatial resolution is also crucial, depending on the study objective. Just as AVIRIS may provide more bands than needed, the new generation of sensors may provide more spatial detail than needed, depending on study objectives. Figure 4.7 shows that for a given test, accuracies of classification improved from IKONOS (4 bands at 4 m), to TM (6 bands at 30 m), to AVIRIS (14 selected bands at 10 m). The authors of Chapters 4 and 5 make the point that there are times when spectral sensitivity outranks spatial resolution in achieving suitable classification results.

In the following sections, a number of studies in V-I-S are summarized to illustrate the utility of different datasets for various applications with various sensors of different spectral and spatial resolutions. Also illustrated are various analytical techniques.

6.4   Morphological Application of the V-I-S Model

This section applies the model to the morphology of cities. Five examples are given with increasing spectral resolution (Table 6.3). Spatial resolution varies from 30 m to 10 m. Four examples employ sub-pixel analysis and one example employs artificial neural network methods.

Table 6.3 Selected morphological applications of the V-I-S Model

Author(s)	Location	Year	Data	Spectral bands	Spatial resolution (m)	Techniques
Chung	Salt Lake City	1989	SPOT-X	3	20	Subpixel, visual
Rashed et al.	Cairo	2001	IRC-1C	4	24/10	Subpixel, SMA
Card	Salt Lake City	1993	TM	6	30	Subpixel, visual
Phinn et al.	Brisbane	2002	TM, photog	6	30	Subpixel, SMA
Ridd et al.	Pasadena	1997	AVIRIS	224	20	Per pixel, ANN

6.4.1   SPOT-X Data: Spatial Variation in Urban Communities

Chung (1989) completed the first study using the V-I-S model in Salt Lake City. An early use of SPOT-X data with its then-new 20 m resolution required precise spatial rectification, especially to achieve sub-pixel identification. Key features of this investigation are:

- Large scale CIR photography for land cover calibration
- Land survey monuments for ground control

Large scale CIR (color infrared) photography (1:4,600) provided very high quality ground information. To fix the photos to ground control points, state plane coordinates of road intersections were obtained and converted to a UTM projection. Beginning with a 60-group unsupervised classification, print character maps in transparency form were overlaid on the CIR photos. Each pixel was large enough (about 1 cm) that the cover type could be accurately determined at the center of each pixel quadrant.

Seven V-I-S cover types, plus water, were sought: Vts – trees and shrubs, Vg – green grass, Ia – asphalt, Ic – concrete, Ir – roofs, Sl – light soil, Sd – dark soil, and W – water. From 776 random pixels a total of 3,104 point observations were recorded. Table 6.4 displays the composition of four sections of the city for the cover types. Note the transition from industrial, to commercial (CBD), to a near town residential area (Liberty), to a low density suburban area (Sugar). Vegetation increases from 4%, to 10%, 47%, and 57% respectively, while impervious surface is essentially a reciprocal: 96%, 90%, 50%, and 37%, respectively. Soil varied between 0 and 5%.

Figure 6.9 displays a Red vs. NIR (x,y) spectral feature space plot (excluding the brightest signatures), indicating the V-I-S composition of each spectral cluster. Note the transition from 100% non-vegetation, mostly impervious, to 100% vegetation along each trend line. Note also that class 24 is 80% water, 20% impervious. Figure 6.10 employs colors to group the original 60 clusters into V-I-S related

Table 6.4 Pixel counts and percent cover of V-I-S components in four study sites (Chung 1989)

Study Site		Vegetation		Impervious			Soil	Water	
		Vts	Vg	Ia	Ic	Ir	S	W	Totals
Industrial	Count	40	5	709	0	310	0	0	1,064
	%	4	0	67	0	29	0	0	100%
		4%		96%			0%	0%	
CBD	Count	53	12	370	0	245	0	0	684
	%	8	2	54	0	36	0	0	100%
		10%		90%			0%	0%	
Liberty	Count	232	118	160	5	205	10	14	744
	%	31	16	22	1	28	1	2	100%
		47%		50%[a]			1%	2%	
Sugar	Count	173	173	87	11	129	33	6	612
	%	28	28	14	2	21	5	1	100%
		57%[a]		37%			5%	1%	

[a]Rounding causes irregularities in some totals

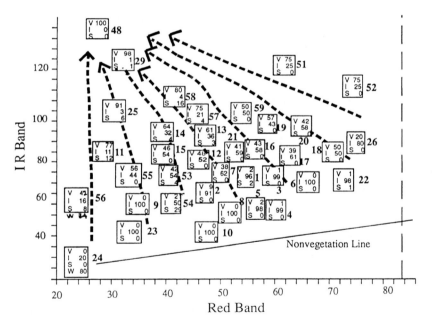

Fig. 6.9 Enlargement of a portion of a 60-class unsupervised plot illustrating V-I-S components of spectral classes. Arrows demonstrate an increasing V component from the non-vegetation line (0%) to the green corner, class 48, at 100% (Chung 1989)

environmental character. Those colors are then applied to a per-pixel classification to illustrate the V-I-S composition of Salt Lake City (see Fig. 6.11). Those familiar with the city readily identify with this environmental display.

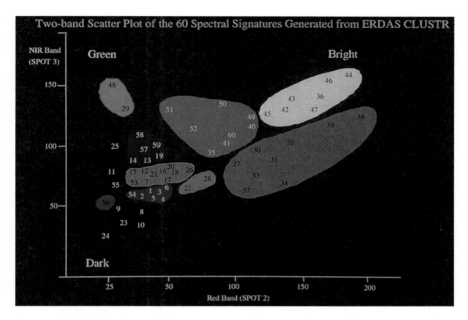

Fig. 6.10 Color applied to groups of spectral signatures from Fig. 6.9, and then applied to Fig. 6.11 (Chung 1989)

Fig. 6.11 V-I-S classified image/map of Salt Lake City and environs, according to the color scheme represented in Fig. 6.10. Overall, the image is a quite faithful representation of the urban/peri-urban environment (Chung 1989)

6.4.2   IRS-1C Data: Anatomy of a Metropolis

Spectral mixture analysis, as applied to the V-I-S model, is beginning to emerge in the literature (Rashed et al. 2001; Wu and Murray 2003). Rashed et al. (2001) employed four-spectral-band Indian IRS-1C data to study the anatomy of Greater Cairo in terms of end-member fractions through spectral mixture analysis (SMA). Key features of this investigation are:

- Spectral mixture analysis
- Decision tree analysis
- Comparative accuracy assessment

Four end-members were selected: vegetation, impervious surface, soil, and shade from pure pixel end-member samples. The resulting fractions were used to classify the urban scene through a decision tree classifier. The V-I-S model provided the link to the spectral mixing procedure because the spectral contribution of each of the three V-I-S components, and shade, can be resolved at a sub-pixel level through SMA. Shade was added as an end-member to accommodate the shadowing of tall buildings, a common feature in large cities, and grouped spectrally with water as dark objects. The end-member spectra were derived from the image. Two different soils were tested, one lighter, one darker. A four end-member SMA with the darker soil produced the best results.

Figure 6.12a–d displays the four fractions. The shade/water fraction highlights the through-flowing Nile River. A binary decision tree was used to further classify the results into eight cover and land use types. The results were tested against two traditional per pixel classifiers, maximum likelihood (ML) and minimum distance to means (MDM). Accuracies were determined through use of the co-registered 10 m IRS panchromatic band and field familiarity. Table 6.5 shows that, as hypothesized, the SMA/decision tree fractions are more accurate than either the ML or MDM classifications.

This study shows the value of the V-I-S model in providing a direct identification of biophysical features so important in urban area remote sensing. The study also demonstrates the flexibility of the model in sub-pixel analysis, SMA endmember determination, and adding shadowing as an important element of urban features of varying heights, although spectrally virtually indistinguishable from water.

6.4.3   Thematic Mapper Data: Multivariate Calibration
and Color Display

Card (1993) used the six reflective bands of TM data to distinguish four V-I-S classes: green vegetation, dry vegetation, soil, and impervious surface, in Salt Lake City. Enlarged black and white photography (1:7,920 scale) was used for calibration and accuracy assessment. Key features of this investigation are:

- Transforming raw TM data to radiance values
- Dark object subtraction for atmospheric correction

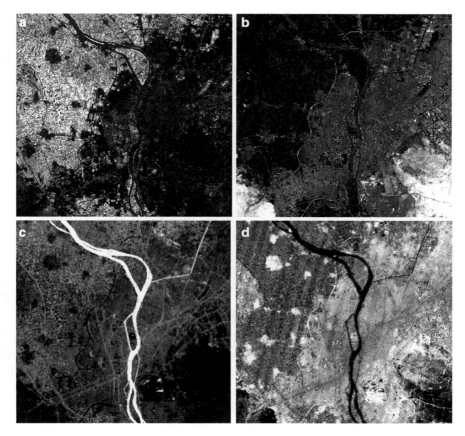

Fig. 6.12 Fraction images produced by the 4-endmember model utilizing vegetation, impervious, dark soil, and shade. Brighter areas indicate higher abundance while darker areas indicate lower abundance (Rashed et al. 2001)

Table 6.5 A comparison between overall classification accuracy and kappa accuracy for the three classifications (Rashed et al. 2001)

	Decision tree based on SMA	Maximum likelihood	Minimum distance
Overall accuracy	89.52	64.51	52.69
Kappa coefficient	0.88	0.59	0.45

- Multivariate calibration by an Lwin and Maritz (1980) technique which uses all six reflective bands for sub-pixel identification that sum to 100 for each pixel
- Display of four fractions in a single color image
- A challenge to the conventional wisdom of "ground truth" wherein the burden of RS classification is to "match" some other source of presumed correctness

Fig. 6.13 Color image of classification applied to Salt Lake Valley: green vegetation applied to green color gun, dry vegetation to blue color gun, and bare soil to red color gun. Black areas are unassigned imperious surface, but may include small amounts of other cover types, including water (Card 1993)

Figure 6.13 displays the four fractions on three color guns as: green vegetation to green, dry vegetation to blue, and bare soil to red. Pure, bright colors of green, blue, and red represent 100% of the cover type. Mixes of green and dry vegetation appear as hues from green to cyan to blue; mixes of green vegetation and soil appear as hues from green to yellow to red; mixes of dry vegetation to soil appear as hues from blue to magenta to red. By "default" impervious surface is basically black where the other three fractions are minimal. The black or near black also represents shade, prominent in the commercial and high density residential high-rise district, as well as a few small water bodies. The advantage of the display is that all four fractions are discernable in one color image, and mixes are integrated into the same image.

6.4.4 Thematic Mapper Data: Radial Transects for Environmental Patterns

Phinn et al. (2002) applied the V-I-S model to TM imagery of Brisbane to determine urban environmental patterns of the city. The investigators claim that the model "exhibits potential to be implemented as a standard measure applicable to urban ecology." Special features of the study are:

- Per pixel V-I-S plus water classification from 1:5,000 scale CIR photography along transects outward from city center
- TM per pixel classification of forest/woodland, sparse vegetation/grass, cleared land/exposed soil, developed (impervious), and water
- Constrained spectral unmixing of TM data

Several useful results came from the investigation: (a) distinctive zones from the inner city to the urban fringe were quantified, as impervious surface gives way to vegetation, (b) a "density of development index" was suggested based upon the impervious fraction, and (c) a "stability index" was suggested based upon the soil fraction and the notion that in a built environment, exposed soil indicates land conversion.

6.4.5   AVIRIS Data: Neural Network Analysis

Ridd et al. (1997) employed a neural network analysis on AVIRIS data of the Pasadena area to detect and map eight V-I-S classes: green trees and shrubs (Vts), green grass (Vgg), dry grass (Vgd), dark impervious (Id), gray impervious (Ig), bright impervious (Ib), soil (S), and water (Wtr). The study area includes a portion of the San Gabriel Mountains. The key elements of this investigation are:

- Two hundred and twenty-four narrow spectral bands (10 nm) to improve sensitivity to the nuances of environmental variety with 20 m spatial resolution
- Several sub-classes of V-I-S composition felt to be significant features of the area of study
- A directed effort to identify the most diagnostic bands specific to the eight cover types from AVIRIS data
- An artificial neural network (ANN) classification technique to maximize separation of sub-classes

The first step was to select several optimal training sites in the field to represent each cover type in order to generate a signature. Using GenIsis® software the 224-channel signature for each of the eight classes was displayed. Using the Similarity Index Map module of GenIsis® the field site selections were refined and final signatures set. Then, the eight most diagnostic AVIRIS channels were selected to separate the classes. Table 6.6 shows the eight neural network channels and the associated AVIRIS channel numbers. (Although AVIRIS channels are nominally 10 nm band width, it was determined that they be doubled or even tripled for best discrimination.) The "color" range, band range, and band width are displayed. To reduce massive processing time bands 5 and 6, and bands 7 and 8 were ratioed to result in six input bands.

    The neural network classification utilizes a "hidden layer" which allows multiple combinations of linkages (synapses) between input vectors and output classes,

Table 6.6 Selection of AVIRIS channels resulting in eight ANN channels

NN channel	AVIRIS channels	"Color" range	Band range (nm)	Band width (nm)
1	12–13	Blue	508–518	20
2	29–31	Red	675–694	20
3	50–52	NIR	843–863	30
4	99–100	NIR	1,253–1,275	31
5	128–130	MIR	1,550–1,569	29
6	145–146	MIR	1,718–1,728	20
7	181–182	MIR	2,030–2,040	20
8	200–202	MIR	2,219–2,238	29

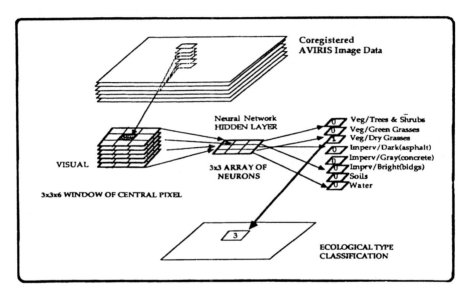

Fig. 6.14 Schematic neural classification from six spectral bands to an eight class map

until the optimal separability is reached. Figure 6.14 is a schematic diagram depicting the role of the hidden layer leading to final classification. A pixel centered in a $3 \times 3 \times 6$ matrix has multiple opportunities to be added to the hidden layer and ultimately assigned to one of the eight V-I-S classes. Once assigned, the pixel takes its place in the classification, in this case, class 3 – vegetation dry grass.

Figure 6.15 shows the overall accuracy in the training set at 86.8%, with Vgg and Ib registering 100%, no errors of omission. Ig shows several errors of commission, confused with Id and soils. In the test set, overall accuracy was 89.95% with Ig and soil registering 100% hits while Id missed about half of the actual Id pixels, mistaking a large number as Ig. Because so little of the scene was water and in the training set much of it remained unknown, it was omitted from the test set. It appears from this study that AVIRIS 224-channel data coupled with ANN classification is

Training Set

	unkn	vts	vgg	vgd	ld	lg	lb	soil	wtr	Total	%Hit
vts	20.00	141.00	0.00	6.00	3.00	0.00	0.00	0.00	0.00	170.00	82.94
vgg	0.00	0.00	137.00	0.00	0.00	0.00	0.00	0.00	0.00	137.00	100.00
vgd	1.00	0.00	0.00	167.00	2.00	0.00	0.00	0.00	0.00	170.00	98.24
ld	2.00	0.00	0.00	0.00	64.00	18.00	0.00	2.00	0.00	86.00	74.42
lg	0.00	0.00	0.00	0.00	2.00	170.00	0.00	0.00	0.00	172.00	98.84
lb	0.00	0.00	0.00	0.00	0.00	0.00	40.00	0.00	0.00	40.00	100.00
soil	12.00	0.00	0.00	7.00	4.00	23.00	2.00	86.00	5.00	139.00	61.87
wtr	15.00	0.00	1.00	0.00	2.00	0.00	0.00	0.00	32.00	50.00	64.00

Overall: 86.8%

Test Set

	unkn	vts	vgg	vgd	ld	lg	lb	soil	wtr	Total	%Hit
vts	4.00	117.00	0.00	0.00	0.00	0.00	0.00	0.00	0.00	121.00	96.69
vgg	1.00	0.00	119.00	1.00	0.00	0.00	0.00	0.00	0.00	121.00	98.35
vgd	4.00	6.00	0.00	98.00	0.00	0.00	0.00	4.00	0.00	112.00	87.50
ld	5.00	2.00	0.00	0.00	29.00	18.00	0.00	5.00	0.00	59.00	49.15
lg	0.00	0.00	0.00	0.00	0.00	44.00	0.00	0.00	0.00	44.00	100.00
lb	1.00	2.00	0.00	0.00	6.00	7.00	72.00	1.00	0.00	89.00	80.90
soil	0.00	0.00	0.00	0.00	0.00	0.00	0.00	121.00	0.00	121.00	100.00
wtr											

Overall: 89.5%

Fig. 6.15   Accuracy of the neural network classification of eight V-I-S subtypes (Ridd et al. 1997)

quite successful in characterizing urban/peri-urban environments even where nearby mountains are in the data set. What is not known is the degree to which the success stems from the hyperspectral data with a narrow band width and 20 m cell size or the neural network classification.

6.5   Environmental Process Application of the V-I-S Model

The V-I-S model has been shown to be a suitable foundation for physical process models. The two examples summarized here represent energy and moisture flux, the two primary drivers in the urban ecosystem (and ecosystems generally).

6.5.1   ATLAS Data: Thermal Emittance per Land Cover Class

Gluch et al. (2006) utilized ATLAS (Advanced Thermal Land Applications Sensor) 10 m airborne data to determine daytime thermal emittance from eight V-I-S cover types over Salt Lake Valley, Utah: two vegetation types, two impervious types, two soil types, shade, and water. Clearly the V-I-S model of *land cover* is superior to models or classifications of *land use* for energy analysis. Energy dynamics are tied

Table 6.7 Thermal statistics by
land cover (Wm⁻²) (Gluch et al.
2006)

	Min	Max	Mean	SD
imp_dk	27.82	38.78	31.77	3.43
soil_dk	28.98	32.94	30.37	0.58
soil_lt	27.58	30.70	28.96	0.64
imp_lt	26.07	32.54	28.53	1.69
veg_grass	23.06	27.41	26.08	0.64
veg_trees	22.31	27.54	23.78	0.89
Shadow	20.64	24.88	23.04	1.08
Water	20.65	24.60	21.96	0.72

to land cover types (as in V-I-S), which can be directly measured through remote sensing. Furthermore, energy flux responds very differently among the three V, I, and S urban cover types, as well as water (see Table 6.1). Consequently, values derived from thermal sensors of vegetation, impervious surface, soil, and water are directly related to such models as the urban heat island (UHI), evapotranspiration rates, and other physical phenomena.

Table 6.7 shows the mid-afternoon thermal emittance values by land cover class according to ATLAS channel 13 expressed in watts per square meter (Wm⁻²). Note that dark impervious surface has the highest mean, followed by dark soil. Vegetation is coolest (emitting less energy) of the three terrestrial cover types with trees and shrubs cooler than grass. Shadows are still cooler and water is the least emissive in mid-afternoon. It follows that ambient air temperature, as sensed by human populations, run in the same order.

6.5.2 Thematic Mapper/SPOT-P Data: Storm Runoff Prediction

Chen (1996) examined the effect of V-I-S cover types as input to storm runoff models for runoff prediction of storms of different intensities – using merged Landsat TM and SPOT-P data re-sampled to 10 m. Widespread conversion of land cover attendant to urbanization typically increases storm water runoff in amount, intensity, and in routing. However, there is significant variability across the urban landscape due to the spatial heterogeneity of cover composition. To be able to predict runoff rates and patterns would significantly assist in storm water drainage design and engineering, and other environmental planning.

To perform effective research on the process it is necessary to have empirical data for both precipitation records and concurrent runoff records. In Salt Lake City and County runoff records were available at outfall points from specifically engineered drainage basins. Nearby weather stations provided precipitation rates for the study period. The Soil Conservation Service (SCS) storm runoff model provided the appropriate input for various sub-classes of V, I, and S. Table 6.8 displays the results of the SCS model applied to 14 rainfall events across five urban basins over a 4-year time span. Rainfall amount per event is shown by date. Measured runoff

Table 6.8   Measured and estimated runoff amounts from the SCS Model (Chen 1996)

	Date	Precip. (in.)	Measured runoff (ft³)	Estimated runoff (ft³)	Error (%)
Basin 1	5/21/92	0.32	21,081	20,501	−2.8
	4/24/93	0.46	47,574	45,958	3.4
Basin 2	5/21–22/93	0.51	175,478	168,447	4.0
	6/17/93	0.34	58,988	48,621	17.6
Basin 3	5/08/93	0.97	298,174	370,230	24.2
Basin 4	5/21/92	0.32	17.705	14,631	−17.4
	4/01/93	0.50	61,628	54,947	−4.1
	5/05/93	0.56	76,830	72,217	−6.0
	5/06–07/93	0.86	145,866	184,371	26.0
	5/08/93	0.85	213,671	172,787	−25.4
Basin 5	3/21/95	0.63	146,406	61,266	−58.2
	3/23/95	0.99	215,290	326,322	51.6
	4/08/95	0.77	136,607	143,693	5.2
	5/25/95	1.02	328,555	355,778	8.0

is compared to estimated runoff in cubic feet per storm event. Under-estimates are represented by a minus sign, and over-estimates by a plus sign. Seven of the errors are less than 10% with five between 17% and 26%, still within the practical working range according to collaborating practitioners. Estimates for mid-valley basins 1, 2, and 4 performed best. Basins 3 and 5 are foothill sites with granular soils and considerable open space.

Remote sensing applied to estimating/predicting urban storm runoff is of great scientific and practical planning value. Yet, it is virtually unexplored at this point in time due to rare coincidence of empirical rainfall and runoff data at specified urban drainage basins. While this study dealt only with water runoff quantity, similar remote sensing opportunities are pursued for water quality.

6.6   Temporal Application of the V-I-S Model

The V-I-S model has been applied to several change detection studies (Ward et al. 2000; Madhaven et al. 2001; Hung 2003; Kaya et al. 2004). This section highlights two examples, one in Turkey and one in Utah.

6.6.1   SPOT-X Data: Pre- and Post-earthquake Analysis

Kaya et al. (2004) employed SPOT-X to record earthquake-induced land cover change in northwest Turkey in consequence of the Izmit earthquake of August 1999. The trend and relative magnitude of land cover change were displayed using the

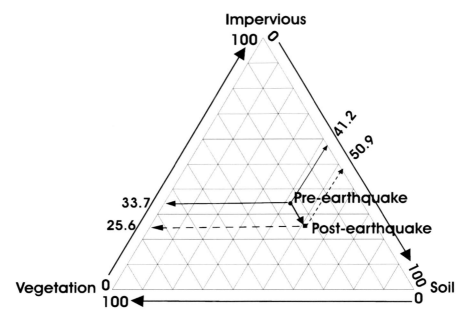

Fig. 6.16 Trend of V-I-S from pre-earthquake to post-earthquake conditions, Izmit, Turkey (Kaya et al. 2004)

V-I-S model. SPOT-X (20m) data from 25 July 1999, and 4 October 1999, were geometrically corrected to 26 ground control points and transformed to UMT coordinates. Using the ISODATA (Interactive Self Organizing Data) classifier, 25 spectral classes were combined to form four classes: vegetation, impervious I (urban), impervious II (collapsed buildings), and soil (neither vegetation nor urban). The concept adopted by the investigators is that the collapsed buildings no longer functioned as urban, where the term, impervious, implies buildings and roads with urban functions. Rather they are treated as rubble and grouped with the soil class.

From the SPOT-X classification there was a slight loss in vegetation (−1.6%) and a greater loss in impervious surface (−8.1%), and a significant gain in soil with the addition of collapsed rubble (+9.7%). Some 7.1% of the original impervious surface was reduced to "non-urban" function and added to the soil class. Figure 6.16 displays the results plotted on the V-I-S ternary diagram.

6.6.2   MSS/TM/ETM+ Data: Urban Growth, Expert System Analysis

Hung (2003) employed Landsat MSS, TM, and ETM+ data to monitor land cover change in Salt Lake Valley, Utah, 1972–1999. He used an expert system developed earlier (Hung and Ridd 2002) to determine sub-pixel composition. Seven cover types were targeted: green grass, trees and shrubs, three classes of impervious, soil/ dry vegetation, and water. Water was separately treated, first by identifying all water features, then masking them out of the terrestrial classification.

The green grass component is usually found in golf courses, baseball fields, cemeteries, parks, and some residences. Green pastures and alfalfa fields are also included in this class, representing peripheral active irrigated farmlands. The trees/ shrubs vegetation class is found in mountain foothills, residential areas, and parks. The bright impervious class includes materials of high reflectance such as rooftops typically of commercial/industrial buildings. The medium impervious class includes many rooftops as well as concrete and weathered asphalt common in commercial/industrial areas as well as residential areas. Dark impervious surfaces of low reflectance include recently placed asphalt on roads and parking lots, common in commercial and industrial areas. This class also includes shadowed areas occurring in all areas but especially in high rise downtown commercial and apartment districts. In the V-I-S concept it is important to separate soil from impervious surfaces because of their very different contributions to environmental processes (see Fig. 6.1). In this case, dry vegetation is included in the soil class as this is the common condition of soil in the study area, and does not alter the energy/moisture regime appreciably.

The six fractions show significant change across Salt Lake Valley in the 27-year time span. Farm fields are substantially reduced under urban sprawl. Trees and shrubs increased significantly in fringe area subdivisions during the period. Impervious surfaces show an expansion of the downtown district, as well as an extensive expansion of large industrial tracts to the west, replacing arid land soil and dry vegetation. Many shopping centers also emerged in the outlying suburban and farm areas undergoing transition. Through regression analysis, accuracies are expressed in correlation coefficients. Most of the six classes show significance at the 0.01 level, and the rest at 0.05.

Figure 6.17 shows change over time for the three principal V-I-S components in a linear-time plot (1972–1999), with soil diminishing while impervious surface increases from 40% to 65%. Vegetation also increases for a time, and then weakens. These data represent the University of Utah campus, as the medical center complex and research park extend up into barren benches and foothills.

Figure 6.18 shows three methods of displaying change over time for an alfalfa farm field being converted to a residential subdivision. From 1972, the green field was cleared in preparation for urban development. In the upper chart, vegetation drops sharply from near 80% to 25% in the first 3-year interval and then to 16% by 1981. From that point on, the balance between soil and impervious surface is interchanged as a sub-division takes shape. The last stage (1996–1999) shows vegetation beginning to rise as landscaping commences. The other two plots show the same trend over the 27-year period. The lower left displays the same transition in a Red/NIR feature space plot as a field of green vegetation suddenly approaches the non-vegetation line and then proceeds generally toward the bright point to the right as rooftops and concrete increase. The ternary V-I-S plot illustrates the same land conversion. In due time as landscaping matures, a transition toward the vegetation corner will ensue.

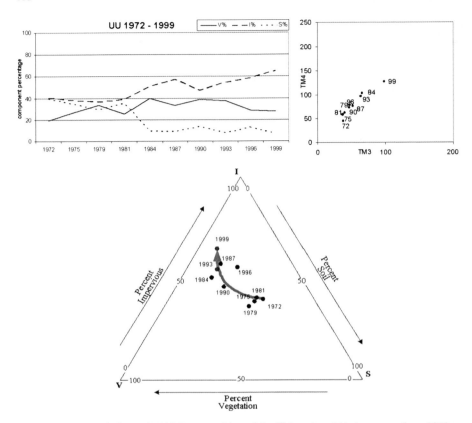

Fig. 6.17 Temporal change in V-I-S composition of the University of Utah campus from 1972 to 1999 (Hung 2003)

Chapter Summary

For urban environmental/ecological/planning purposes it is important to distinguish the three general V-I-S classes – vegetation, impervious surface, and soil – because their influences upon both biophysical processes and human experience are so different. To subdivide them further also helps to understand and map variations within class. The model readily characterizes the morphological nature of an urban area, and distinguishes it from surrounding environments, whether farm, forest, desert, grassland, wetland, or tundra. The model also serves to monitor change in urban morphology over time, so important in urban and environmental planning and management. The figures in Section 6.6 show in interrelated graphic form the environmental vitality of the V-I-S model for an urban peri-urban landscape undergoing change. Of particular significance is the relationship between the spectral signatures and the ternary plot. It is of course the spectral signatures, which define the classes.

Quantifying and mapping the V-I-S urban/peri-urban environmental complex serves as a foundation for many physical models in science and engineering, such as the two exhibited herein. Many others such as evaporation rates, water uptake, air quality, and soil condition find a connection to and from the V-I-S foundation and its changing patterns. Human factors such as housing density, population, and income levels may also be related. Much research lies ahead. Improving sensor resolution, interfaced sensor data (VISNIR, SWIR, thermal, radar, lidar, etc.), and more robust information extraction techniques provide continued opportunities for remote sensing applications of the V-I-S model.

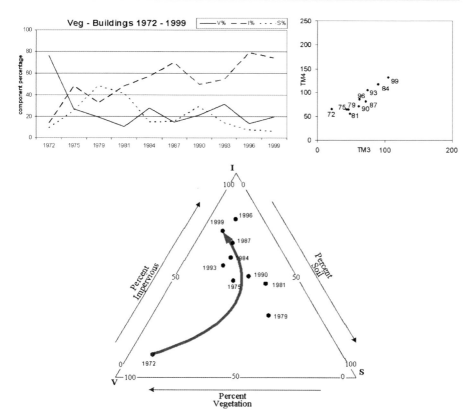

Fig. 6.18 The three methods of displaying another "micro-scale" site, a productive farmland suddenly thrust into new urban development, with much bare soil, and gradually increasing impervious (roads and buildings) and vegetation (landscaping) (Hung 2003)

LEARNING ACTIVITIES

Outdoor Activities

- Select a plot, such as the student's home lot, for the following activities:
  - Observe and list several types of vegetation, impervious surface, and exposed soil.
  - Group the subsets together as V, I, and S, and estimate the percent cover of each, summing to 100%.
  - Place the results on the V-I-S ternary diagram.

- Repeat the above steps for a nearby school ground and compare the two sites:
  - Which produces greater storm runoff and why?
  - Which cover types change temperatures the most, the least, from day to night and why?

- Select a routing for a driving transect from the city center to the outlying un-urbanized environment. With a data sheet (constructed in tabular form) in hand, beginning at city center, drive the route, stopping at one-kilometer intervals and record the V-I-S composition (summing to %) as visible from the roadside. At each stop also state the dominant land use. Mark and number the stopping points on a map and your data sheet.

  - Observing your table of data, write a brief statement about changing V-I-S composition along the route, considering the general trend and any anomalies.
  - Write a paragraph explaining the association of changing V-I-S values with land use observed along the way.

- Back in the classroom or lab, compare your transect with that of other students who took a different route. Together create a schematic of zones of land use around the city center. Generalize the V-I-S proportions per zone and record them on the schematic diagram, and on a map of the city.

Study Questions

- Define an ecosystem and explain the urban/peri-urban environment as an ecosystem with examples from personal experience.
- Identify and list 12–15 urban/peri-urban land cover types and group them into V-I-S categories.
- Explain how spatial resolution affects mapping of V-I-S patterns in urban/peri-urban areas.

- Considering the local neighborhood, discuss what might be an optimal spatial resolution (pixel size) and which current sensors might provide for such a V-I-S analysis.
- Explain in general how spectral resolution affects detection of various kinds of V, I, and S, and how spatial resolution would relate to such detection.
- Sketch typical spectral reflectance curves of V, I, S, and water, across visible, near-infrared, and mid-infrared wave bands.
- Explain the spectral change when a green field (A) is converted to a shopping center (B), and place A and B on the V-I-S ternary diagram.
- Discuss how ambient air temperatures likely vary with changing V-I-S composition from the urban core to the urban periphery.
- Explain how storm water runoff would likely vary from those same sites.

References

Burgess EW (1925) The growth of the city. In: Park RE, Burgess EW, McKenzie RD (eds) The city. University of Chicago Press, Chicago, IL

Card DH (1993) Examination of a simple surface composition model of the urban environment using remote sensing. Doctoral dissertation, The Department of Geography, The University of Utah, Salt Lake City, Utah

Chen J (1996) Satellite image processing for urban land cover composition analysis and runoff estimation. Doctoral dissertation, The Department of Geography, University of Utah, Salt Lake City, Utah

Chung JM (1989) SPOT Pixel analysis for urban ecosystem study in Salt Lake City, Utah. Master's thesis, The Department of Geography, The University of Utah, Salt Lake City, Utah

Dorney RS, McLellan PW (1984) The urban ecosystem: its spatial structure, its scale relationships, and its subsystem attributes. Environment 16:9–20

Douglas I (1983) The urban environment. Edward Arnold, London

Forster BC (1985) An examination of some problems and solutions in monitoring urban areas from satellite platforms. Int J Remote Sens 6:139–151

Gluch RM, Quattrochi DA, Luvall JC (2006) A multi-scale approach to urban thermal analysis. Remote Sens Environ 104:123–132

Harris CO, Ullman EL (1945) The nature of cities. Ann Am Acad Pol Soc Sci 242:7–17

Herold M, Gardner ME, Roberts DA (2003) Spectral resolution requirements for mapping urban areas. IEEE Trans Geosci Remote Sens 41:1907–1919

Hoyt H (1939) The structure and growth of residential neighborhoods in American cities. Federal Housing Administration, Washington, DC

Hung MC (2003) Remote sensing and GIS for urban environmental modeling monitoring and visualization. Doctoral dissertation, Department of Geography, University of Utah, Salt Lake City, Utah

Hung MC, Ridd MK (2002) A subpixel classifier for urban land-cover mapping based on a maximum-likelihood approach and expert system rules. Photogramm Eng Remote Sens 68:1173–1180

Kaya, S, Llewellyn G, Curran PJ (2004) Displaying earthquake damage in urban area using a vegetation-impervious-soil model and remotely sensed data. In: Proceedings of the XX congress international society of photogrammetry and remote sensing, Istanbul, Turkey, 12–23 July 2004

Lwin T, Maritz JS (1980) Note on the problem of statistical calibration. J R Stat Soc Ser C 29:135–141

Madhaven BB, Kubo S, Kurisaki N, Sivakumar TVLN (2001) Appraising the anatomy and spatial growth of the Bangkok Metropolitan area using a vegetation-impervious-soil model through remote sensing. Int J Remote Sens 22:789–806

Phinn SR, Stanford M, Scarth PF, Shyy T, Murray A (2002) Monitoring the composition and form of urban environments based on the vegetation-impervious surface-soil (VIS) model by sub-pixel analysis techniques. Int J Remote Sens 23(20):4131–4153

Quattrochi DA (1996) Cities as urban ecosystems: a remote sensing perspective. In: Proceedings, Pecora Thirteen, Sioux Falls, SD, 20–22 Aug 1996, pp 470–479 (on CD)

Rashed T, Weeks JR, Gadalla MA, Hill AG (2001) Revealing the anatomy of cities through spectral mixture analysis of multispectral satellite imagery: a case study of the greater Cairo region, Egypt. Geocarto Int 16(4):5–15

Ridd MK (1995) Exploring a V-I-S (Vegetation-Impervious Surface-Soil) model for urban ecosystem analysis through remote sensing: comparative anatomy for cities. Int J Remote Sens 16:2165–2185

Ridd MK, Hipple J (eds) (2006) Remote sensing of human settlements. Manual for remote sensing, 3rd edn. ASPRS, Washington DC

Ridd MK, Merola JA, Jaynes RA (1983) Detecting agricultural to urban land use change from multispectral MSS digital data. In: Proceedings of the ASP-ACSM fall convention, Salt Lake City, Utah, pp 473–482

Ridd MK, Ritter ND, Green RO (1997) Neural network analysis of urban environments with airborne AVIRIS data. In: Proceedings of the 3rd international airborne remote sensing conference and exhibition, Copenhagen, Denmark, 7–10 July 1997

UNFPA (2001) The state of world population 2001. United Nations Population Fund, United Nations Publications, New York. Available at http://www.unfpa.org/swp/2001/english

Ward D, Phinn SR, Murray AT (2000) Monitoring growth in rapidly urbanizing areas using remotely sensed data. Prof Geographer 52:371–386

Wu CS, Murray AT (2003) Estimating impervious surface distribution by spectral mixture analysis. Remote Sens Environ 84:493–505

Part II
Techniques and Applications

Chapter 7
A Survey of the Evolution of Remote Sensing Imaging Systems and Urban Remote Sensing Applications

Debbie Fugate, Elena Tarnavsky, and Douglas Stow

The increasingly diverse nature of sensor systems and imagery products, as well as their commercial availability, have led to a broad set of applications resulting in rich, interdisciplinary topics that come under the umbrella of urban remote sensing. This chapter reviews the development of remote sensing systems, their contribution to the emergence of urban remote sensing, and how they have given rise to the pursuit of novel approaches to the study of urban environments.

Learning Objectives

Upon completion of this chapter, you should be able to:

❶ Describe the relationships between the technological advancements in sensor systems and evolving urban remote sensing applications

❷ Speculate on the diversity and interdisciplinary nature of the topics addressed under the umbrella of urban remote sensing

---

D. Fugate (✉)
Department of Geography, San Diego State University, 5500 Campanile Dr., San Diego, CA 92182-4493, USA
e-mail: fugate.debbie@gmail.com

E. Tarnavsky
Geography Department, King's College London, Strand, London WC2R 2LS, UK
e-mail: elena.tarnavsky@kcl.ac.uk

D. Stow
Department of Geography, San Diego State University, 5500 Campanile Dr., San Diego, CA 92182-4493, USA
e-mail: stow@mail.sdsu.edu

T. Rashed and C. Jürgens (eds.), *Remote Sensing of Urban and Suburban Areas*, Remote Sensing and Digital Image Processing 10, DOI 10.1007/978-1-4020-4385-7_7, © Springer Science+Business Media B.V. 2010

7.1   Introduction

The variety of remote sensing imaging systems and urban remote sensing applications is due in part to the reciprocal exchange of expertise between the domains of information technology and an emergent interest in the structural complexity and dynamics of urban environments. Remote sensing is a domain of interdisciplinary studies encompassing a wide array of applications. Broader and increasingly interdisciplinary applications are driven by the evolving technological characteristics of remote sensing systems, increased availability of remotely sensed imagery, and advances in data processing and analysis techniques. These drivers have inspired new thinking about the urban environment and provided the impetus for urban remote sensing (URS).

In this chapter, we trace the evolution of URS applications as driven by technological developments of imaging instruments. The focus is on satellite systems that have been most commonly used for urban analyses. Through a chronological overview of applications, the development of urban remote sensing is outlined from the early days of aerial photography to recent advances in satellite imaging and image processing technologies. Finally, the chapter concludes with a summary of key trends in the joint evolution of remote sensing systems and urban remote sensing applications.

7.2   Evolution of Remote Sensing Systems

Urban environments are characterized by complex and dynamic physical and socio-economic attributes that vary continuously across space and time. The ability to map, monitor, and analyze such attributes from remotely sensed imagery greatly depends on the characteristics of the imaging instruments (see related discussions in Chapters 4 and 5). Remote sensing instruments include photographic (film-based cameras) and non-photographic sensor systems (radiometers, digital cameras, radar systems, and electro-optical scanners).

Current operational remote sensing systems are designed for specific earth observation missions, and thus have different operational principles and technical characteristics. The focus in this chapter is on imaging systems mounted on earth orbiting satellite platforms, designed to collect imagery of the earth surface at regular time intervals. Airborne imaging systems have been commonly used in the early years of URS but their technical characteristics are not discussed here. Jensen (1996) provides a more detailed review of airborne imaging systems' characteristics. The following major imaging satellites will be covered in this chapter:

current operational remote sensing systems have different operational principles and technical specifications

- Landsat (originally called Earth Resources Technology Satellite or ERTS) developed by National Aeronautics and Space Administration (NASA) and operated by the United States Geological Survey (USGS)
- The French earth observing series of satellites called Satellite Pour l'Observation de la Terre (SPOT)
- Indian Remote Sensing (IRS) operated by the Indian government
- The U.S. Defense Meteorological Satellite Program (DMSP) series operated by the National Oceanic and Atmospheric Administration (NOAA)
- IKONOS by Space Imaging, Inc.
- The ASTER imaging system mounted on the TERRA earth observing satellite and operated by NASA
- The QuickBird mission by DigitalGlobe (formerly EarthWatch, Inc.)

The first successful launch year for each of the above missions is marked in Fig. 7.1 to illustrate their operational time scale.

Although the operating principles of various imaging instruments have changed over the years, the spectrum of applications and usability of imagery have been largely determined by the spatial, radiometric, spectral, and temporal resolutions of the imaging system. Image resolution characteristics play a major role in determining the size and properties of the features or phenomena that can be discriminated in remotely sensed imagery. The spatial, radiometric, spectral, and temporal resolutions are discussed below for the digital imaging systems and multi-spectral scanners listed in Fig. 7.1. Spatial, radiometric, spectral,

usability of imagery in urban applications is determined by the spatial, radiometric, spectral, and temporal resolutions of the imaging system

and temporal resolutions can each be defined in two different contexts, the users' needs and the technical specifications of the imaging system. In the context of users' needs, definitions depend on the purpose of the study. The definitions of

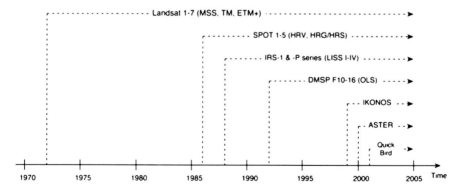

Fig. 7.1 Time scale of operational earth observing missions most commonly used in urban remote sensing applications

spatial, radiometric, spectral, and temporal resolutions discussed below refer primarily to the imaging sensor characteristics.

While the emphasis in this section is on the resolution or precision of remotely sensed measurements, the reader should be aware that the extent, duration, or location of the measurement in space, time, and spectral or energy intervals may be equally important. For instance, the spatial extent of coverage, range of energy sensed, position of wavebands throughout the spectrum, or temporal record length of archived imagery are also important factors when selecting imagery to match a particular urban application.

7.2.1 Spatial Resolution

The spatial resolution of remote sensing instruments is a function of the altitude of the platform relative to the earth surface and the resolving power of the sensor, and is often expressed as the ground sampling distance (GSD) of the sensor at nadir. High spatial resolution imagery provide a closer view of smaller objects while low spatial resolution imagery provide a farther view of larger objects.

spatial resolution is expressed in terms of the ground sampling distance of the sensor at nadir

Moderate to high spatial resolution imagery have been used for urban remote sensing applications. Imagery at 80 m spatial resolution have been available from the Landsat Multispectral Scanner (MSS) archive since 1972, and at 30 m since the mid-1980s, from the Landsat Thematic Mapper (TM) sensor. In the late 1980s, the French Centre National d'Etudes Spatiales (CNES) launched the SPOT mission with the first four satellites carrying the High Resolution Visible (HRV) payload acquiring images of 10 m resolution in panchromatic mode and 20 m resolution in visible/near infrared (V/NIR) mode. On May 4, 2002 SPOT 5 was launched with two different payloads, the High Resolution Geometric (HRG) panchromatic instrument and the High Resolution Stereoscopic (HRS) multispectral sensor with spatial resolutions of 2.5 and 10 m, respectively. In addition to the higher spatial resolution than the HRV, the HRS includes a shortwave infrared (SWIR) band. The Indian Department of Space launched a series of IRS satellites equipped with high and moderate spatial resolution versions of the Linear Imaging Spectral Scanner (LISS). The latest IRS RESOURCESAT-1 satellite was launched on October 17, 2003 and provides imagery from three spaceborne sensors, the high resolution LISS sensor (LISS-IV), the medium resolution LISS sensor (LISS-III), and the Advanced Wide Field Sensor (AWiFS) with spatial resolutions of 6, 24, and 60 m, respectively (Rao and Cook 2004). RESOURCESAT-2 is planned for launch in 2006 and will carry the high spatial resolution version of the LISS payload.

Since 1999 the urban remote sensing community has had access to imagery from very high spatial resolution sensors such as the digital imaging payloads onboard IKONOS and DigitalGlobe's QuickBird satellite systems. Although IKONOS 1 and QuickBird 1 both experienced launch failures, the second missions launched in

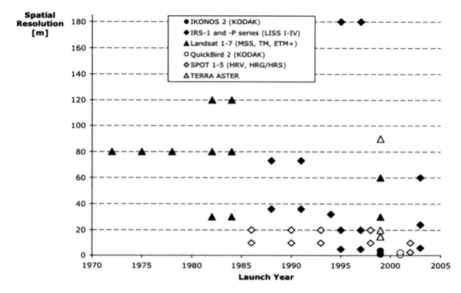

Fig. 7.2 Historical trend of increasing spatial resolution of imaging instruments most commonly used in urban remote sensing applications. The DMSP Operational Linescan System (OLS) is not included in this figure due to its relatively low spatial resolution at 2.7 km low spatial resolution (smooth mode) and at 550 m high spatial resolution (fine mode) (Dickinson et al. 1974). The thermal infrared (TIR) band of Landsat Multispectral Scanner (MSS) is not included in the figure due to its limited use in urban analysis

the end of 1999 and 2000, respectively, are currently operational. IKONOS and QuickBird provide images at 1 m and 0.61 cm spatial resolution in panchromatic mode and 4 and 2.5 m spatial resolution in V/NIR mode respectively. The OrbView-3 satellite system was launched in 2003 and has similar sensor specifications as the IKONOS system. A new IRS mission (CartoSat) designed for cartographic applications was launched on 5 May, 2005 and will provide panchromatic imagery at 2.5 m.

Urban remote sensing applications have substantially benefited from the high spatial resolution characteristics of commercial satellite image data and the ensuing ability to monitor smaller features of interest in complex urban environments (Donnay et al. 2001). The trend of increasing spatial resolution of imaging sensors since the 1970s is illustrated in Fig. 7.2 and includes all satellite imaging systems of pertinence.

7.2.2 Radiometric Resolution

The radiometric resolution (or radiometric precision) of remote sensing data is determined by the detector or film sensitivity and noise level, as well as the quantitative manner in which analog radiance measurements are digitized and stored as digital number (DN) values (called the quantization level). The detector-related effects are often characterized as the noise equivalent difference in radiance (NEΔL)

radiometric resolution refers to the quantization level at which analog radiance measurements are digitized and stored as digital number (DN) values

or reflectance (NEΔρ). Since the recorded digital number is a discrete quantified measure of the originally continuous radiance measurement (i.e., the energy received at the sensor), the greater the quantization level the closer the quantized image output approximates the original continuous data (Schowengerdt 1997). Generally, quantization levels for most systems range between 8-bits (256 gray levels) and 12-bits (4,096 gray levels).

Successful analysis of remotely sensed imagery is dependent on sufficient radiometric resolution, meaning that the NEΔL is sufficiently low and the quantization level high enough to allow discrimination among the reflected or emitted energy levels recorded for various phenomena of interest in urban environments. Conversely, low radiometric resolution (such as 6-bit quantization level) can be prone to saturation, impeding discrimination of urban environment features due to insufficient contrast (Jensen and Cowen 1999). The first space-borne imaging systems, such as the Landsat MSS, used a 6-bit quantization level. Similarly, some of the earlier IRS platform payloads output images at 7-bit quantization level, while other IRS series record images at 8-bit and 16-bit quantization levels. The quantization level of most imaging systems (Landsat TM and ETM+, SPOT, DMSP, IKONOS, ASTER, and QuickBird) is generally between 8-bit and 12-bit.

7.2.3   Spectral Resolution

Spectral resolution refers to the capability of an imaging system to measure and record sensed radiation within discrete ranges of wavelengths (referred to as wavebands). The importance of spectral resolution in urban remote sensing analyses is discussed in detail in Chapter 4 (see also Herold et al. 2004). The historical trend in the evolution of imaging systems has been toward increased spectral sensitivity (i.e., increased number of bands and narrower wavelength ranges).

Imaging spectrometers or hyperspectral imagers are capable of sensing hundreds of narrow wavebands across the UV, visible, NIR, SWIR, and TIR portions of the spectrum (see Chapter 9 for a review of applications). The only civilian spaceborne imaging spectrometer has been the Hyperion instrument on the NASA EO-1 satellite. However, a number of airborne imaging spectrometers have and continue to acquire hyperspectral imaging data. The key advantage of hyperspectral image data (~10 nm spectral resolution) over broadband multispectral data (~100 nm spectral resolution) is the ability to select specific and multiple narrow wavebands that optimally discriminate urban surface materials or quantify urban surface conditions.

7.2.4   *Temporal Resolution*

The temporal resolution of a remote sensing system can be defined as either the theoretical or operational capability for acquiring repetitive imagery over some time interval. Platform characteristics are the primary factors influencing the theoretical temporal resolution. Aircraft are more mobile than spacecraft platforms and provide the potential for temporal resolutions of a few minutes (seconds for rotary-wing aircraft). Most spaceborne systems utilize polar-orbiting satellites, with limited repeat potential from several days to a few weeks. The spatial extent of images depends on the swath width, and influences the resulting temporal resolution.

> **temporal resolution refers to theoretical or actual revisit time intervals of the sensor's platform over the same location of the earth surface**

Larger viewing swath widths allow for monitoring of the Earth's surface in shorter time periods and thus, shorter revisit intervals. Equatorial-orbiting, geo-stationary satellites have the potential for nearly continuous imaging (i.e., infinitely fine temporal resolution). Operationally, even polar-orbiting satellites tend to achieve higher temporal resolution than airborne systems, because of their regular orbits and systematic data collection and archiving capabilities. Also, the pointability of many newer generation satellite sensors enables higher temporal resolution for polar-orbiting satellites that would normally have a long repeat interval.

In the early 1970s temporal resolution tended to vary inversely with spatial resolution. The closer the platform is to the earth surface (i.e., lower the orbit), the higher the spatial resolution and the lower spatial coverage and therefore, the temporal resolution. This relationship is valid for scanning radiometers (such as the Landsat Thematic Mapper) designed to collect images over a fixed swath width centered at nadir. However, as discussed previously, with the advent of pointable

> **a system is called pointable, if off-nadir image capturing is possible, implying an increase in the revisit capability of the satellite**

imaging sensors such as SPOT, IKONOS, and Quickbird, that can capture imagery off-nadir while orbiting the earth, temporal resolutions have been reduced.

Selecting the Appropriate Imagery for Urban Analysis

Given the complexity of urban environments and the high spatial heterogeneity of urban materials, spatial, radiometric, and spectral resolutions of imagery should be considered concurrently, with the objective of achieving highest possible separability of features of interest. Current operational imaging instruments are not fully optimized in their spectral, radiometric, and spatial resolutions to handle urban phenomena. High separability is achieved through proper selection of resolutions according to the level of detail targeted in the information extraction process in a particular study. Spatial properties of features may lead to the use of techniques that also incorporate their spectral properties. Features of interest may be represented as part of a single pixel in a coarse spatial resolution image, thus resulting in spectral mixing of multiple land cover materials in a single image pixel and the use of spectral unmixing models. Similarly, features or objects of interest may be represented as multiple pixels in high spatial resolution imagery. Thus, the improper selection of spectral and spatial resolutions can hinder the ability to extract information from imagery at the necessary level of detail.

Although there is an apparent need for careful selection of resolution characteristics according to the nature and goals of urban studies, there is no unanimous agreement on the sole importance of a specific resolution characteristic for selection of suitable imagery. Some researchers have observed that spatial resolution has generally been the single most important technical characteristic for urban remote sensing applications (Welch 1982). However, Jensen and Cowen (1999) discussed a rating system for determining the usefulness of imagery that emphasized the importance of other technical characteristics and argued that spatial resolution alone cannot determine the usefulness of imagery. While higher resolution generally means greater resolving ability in the spatial, radiometric, spectral, or temporal domains, requirements such as storage, processing time, and data exchange should be taken into account, especially if these are the limiting factors.

Selection of the most appropriate or optimal image resolution should also be based on the image processing and interpretation approach utilized. If visual interpretation and manual mapping techniques are employed, then the highest spatial resolution imagery available should generally be chosen. Extraneous detail is normally filtered out or ignored through the human cognitive element of visual interpretation. Of course, areal coverage and cost considerations may need to be traded-off with high spatial resolution, when making such a choice. If semi-automated, per-pixel image classification approaches are to be implemented, then moderate resolution imagery (e.g., SPOT HRV or Landsat TM) may provide superior land use and land cover mapping results for urban and urbanizing areas, relative to high spatial resolution image data (Chen et al 2004). Conversely, land use and land cover classification or linear feature (e.g., road) extraction based primarily on texture measures will tend to yield superior results when derived from high spatial resolution image data (Chen and Stow 2002).

7.3   Evolution of Urban Remote Sensing and a Gallery of Applications

With the advent of digital imaging and soft-copy photogrammetric techniques, the number, variety, and nature of remote sensing applications have increased tremendously. For example, the commercial availability of high spatial resolution imagery (primarily in panchromatic mode) has led to developments of image processing techniques targeting the improvement of coarse spatial resolution imagery. As detailed in Chapter 11, one way of achieving this is by image fusion techniques allowing the preservation of spectral detail in coarse resolution imagery and improvement of their spatial resolution (Ranchin et al. 2001). Additionally, the complexity and dynamics of urban environments often necessitate the use of multi-sensor, multi-resolution data sets, eventually leading to the use of data compression techniques such as principal component analysis (PCA), spatial filtering, and image transforms in the wavelet domain (Donnay et al. 2001). As a result, a number of image processing and analysis techniques have been tailored for implementation into efficient urban remote sensing analysis, resulting in broader and more diverse applications.

7.3.1   Early Urban Remote Sensing (1950–1970)

This section presents a chronological overview of urban remote sensing applications from 1950 to 1970. This is primarily a review of typical applications that utilized film-based systems (prior to the launch of the Landsat satellites), and were dominated by studies of urban and infrastructure development.

Studies of urban places using remotely sensed data can be found in the journal of Photogrammetric Engineering (now titled Photogrammetric Engineering and Remote Sensing) as far back as the second volume published in 1936. An annotated bibliography compiled by Chardon and Schwertz (1972) and leading remote sensing journals reveals that between 1950 and 1970, remote sensing researchers concerned with urban applications mostly published results from exploratory analysis of the potential of airborne remote sensing for the study **interest in "civilian" urban remote sensing applications began to develop in the 1930s** of urban environments and specific urban area problems. The three key areas of interest were: (1) urban land use and dynamics of the residential and industrial components (starting as early as in the 1950s), (2) mapping, analysis, and planning of the urban transportation infrastructure (late 1950s), and (3) mapping density and quality, of housing units, including environmental health conditions (starting as early as 1960s).

Geographical areas of interest during the 1950–1970 period varied considerably. Within the United States, Los Angeles and Phoenix appear to have been studied more often than other cities, with the objective of determining pollution levels and urban growth boundaries from remotely sensed imagery. In the early urban remote sensing period, mostly film-based cameras (using black and white, color, or color infrared film) were used to collect aerial photographs, but some studies used video and multi-spectral digital cameras. Radar, image spectroscopy, and thermal infrared imaging were also used, but far less frequently. In summary, the scope of early urban remote sensing was limited to urban planning and infrastructure development and was hindered by the limited availability of imagery. Since the 1950s, there has been a noticeable movement from studying urban infrastructure and morphology, to studying the human dimension of urban environments using remote sensing techniques.

7.3.2   *Recent Urban Remote Sensing (1971 to Present)*

This section covers the more recent period in urban remote sensing, 1971 to present, with an emphasis on the current state-of-the-art. Studies of urban land cover/land use, urban structure, and change are reviewed. Then, a variety of studies designed to study urban populations are outlined. Finally, the emergence of a diverse group of studies concerned with the social science aspects of urban places is traced.

7.3.2.1   Land Cover and Land Use, Urban Structure, and Change Analyses

Studies of urban land cover and land use are the backbone of urban remote sensing. Although the specific goals of many studies differ, most begin with the identification and classification of the land cover or land use within the urban scene. Due to the many complexities of urban environments researchers have devised new approaches designed to improve classification accuracies. One of such approaches has been the development of expert systems (see relevant discussions in Chapter 12); essentially logic based systems that allows for the integration of remotely sensed data with ancillary data in an attempt to provide more accurate classification results. Stefanov et al. (2001) developed an expert system to monitor urban land cover change in Phoenix, Arizona. System inputs included Landsat TM imagery, an image derived texture measure and vegetation index, water rights and land use data. A novelty of this research is that it was one of the first applications of such an approach for a semiarid-arid urban environment. The researchers found that improved classification accuracies support the use of an expert system, especially in urban environments with heterogeneous land covers. That same expert system was later modified for use with ASTER data (Stefanov 2002), and used to study land cover diversity across a number of different urban centers (Netzband and Stefanov 2003).

Hybrid approaches to urban land use and land cover mapping have also been developed. Lo and Choi (2004) designed a hybrid method using both unsupervised and supervised approaches as well as hard and soft classifications. They used Landsat 7 ETM+ data to study metropolitan Atlanta, Georgia. The authors found that the hybrid approach performed slightly better (compared to the unsupervised clustering approach), and that the supervised fuzzy classification was efficient in low-density urban use areas characterized by mixed pixels.

"hybrid" approaches refer the methods that combine both supervised and unsupervised classifications

Recently though, researchers have used more advanced sub-pixel methods such as spectral mixture models to quantify and measure the composition of urban land cover (Rashed et al. 2001; Small 2001; Wu and Murray 2003). These "soft" approaches to classification are becoming increasingly more popular as researchers attempt to model the heterogeneity that exists within urban environments. Rashed et al. (2001) used spectral mixture analysis (SMA) to derive physical measures of urban land cover that describe the morphological characteristics of the Greater Cairo region. The authors suggest that SMA may be superior to other standard classification methods, especially when used in an urban context. They applied their model to an Indian Remote Sensing multispectral image (IRS-IC LISS-III) of Cairo, Egypt, in an attempt to produce a replicable procedure to analyze the anatomy of cities. The study of Cairo was followed with a multi-temporal study of change analysis between the years 1987 and 1996 (Rashed et al. 2005). The authors sought to emphasize land cover change within classes (rather than between classes) in an attempt to model urban morphology.

Since urban environments are particularly complex, given the diversity of land cover and land use types, Herold et al. (2004) suggest the use of hyperspectral data as a way to deal with the spectral complexities of urban environments. The authors developed and analyzed a field spectral library in the 350–400 nm spectral range, consisting of approximately 4,500 individual spectra. They also evaluated the most suitable wavelength for the separation of urban land cover and related them to the spectral bands of two sensor systems, IKONOS and Landsat ETM+ (results of this evaluation are reviewed in Chapter 4). This interesting comparison highlights the spectral limitation of most current sensor systems – often the optimal bands lie outside or near the edge of the sensors' spectral range. The results of their study indicate the future potential of hyperspectral data for urban analysis (Chapter 9).

Image processing approaches such as image texture generation and spatial pattern analysis have been applied to the study of urban structure. Application of image texture to urban environments is particularly widespread. A variety of quantitative measures of image texture have been developed with the aim of capturing the underlying spatial structure of the urban scene (Brivio and Zilioli 2001). Barnsley et al (2001) modeled image texture using standard gray level co-occurrence matrices, and provided a good review of image texture methods. High within class variability is a common characteristic of urban regions, particularly in urban–rural transition areas (Pesaresi and Bianchin 2001). In an attempt to increase classification accuracies, texture measures

spatial pattern analysis is a method for quantifying the spatial arrangement of image brightness components in an image

have been utilized as an additional input in the classification process (Kontoes et al. 2000; Orun 2004; Shaban and Dikshit 2001; Stefanov et al. 2001). Stefanov et al. (2001) showed that texture measures of variance, generated from ASTER imagery, could be used to study levels of urbanization in cities (i.e., decentralized, centralized). Spatial pattern analysis is another method for analyzing image structure by quantifying the spatial arrangement of image brightness components in an image. This can be accomplished using geo-statistical techniques such as semi-variogram analysis as discussed by Brivio and Zilioli (2001).

Studies of urban change over both space and time have become increasingly important as the world's urban population continues to grow. The anatomy of spatial growth and change in Bangkok, Thailand was studied using Landsat 5-TM imagery (Madhaven et al. 2001). The researchers utilized two classification methods, a traditional maximum likelihood supervised classification and a biophysical, vegetation-impervious-soil (VIS) approach based on the model developed by Ridd (1995), which was reviewed in detail in Chapter 6. Wilson et al. (2003) developed a geo-spatial model to quantify and map urban growth. The model, which utilized Landsat TM imagery, provided researchers with a powerful visual and quantitative analysis of both the type and extent of urban sprawl.

Other methods have been developed with the aim of improving change analysis studies. Westmoreland and Stow (1992) incorporated ancillary data with fused SPOT panchromatic and Landsat TM multispectral images in an integrated image processing and GIS platform to aid in the visual interpretation of changed land use. A framework was developed to aid in the interactive identification of land use categories and land use updates through the simultaneous display of digital image data with the vector land-use data to be updated. In a study of urban expansion and flood risk assessment in Nouakchott, Mauritania, Wu et al. (2003) utilized image fusion techniques. SPOT panchromatic imagery were fused with multispectral imagery for two study dates, 1995 and 1999. The fused imagery were then incorporated with ancillary demographic data in a GIS to map areas of expansion and flood risk.

7.3.2.2   Urban Population Studies

Humans continuously modify their living environment. Remotely sensed imagery, both day and night-time, can be used to study urban populations, their size, density and distribution, through the characteristics of their environments. Urban populations have been studied using image-derived variables from color infrared photography, Landsat MSS and TM, and SPOT imagery, all captured under daylight conditions.

several methods have been developed since the mid-1970s to estimate urban populations from satellite observations

Ogrosky (1975) suggested the usefulness of satellite imagery for estimating urban populations. He showed how imagery derived variables such as urban area boundaries, transportation links, urban area of nearest largest neighborhood (NLN), and highway distance to the NLN were strongly associated with population count data taken from the census. High-altitude color infrared aerial photography at a scale of 1:135,000 was used to simulate and high-light the potential use of medium spatial resolution satellite imagery. Other research-ers applied similar models to study population density using remotely sensed imagery (Lindgren 1971; Lo and Welch 1977).

Iisaka and Hegedus (1982) introduced a method for estimating populations using Landsat MSS data. They developed a regression analysis model to test the relationship between population density and the reflected electromagnetic energy of their study area, Tokyo, Japan. The model provided a reliable estimate of popula-tion density. A similar method using SPOT imagery was developed by Lo (1995) and applied in Kowloon, Hong Kong, to estimate both population and dwelling units. The author showed that overall estimates were reasonable at the macro level, yet individual estimates of population and dwelling units were not reliable in areas of very high population density characterized by multi-use buildings. Harvey (2002) explored various refinements to the previously described methods for esti-mating population and achieved reasonably accurate results. He noted that accuracy was generally greater in suburban areas and that high densities were more often under-estimated while low densities over-estimated.

More recently some researchers have developed ways to integrate remotely sensed imagery with census (or survey) data to estimate populations. Chen (2002) developed a method to correlate areal census data with residential densities. The residential densities were classified from remotely sensed data using TM bands 1–5 and 7, together with a texture measure (derived from the TM imagery). Disaggregating areal census data via this approach was successful in a suburban environment.

Urban places have also been examined extensively through the use of night-time satellite imagery (see Chapter 17 for a discussion of the techniques). Human activ-ity and levels of development associated with urban places can be assessed using night-time imagery, allowing researchers not only to detect urban places, but also to monitor urban growth and to compare urban places via their relative light levels.

Night-time imagery from the DMSP OLS has proven extremely useful in the evaluation of urban areas in a variety of ways at both the global and regional scales. Although, available in analog format since 1972, the availability of OLS imagery in digital format beginning in 1992 spurred its use as a source for analyzing the dynamics and temporal aspects of urban places.

In order to explore the variation of population density within cities, Sutton et al. (1997) used DMSP OLS data to measure urban extent and estimate population within the United States. Although the model accounted for just 25% of the varia-tion in population densities, the authors emphasized the usefulness of OLS imagery as a foundation for further development. Later, Sutton et al. (2001) used DMSP OLS imagery to estimate both the urban population of every nation in the world, as well as national and global populations. Their quantitative analysis of human popu-lations was based on the known relationship between urban areal extent and popula-tion size established by Tobler (1969) and Nordbeck (1965). This type of analysis is particularly useful considering the large number of 'data poor' nations, places where either recent or reliable estimates of population size and distribution are not available. Given that estimating a country's population through a national population census is an enormously expensive and time-consuming undertaking, perhaps in the future satellite imagery might prove a dependable method for inter-censal estimates of national populations and urban growth.

Night-time lights imagery have also been useful as a data source for generating urban indicators that may enable better understanding of urban populations than demographic data alone. Lo (2002) used zones of radiance extracted from OLS data and a Triangulated Irregular Network (TIN). Surface area and volume were also generated and input into an allometric growth model used to estimate urban indicators for 35 cities in China. This method proved to be very accurate when used to estimate total non-agricultural population, a commonly used demographic measure of urbanization.

Gallo et al. (2004) also used DMSP OLS data of the United States to study changes in levels of night-time light emitted between 1992/1993 and the year 2000. In concert with the night-time lights data, the researchers also computed a time-integrated Normalized Difference Vegetation Index (TINDVI) for the same period across time and space. Those places where increased levels of emitted night-lights occurred also showed a decrease in the TINDVI. The authors suggest that areas of land cover change and increased night-light brightness have likely undergone increased urbanization.

DMSP OLS data have also been used to map human settlements (Tobler 1969; Elvidge et al. 1997) and delineate urban centers (Imhoff et al. 1997; Henderson et al. 2003). In an effort to create a scale-adjusted index of urban sprawl, Sutton (2003) used DMSP OLS radiance calibrated night-time lights imagery of the United States as a proxy for the extent of urban areas. Urban population counts from the 1990 census were then combined with the information on urban area in a regression model to allow for city-to-city comparisons of urban sprawl.

There have been some technical challenges to using the DMSP OLS data. Concern about the overestimation of urban extent and shift in locational accuracy is due in part to the difficulty in selecting light thresholds. One approach has been to use an alternative higher resolution image, i.e., Landsat TM, as a calibration data set (Henderson et al. 2003). The authors found that cities at different levels of development require different light (both stable and radiance) thresholds in order to assess urban size and location most accurately. Another issue has been that of pixel saturation due to the 6-bit radiometric resolution of the DMSP OLS data (Sutton 1997). Finally, an obvious limitation is the low spatial resolution of the imagery. The use of DMSP OLS data has been successful at the global, regional and city scale; however their effectiveness has not been shown at any finer scale.

7.3.2.3 Social Science Applications

Many forms of social data are increasingly being georeferenced, distributed, and archived. This opens up new avenues for the exploration of these data and allows for new ways of analyzing them, when combined with remotely sensed data.

The vast array of remotely sensed data now available to researchers allows for the analysis of the physical environment in ways not always possible through other methodologies. Remote sensing can serve as a link between the physical environment and the social lives of urban residents, and perhaps lead to improved understanding of the relationship between the built

> **remote sensing can serve as a link between the physical environment and the social lives of urban residents**

environment and numerous urban demographic and social processes.

Sociologists in the past seemed to overlook the potential of remote sensing for use in deriving indicators of social change (Blumberg and Jacobson 1997). However, a number of recent studies described here have integrated remote sensing data with urban population, quality of life, health, poverty, crime, and other similar socio-economic data types. These studies do much to promote the utility and inter-disciplinary attractiveness of remotely sensed data.

Many regions of the world are rapidly becoming more urbanized and the analysis of urban areas along with a greater understanding of quality of life is crucial to the understanding of what the future may hold for urban populations (Vlahov and Galea 2003). In a study that shows the utility of remote sensing for health and quality of life related research, Lo (1998) integrated environmental variables derived from satellite imagery, such as land use/land cover, NDVI, and surface temperature, with census data on population density, house value, education, and income. Using this combined data set he assessed quality of life in Athens-Clarke County, Georgia and found that remotely sensed data adds value to the study of urban quality of life.

The utility of remote sensing in famine early warning systems has also been explored (Hutchinson 1998). The author showed how famine is a function of many factors including economics, access to food, and environmental effects. Through the use of remote sensing, environmental effects are quantifiable and can be input into models designed to provide early warnings on famine. Although typically applied to rural areas at the regional or national scale, some work has focused on famine in urban areas and methods for adopting rural-centric models to urban environments (Bonnard 2000) in recognition of the growth in urban poverty and under-nutrition (Haddad et al. 1999).

The current pace of urbanization around the world calls for an appreciation of the demands, quality, and affordability of housing. Another study in the socio-economic realm sought to show the utility of remote sensing and GIS to identify potential low-income housing sites (Thomson and Hardin 2000). The authors used TM imagery of Bangkok, Thailand, to identify land cover and land use classes using a two-stage hybrid classification technique. Through a set of criteria (i.e., flood risk, density) input into a GIS, they were able to identify land area most suitable for low-income housing. The authors note that planners need to consider more than just the socio-economic factors of possible housing sites; they must also consider the physical characteristics of the sites.

Walsh and Welsh (2003) describe a variety of approaches for linking data across thematic domains in order to create data sets that extend across social, biophysical, and geographical fields. Although they applied their techniques in a rural setting, some of the methods they highlighted for linking people, place, and the environment are applicable in an urban or suburban setting.

Other types of health and well-being applications have also been explored. Fugate (2003) utilized quantitative measures of the physical environment derived from remotely sensed data to investigate the role of environmental context in relation to child health in Cairo, Egypt. In a study of intra-urban variability of fertility in Cairo, Egypt, Weeks et al. (2004) used imagery from the IRS sensor and processing methods such as spectral mixture analysis to obtain quantitative measures of environmental context, such as the amount of green space and the percentage of impervious materials in urban neighborhoods. They were then able to characterize in a quantitative manner the built and natural environment in each neighborhood. This is a good example of how, through the use of remotely sensed imagery, we can extract measures or details about the physical environment that can then be linked to the socio-economic environment.

In a novel approach to create an alternative to the urban/rural dichotomy for social science research, Weeks et al. (2005) developed an urbanness index derived from a combination of remotely sensed imagery and census data. As discussed in detail in Chapter 3, the urbanness index was developed based on the hypothesis that some variability in human behavior may be captured in surrogate form by knowledge of the built environment (gained through the use of remotely sensed data) and census measures, which act as proxy measures of the social environment.

Public health applications of remote sensing show much promise, and as such are worth reviewing here for their utility in quantifying features of the landscape and how they may affect health. This can of course be applied to both urban and rural places. The use of remote sensing has shown most success in the identification and monitoring of vector habitats, and thus allowing epidemiologists to make some level of disease predictions. Through the use of remotely sensed imagery the presence of specific vegetative cover, water, specific temperatures ranges, and elevation can all be used to characterize potential vector habitats. Tatem and Hay (2004) provide an up to date review of remote sensing and health in relation to urban environments. They also explore the potential of remote sensing for malaria research in an urban context. The use of remote sensing (and GIS) in the field of epidemiology has been most successful due to the capability to monitor environmental aspects of the earth's surface that are related to vector habitat. Considering the pace of urbanization, especially in developing countries, the public health community must refocus some of its efforts towards urban places.

Chapter Summary

In an attempt to synthesize the development and growth of urban remote sensing, this chapter presented the most prominent aspects of the evolution of imaging sensor systems and urban remote sensing applications over the past

50 years. Many other aspects of urban remote sensing have only been addressed briefly here, yet are presented in more detail in Chapters 4 and 5, which are devoted to spectral and temporal properties of urban environments. In addition to relative discussions in this book, the paper by Jensen and Cowen (1999) provides a comprehensive review of the spatial, spectral, and temporal characteristics of urban attributes and remote sensing systems. The paper covers data availability and requirements for a number of the most common earlier remote sensing applications such as land use and land cover classification, building and cadastral infrastructure mapping and planning, and utility and transportation system analysis.

Further developments within the field of remote sensing and the allied mapping sciences of GIS and cartography, including the increasingly available commercial, high spatial resolution imagery, should enhance the ability to learn and understand more aspects of urban places and populations. Besides some challenges and inconclusive results, the first attempts have already been made to link the physical and social attributes of urban environments. Morphing physical remote sensing with social sciences appears to be the current trend in studies of urban environments. Here, we have attempted to emphasize that the joint evolution of remote sensing technologies and urban remote sensing applications is what keeps this research theme alive.

LEARNING ACTIVITIES

Exercises

- Conduct an Internet search for the websites of the following remote sensing imaging systems (LANDSAT [MSS, TM, ETM+]; NOAA [AVHRR, DMSP-OLS]; SPOT [HRV, HRVIR, HRG]; IRS [LISS-II; LISS-III; PAN]; ERS and JERS [SAR-AMI; SAR-OPS]; RADARSAT [SAR]; GOES; AVIRIS, EarlyBird; OrbView; EOS [ASTER; MISR; MODIS]; IKONOS; QuickBird; EO [HYPERION; ALI]; EnviSat [MERIS; ASAR]). Use the technical documentation available on these websites to construct a comparative matrix for some or all of the above-listed imaging sensors detailing the following information: sensor name, sensor operational country and/or organization, operational period (mission's start and end dates, if applicable), swath, spatial resolution, spectral coverage, spectral resolution and number of spectral bands, radiometric resolution, revisit time-intervals (if applicable); and mode of sensing (i.e., passive/active).

- Use the matrix you constructed in the above exercise to carryout a feasibility study for the use of remote sensing in the investigation of one or more of the following urban problems. Your feasibility study(ies) should articulate the nature of information that needs to be extracted from the remotely sensed imagery in a given investigation; the spatial, spectral, radiometric, and temporal resolution requirements for the investigation; a list and justification of the type, date, and number of images needed (you can use more than one sensor type in one investigation); and a brief highlight or workflow of how you will use the image(s) to fulfill the purpose of the investigation.

  - The changing distribution of the world's urban population between 1975 and 1995 (you can choose to keep the scope of your investigation global or limit it to some regions of the world)
  - Urban gentrification process in a city of your choice
  - Post-war urban planning in a city located either in Afghanistan or Bosnia
  - Environmental impact assessment in a city affected by a storm-surge or flash floods (e.g., New Orleans after hurricane Katrina)
  - Comparative study of the urban heat islands effects in Manchester, UK (an industrial city), and Elenite, Bulgaria (a Black Sea resort)
  - Planning for a new highway around Mexico City
  - Comparative study of the nature of spatial form of ethnically segregated areas in Pretoria, South Africa, in 1988
  - Relationship between environmental stressors (e.g., air pollution, environmental degradation, deforestation, etc.), socioeconomic income distribution, and rates of asthma hospitalizations in the city of Singapore

References

Barnsley MJ, Moller-Jensen L, Barr AL (2001) Inferring urban land use by spatial and structural pattern recognition. In: Donnay JP, Barnsley MJ, Longley PA (eds) Remote sensing and urban analysis. Taylor & Francis, London, pp 115–144

Blumberg D, Jacobson D (1997) New frontiers: remote sensing in social science research. Am Sociol 28(3):62–68

Bonnard P (2000) Assessing urban food security, Famine Early Warning System (FEWS) Project Report. Online document: http://www.eldis.org/static/DOC1747.htm. Accessed 20 Feb 2009

Brivio PA, Zilioli E (2001) Urban pattern characterization through geostatistical analysis of satellite images. In: Donnay JP, Barnsley MJ, Longley PA (eds) Remote sensing and urban analysis. Taylor & Francis, London, pp 39–54

Chardon RE, Schwertz EL (1972) An annotated bibliography of remote sensing applied to urban areas, 1950–1971. School of Geoscience Louisiana State University, Baton Rouge, LA

Chen K (2002) An approach to linking remotely sensed data and areal census data. Int J Remote Sens 23:37–48

Chen D, Stow D (2002) The effect of training strategies on supervised classification at different spatial resolutions. Photogramm Eng Remote Sens 68:1155–1161

Chen D, Stow D, Gong P (2004) Examining the effect of spatial resolution on classification accuracy: an urban environmental case. Int J Remote Sens 25:2177–2192

Dickinson LC, Boselly SE, Burgmann WW (1974) Defense meteorological satellite program: User's guide. Air Weather Service (MAC), US Air Force

Donnay JP, Barnsley MJ, Longley PA (eds) (2001) Remote sensing and urban analysis. Taylor & Francis, London

Elvidge C, Baugh KE, Hibsin VR, Kihn EH, Kroehl HW, Davis ER, Cocero D (1997) Satellite inventory of human settlements using nocturnal radiation emissions: a contribution for the global tool chest. Global Change Biol 3:397–395

Fugate D (2003) Environmental context and health: an intra-urban level analysis in Cairo, Egypt. M.A. thesis, San Diego State University, San Diego, CA

Gallo KP, Elvidge CD, Yang L, Reed C (2004) Trends in night-time city lights and vegetation indices associated with urbanization within the Conterminous USA. Int J Remote Sens 25:2003–2007

Haddad L, Ruel MT, Garrett JL (1999) Are urban poverty and undernutrition growing? Some newly assembled evidence. World Dev 27:1891–1904

Harvey JT (2002) Estimating census district populations from satellite imagery: some approaches and limitations. Int J Remote Sens 23:2071–2095

Henderson M, Yeh ET, Gong P, Elvidge C, Baugh K (2003) Validation of urban boundaries derived from global night-time satellite imagery. Int J Remote Sens 24:595–609

Herold M, Roberts DA, Gardner M, Dennison P (2004) Spectrometry for urban area remote sensing: development and analysis of a spectral library from 350 to 2400 nm. Remote Sens Environ 91:304–319

Hutchinson CF (1998) Social science and remote sensing in famine early warning. In: Liverman DM, Moran EF, Rindfuss EE, Stern PC (eds) People and pixels. National Academy Press, Washington, DC, pp 189–196

Iisaka J, Hegedus E (1982) Population estimation from Landsat imagery. Remote Sens Environ 12:259–272

Imhoff ML, Lawrence WT, Stutzer DC, Elvidge CD (1997) A technique for using composite DMSP/OLS 'city lights' satellite data to map urban area. Remote Sens Environ 61:361–370

Jensen JR (1996) Introductory digital image processing. Prentice-Hall, Upper Saddle River, NJ

Jensen JR, Cowen DC (1999) Remote sensing of urban/suburban infrastructure and socio-economic attributes. Photogramm Eng Remote Sens 65:611–622

Kontoes CC, Raptis V, Lautner M, Oberstadler R (2000) The potential of kernal classification techniques for land use mapping in urban areas using 5m spatial resolution IRS-1C imagery. Int J Remote Sens 21:3145–3151

Lindgren DT (1971) Dwelling unit estimation with color-IR photos. Photogramm Eng 37:373–377

Lo CP (1995) Automated population and dwelling unit estimation from high-resolution satellite images: a GIS approach. Int J Remote Sens 16:17–34

Lo CP (1998) Application of Landsat TM data for quality of life assessment in an urban environment. Comp Environ Urban Syst 21:259–276

Lo CP (2002) Urban indicators of China from radiance-calibrated digital DMSP-OLS nighttime images. Ann Assoc Am Geographers 92:225–240

Lo CP, Choi J (2004) A hybrid approach to urban land use/cover mapping using Landsat 7 Enhanced Thematic Mapper Plus (ETM+) images. Int J Remote Sens 25:2687–2700

Lo CP, Welch R (1977) Chinese urban population estimates. Ann Assoc Am Geographers 67:246–253

Madhaven BB, Kubo S, Kurisaki N, Sivakumar TVLN (2001) Appraising the anatomy and spatial growth of the Bangkok Metropolitan area using a vegetation-impervious-soil model through remote sensing. Int J Remote Sens 22:789–806

Netzband M, Stefanov WL (2003) Assessment of urban spatial variation using ASTER data. Int Arch Photogramm Remote Sens Spat Inf Sci 34(7/W9):138–143

Nordbeck S (1965) The law of allometric growth. Michigan Inter-University Community of Mathematical Geographers, Ann Arbor, Paper 7

Ogrosky C (1975) Population estimates from satellite imagery. Photogramm Eng Remote Sens 41:707–712

Orun AB (2004) Automated identification of man-made textural features on satellite imagery by bayesian networks. Photogramm Eng Remote Sens 70:211–216

Pesaresi M, Bianchin A (2001) Recognizing settlement structure using mathematical morphology and image texture. In: Donnay JP, Barnsley MJ, Longley PA (eds) Remote sensing and urban analysis. Taylor & Francis, London, pp 55–68

Ramsey SWE, MS CPR (2001) Monitoring urban land cover change: an expert system approach to land cover classification of semiarid to arid regions. Remote Sens Environ 77:173–185

Ranchin T, Wald L, Mangolini M (2001) Improving the spatial resolution of remotely-sensed images by sensor fusion: a general solution using the ARSIS method. In: Donnay JP, Barnsley MJ, Longley PA (eds) Remote sensing and urban analysis. Taylor & Francis, London, pp 21–38

Rao RS, Cook AM (2004) New eye in the sky. RESOURCESAT-1, Imaging Notes, Winter 2004, pp 9–30

Rashed T, Weeks JR, Gadalla M, Hill A (2001) Revealing the anatomy of cities through spectral mixture analysis of multispectral satellite imagery: a case study of the greater Cairo region, Egypt. Geocarto Int 16(4):5–16

Rashed T, Weeks JR, Stow DA, Fugate D (2005) Measuring temporal compositions of urban morphology through spectral mixture analysis: toward a soft approach to change analysis in crowded cities. Int J Remote Sens 26:699–718

Ridd M (1995) Exploring a V-I-S (Vegetation-Impervious Surface-soil) model for urban ecosystem analysis through remote sensing: comparative anatomy of cities. Int J Remote Sens 16:2165–2185

Schowengerdt RA (1997) Remote sensing: models and methods for image processing. Academic, San Diego, CA

Shaban MA, Dikshit O (2001) Improvement of classification in urban areas by the use of textural features: the case stidy of Lucknow city, Uttar Pradesh. Int J Remote Sens 22:565–593

Small C (2001) Estimation of urban vegetation abundance by spectral mixture analysis. Int J Remote Sens 22:1305–1334

Stefanov WL (2002) Assessment of landscape fragmentation associated with urban centers using ASTER data. Am Geophys Union EOS Trans 83(47). Abstract B61C-0739

Stefanov WL, Christensen PR, Ramsey MS (2001) Remote sensing of urban ecology at regional and global scales: results from the Central Arizona-Phoenix LTER site and ASTER Urban Environmental Monitoring program, In: Juergens C (ed) Remote sensing of urban areas – Fernerkundung in urbanen Räumen. In: Proceedings (Abstracts and Full papers on Supplement CD-ROM) of the 2nd international symposium held in Regensburg/Germany, 22–23 June 2001. Regensburger Geographische Schriften 35, pp 313–321 (on supplemental CD ROM)

Sutton P (1997) Modeling population density with night-time satellite imagery and GIS. Comp Environ Urban Syst 21:227–244

Sutton P (2003) A scale-adjusted measure of "urban sprawl" using nighttime satellite imagery. Remote Sens Environ 86:353–369

Sutton P, Roberts DA, Elvidge C, Meij H (1997) A comparison of nighttime satellite imagery and population density for the Continental United States. Photogrammetric Engineering & Remote Sensing 63:1303–1313

Sutton P, Roberts DA, Elvidge C, Baugh K (2001) Census from Heaven: an estimate of the global human population using night-time satellite imagery. Int J Remote Sens 22:3061–3076

Tatem AJ, Hay SA (2004) Measuring urbanization pattern and extent for Malaria research: a review of remote sensing approaches. J Urban Health 81:363–376

Thomson CN, Hardin N (2000) Remote sensing/GIS integration to identify potential low-income housing sites. Cities 17:97–109

Tobler W (1969) Satellite confirmation of settlement size coefficients. Area 1:30–34

Vlahov D, Galea S (2003) Urban health: a new discipline. Lancet 362:1091–1093

Walsh SJ, Welsh WF (2003) Approaches for linking people, place, and environment for human dimensions research. Geocarto Int 18(3):51–61

Weeks JR, Getis A, Hill AG, Gadalla MS, Rashed T (2004) The fertility transition in Egypt: intra-urban patterns in Cairo. Ann Assoc Am Geographers 94:74–93

Weeks JR, Larson D, Fugate D (2005) Patterns of urban land use as assessed by satellite imagery: an application to Cairo, Egypt. In: Entwisle B, Rindfuss R, Stern P (eds) Population, land use, and environment: research directions. National Academies Press, Washington, DC, pp 265–286

Welch R (1982) Spatial resolution requirements for urban studies. Int J Remote Sens 3:139–146

Westmoreland S, Stow DA (1992) Category identification of changed land-use polygons in an integrated image processing/geographic information system. Photogramm Eng Remote Sens 58:1593–1599

Wilson EH, Hurd JD, Civco DL, Prisloe MP, Arnold C (2003) Development of a geospatial model to quantify, describe and map urban growth. Remote Sens Environ 86:273–285

Wu CS, Murray AT (2003) Estimating impervious surface distribution by spectral mixture analysis. Remote Sens Environ 84:493–505

Wu W, Courel MF, Rhun KL (2003) Application of remote sensing to the urban expansion analysis for Nouakchott, Mauritania. Geocarto Int 18(1):17–24

Chapter 8
Classification of Urban Areas: Inferring Land Use from the Interpretation of Land Cover

Victor Mesev

Ancillary data are vital for successful image classification of urban areas. This chapter explores the role of ancillary data (information from beyond remote sensing) for improving the contextual interpretation of satellite sensor imagery during spectral-based and spatial-based classification. In addition, careful consideration is given to the crucial distinctions between urban land cover and urban land use, and how the inherent heterogeneous structure of urban morphologies is statistically represented between hard and soft classifications.

Learning Objectives

Upon completion of this chapter, you should be able to:

❶ Distinguish between land use and land cover
❷ Explain how between hard and soft classification
❸ Speculate on the role of ancillary data for spectral-based and spatial-based classification

---

8.1 Introduction

We live in a multi-faceted world where our cities are composed of a complex assemblage of both tangible substances and communicable interaction. In order to avoid sensory overload the human brain is designed to focus on objects in a systematic order while at the same time filtering out excess noise. Similar principles apply to the classification of digital remotely sensed data where, instead of the brain, a computer algorithm is employed to make sense of pixels by identifying intrinsic spectral

V. Mesev (✉)
Department of Geography, Florida State University, Tallahassee, FL 32306-2190, USA
e-mail: vmesev@fsu.edu

T. Rashed and C. Jürgens (eds.), *Remote Sensing of Urban and Suburban Areas*,
Remote Sensing and Digital Image Processing 10,
DOI 10.1007/978-1-4020-4385-7_8, © Springer Science+Business Media B.V. 2010

classification algorithms seek to reveal the intrinsic spectral and spatial properties of imagery that are indicative of real patterns and processes

and spatial properties that are indicative of real urban patterns. Of course, as discussed in Chapters 4, 5 and 7, the representation of reality is heavily simplified by the inherent limitations of the sensing instrumentation, which is further hampered by the elevated platform of satellite sensors high above the earth. Nevertheless, remote sensing is rapidly gaining a crucial role in many urban applications, from monitoring urban growth and density attenuation to models that predict quality of life and sustainable development (see Mesev 2003). The interest in urban remote sensing is further fuelled by the availability of satellite sensor imagery at a scale that begins to replicate that of aerial photography. It is worth noting at this stage that although aerial photographic interpretation has revolutionized the mapping of urban areas, all reference to remotely sensed data in this chapter will imply digital representations from multispectral satellite sensors. The reason is that satellite sensors provide a much more flexible data source, capable of comprehensive coverage at multispectral wavelengths, captured across multitemporal intervals, and globally available at relatively lower costs than aerial photography.

The vast majority of contemporary research into urban classification involves the

urban classification can be considered in different ways, whether it is spectral based or spatial based; hard or soft; per-pixel or sub-pixel

manipulation of image pixels to represent either single (hard/crisp) or multiple (soft) thematic classes, and to be treated as either individual entities (spectral based classification) or as groups and objects (spatial based classification). This chapter will outline the main technological and methodological developments in each by demonstrating modifications to the popular per-pixel spectral maximum likelihood decision rule (Strahler 1980; Mesev 1998, 2001) and a spatial method based on nearest-neighbor calculations (Mesev 2005; Mesev and McKenzie 2005). In both, strong cases will be made for the incorporation of ancillary data from beyond the spectral domain for improving classification accuracy – where ancillary data are most conveniently handled by geographical information systems (GIS). In addition, coverage will be given to the conceptual distinctions between urban land cover and urban land use (Dobson 1993); a sensitive dichotomy that underpins all urban remote sensing applications (Mesev 2003).

8.2   Urban Image Classification

In simple terms, the classification of remotely sensed data is the process of generating thematic interpretations from digital signals that represent the world. In even stricter terms, thematic image classification is little more than a statistical data reduction procedure, generalizing from continuous to categorical, or can be visualized as a conversion of data, usually from interval to nominal levels of measurement. For example, an image with an 8-bit radiometric resolution of 256 graylevels at seven spectral bands, when classified, would result in typically 10–20 thematic classes. Moreover, regardless of the degree of sophistication in the classification methodology, the grouping of "similar" pixels, either based on spectral or spatial rules, invariably will be highly subjective, strongly dependent on training samples and class choices, as well as heavily biased by local and scene-specific conditions (Forster 1985; Webster 1995; Cowan and Jensen 1998). In all, the resulting classified image will be a model vastly removed from reality, even more so when representing urban areas. Why, then, should multidimensional image data be classified in the first place? Well, despite wide-ranging caveats, the whole premise for classification can be distilled to matters related with communication. More specifically, communication not only between image data and application objectives but also between data within integrated GIS databases. Unclassified image data may be more objective and have a greater range of measurement but only classified thematic interpretations are acceptable for most applications. Few integrated projects are capable of handling, let alone communicating, graylevel values or digital numbers; they instead demand more pragmatic and user-friendly semantic interpretations that are also consistent with data already present in the GIS (Mesev 1997).

classification usually entails conversion of data from interval to nominal levels of measurements based on spectral or spatial rules

Classification of multispectral satellite sensor imagery, therefore, involves the identification and statistical grouping of pixels with similar digital numbers and/or similar spatial orientation of pixels into meaningful geographical features. In urban examples, this would entail the recognition and conversion of pixels with similar multispectral values and/or positional location into thematic categories with meaningful labels, such as buildings, roads, parks, gardens, etc. The distinction between classifications based purely on spectral signatures and those that also exploit the spatial arrangement of pixels is a particularly important development in urban applications. Unlike the natural environment, urban surfaces are typically a complex mix and intricate arrangement of artificial and natural objects of irregular size, frequently angular in shape, and exhibiting variable density. Such unpredictable compositions and configurations are difficult to replicate from remotely sensed data if the classification relies solely on spectral information (Baraldi and Parmiggiani 1990; Couloigner and Ranchin 2000).

the principal difference between spectral and spatial classification is a matter of whether similar pixels are collected in isolation (spectral) or as part of contiguous groups (spatial)

8.2.1   Urban Land Use from Land Cover

The statistical mechanisms for the classification of images representing urban areas are almost identical to those used for the natural environment. The fundamental exception is the inference of land use from land cover (Chapter 6, also see Dobson

1993; Heikkonen and Varfis 1998). Remote sensing by definition involves sensors that are not in contact with their objects and as such remotely sensed data are capable of only measuring energy deflected off, or emitted from, physical substances from a distance. Environmental applications, such as those focusing on the monitoring of deforestation, changes in meteorological and hydrological cycles are closely amenable to such types of measurements. However, urban areas, although definable by tangible substances such as brick, concrete, metal, vegetation and so on, which are in themselves capable of being detected from a distance by remote sensors, are also collections of human habitation and social interaction, which are not directly tangible and therefore impossible to measure remotely. Some purists would like to limit the range of urban thematic categorization from remote sensing to just land cover, and indeed as illustrated in Chapter 6, the Ridd (1995) Vegetation-Impervious-Soil three-way model comfortably fits within the technical capabilities of remote sensor instrumentation (Fisher 1997). However, and as alluded earlier, urban, unlike most environmental, remote sensing classifications underpin a variety of pragmatic planning and policy-making applications which demand far more meaningful labels (Donnay 1999). They demand semantics that describe how physical and socio-economic characteristics are not only used but how they also function. Fortunately, there are conceptual and mechanical rationales between urban land cover and land use, where one is dependent on the other (Dobson 1993). The process of inferring anthropogenic land use from the physical configuration of land cover is an extricable link between form and function (Geoghegan et al. 1998). For example, changes in the physical shape, size and density of urban built morphologies are an indication of urban growth, which in turn is a proxy for inferring processes such as a suburbanization of the population and a decentralization of commerce (see relative discussions in Chapters 2 and 3). Similarly, measures of ratios between built and biomass may be the outcome of quality of life indicators that can be validated by reference to census parameters (inter alia Brugioni 1983; Ogrosky 1975; Donnay 1999; Chen 2002; Harvey 2002). Nevertheless, all inferences between land cover and land use should be treated as tenuous and should be attempted always with respect to application objectives and the scale of investigation (Woodcock and Strahler 1987).

> **land cover refers to physical properties, readily measurable by remote sensor data**

> **land use, which is not directly measurable by remote sensor data, is an inference of anthropogenic activities from land cover composition and configuration**

Hard Versus Soft Classification: Is Refining Scale the Solution?

Whether spectral or spatial, the inherent geometrical variability of the urban landscape dictates that the scale of measurement is a major factor in both the quality and detail of the resulting classification (Welch 1982). Recent launches of satellites with sensors capable of super fine spatial resolutions of 4 m and less (for example, IKONOS and QuickBird) may have improved visual clarity of urban features but at the same time have also compounded classification ambiguity (Welch 1982; Corbley 1996). The reason is that because of the severe spatial heterogeneity in the composition of urban surfaces, pixels at fine spatial resolution will represent variable mixtures of similar urban objects. Any divergence in the proportion of reflected energy representing urban surfaces (no matter how small) will result in dissimilar pixel values, which in turn, will increase the range of thematic classes. Even adjacent pixels may represent slightly different spectral signatures. This is known as spectral "noise" but is less of a problem for sensors of coarser spatial resolution as variations in urban reflectance are averaged to more similar pixel values and hence a narrower range of possible thematic classes. The use of finer spatial resolution data is further complicated by the degradation of the radiometric resolution. Increasing the spatial resolution capabilities of a sensor essentially narrows the instantaneous field of view (IFOV), which means less energy is collected by the sensor in a shorter period of time (Fisher 1997). In both fine and coarse spatial resolutions the standard classification assumes that only one thematic class is to be allocated to each individual pixel. This perpetuates the aggregated nature of the pixel; where it is conceivable that two pixels of identical multispectral values may represent quite variable combinations of urban land cover properties. In response, rather than deterministically allocating each pixel one exclusive class ('hard' or 'crisp' classification) some techniques employ stochastic mechanisms capable of estimating multiple class memberships within individual pixels ('soft' or 'fuzzy' classification). Hard, or crisp, classification assumes each pixel is assigned one, and only one, thematic class. Soft classification, on the other hand, lifts the restriction and allows each output pixel to be labeled with multiple classes. In a sense, the principles of soft classification are more in tune with the continuous nature of the earth's surface, especially land defined as urban. However, from a pragmatic standpoint, single classes produce simpler maps and are still deemed to be more aesthetically communicable, especially to users with little or no knowledge of remote sensing.

8.2.2   Spectral Classifications

As noted already accurate classifications of remotely sensed data representing urban areas are difficult, and are, on the whole, dependent on the scale of the image data and the scope of the application. Even a coarse classification of built-up land cover will frequently fail the 85% minimum level of accuracy set out by the Anderson criteria (Anderson et al. 1976). The spatial configuration of urban areas

the "targeted" level of image classification depends on the scale of the image data and the scope of the application

is intrinsically complex and typically unpredictable, regardless of the spatial resolution of the data. Logically, any spectral classification, either hard or soft, will generally suffer lower accuracy as a result of mixed pixels (mixels). Inherent spatial variability of urban land cover and acute spectral heterogeneity between pixel values has typically led to lower interpretation accuracy (Baraldi and Parmiggiani 1990). To increase accuracy, methodologies have attempted to incorporate ancillary information (Mesev 1997), usually during the classification process (for example, fuzzy sets, neural nets (Dreyer 1993), and Bayesian modifications (Mesev 1998, 2001)). Other approaches have focused on developing the three-dimensional perspective of urban remote sensing by using, for example, LIDAR data (Laser-induced Detection and Ranging) (Barnsley et al. 2003), interferometric SAR (Synthetic Aperture Radar) data (Grey et al. 2003) and hyperspectral image interpretation (see Chapter 9). The creation of soft classification algorithms, where pixels are allowed to represent more than one class, was a direct response to the inflexibility of hard classifications to adequately represent the continuum of reality (Ji and Jensen 1999). Fuzzy set theory, multiple endmember spectral mixture models (Rashed et al. 2003) and decision rules based on artificial neural networks were employed to fit spectral values using complex recursive mathematical relationships. Output is multivariate, and as such is closer to a more representative interpretation of the continuum of the natural landscape than output from hard classifications. However, soft classified images are also more difficult to understand and their multiple class schemes are less comparable with GIS data.

8.2.3   Spatial Classifications

spatial classifications are separated into those that focus on traditional tonal contrasts among pixels, those that examine the broader inter-relationships between pre-defined groups of pixels representing distinct objects

Given the limitations of spectral-based classifications, there has been a recent re-focus on techniques that are more responsive to the inherent complex spatial configuration of urban areas. Spatial classifications can be loosely separated into those that focus on traditional tonal contrasts by measuring grayscale variations of adjacent pixels, and those that examine the broader inter-relationships between pre-defined groups of pixels representing complete objects. Research on the spatial characterization of urban land cover has gained momentum in recent years bringing about the development of a number of quantitative indices. Amongst the most widely applied have been the scale invariant properties of fractal geometry (see relevant discussion in Chapter 12), which many proponents have argued are capable of measuring the structural complexity and fragmentation of highly heterogeneous urban land cover (Batty and Longley 1994;

Lam et al. 1998). These same objectives were also tested using syntactic pattern recognition systems employing graph-based methodologies (Barnsley et al. 2003), object-oriented algorithms (Chapter 10), and automated expert systems. More recently, spatial metrics have revitalized fractal geometry as part of a suite of indices, including the contagion and patch density measurements (Herold et al. 2002), as well as spatial metrics directly related to urban sprawl (Hasse and Lathrop 2003). Central to the formulation of these metrics is the concept of photomorphic regions, or homogeneous urban patches, which are routinely extracted from aerial photographs but are more of a challenge from satellite sensor imagery. The two principal criticisms of spatial metrics are that their functionality is completely dependent on an initial spectral characterization of the satellite sensor imagery, and that they are conspicuously absent from the actual process of the characterization of homogeneous classes. In this sense they are merely measuring the outcome of the classification regardless of the accuracy. Nevertheless, spatial metrics have channeled contemporary remote sensing research towards group and object based classification, including modifications of established texture analysis based on the spatial co-occurrence matrix (Haralick et al. 1973), geostatistics (Carr 1999; Pesaresi and Bianchin 2001) and wavelet theory (Myint 2003).

8.3   Modified Maximum Likelihood Classification

There are many hard classifications, some statistically deterministic (minimum distance, parallelepiped), others, like the popular maximum likelihood (ML), are based on stochastic mechanisms. The objective is to assign the most likely class w_j, from a set of N classes, $w_1, \ldots, w_N$, to any feature vector $\mathbf{x}$ in the image. A feature vector $\mathbf{x}$ is the vector $(\mathbf{x}_1, \mathbf{x}_2, \ldots, \mathbf{x}_M)$, composed of pixel values in M features (in most cases, spectral bands). The most likely class w_j for a given feature vector $\mathbf{x}$ is the one with the highest posterior probability $\Pr(w_j|\mathbf{x})$. Therefore, all $\Pr(w_j|\mathbf{x})$, $j \in [1, \ldots, N]$ are calculated, and w_j with the highest value is selected. The calculation of $\Pr(w_j|\mathbf{x})$ is based on Bayes' theorem,

$$\Pr(w_j|\mathbf{x}) = \frac{\Pr(\mathbf{x}|w_j) \times \Pr(w_j)}{\Pr(\mathbf{x})} \tag{8.1}$$

On the left hand side is the posterior probability that a pixel with feature vector $\mathbf{x}$ should be classified as belonging to class w_j. The right hand side is based on Bayes' theorem, where $\Pr(\mathbf{x}|w_j)$ is the conditional probability that some feature vector $\mathbf{x}$ occurs in a given class: in other words, the probability density of w_j as a function of $\mathbf{x}$. Supervised classifications, such as the ML, derive this information from training samples. Often, this is done parametrically by assuming normal class probability densities and estimating the mean vector and covariance matrix. Alternatively, it is possible to use Markov random fields (Berthod et al. 1996), or non-parametric methods, such as k-Nearest Neighbor (kNN). The "standard" kNN methods directly implement a decision function based on the number of training pixels per class proportional to the prior probability. This is the prior probability of the occurrence of w_j irrespective of its feature vector, and as such is open to estimation by prior knowledge external to the remotely sensed image. Typically, ML classifiers assume prior probabilities to be equal and assign each $Pr(w_j)$ a value of 1.0 (Strahler 1980). However, variations in prior probabilities can be an important step forward for reducing the detrimental effects of spectrally overlapping classes. If a feature vector **x** has probability density values that are significantly different from zero for several classes, it is not inconceivable for that pixel to belong to any of these classes. When selecting a class solely on the basis of its spectral characteristics, a large probability of error frequently results. The use of appropriate prior probabilities, based on reliable ancillary information, is one way to reduce this error in class assignments. Moreover, it would seem intuitively more sensible to suggest that some classes are more likely to occur than others.

Many proprietary software packages (such as ERDAS Imagine, ENVI and Idrisi) allow the use of prior probabilities, where the user is expected to make estimates by using information on the anticipated (relative) class areas. The increase of classification accuracy from these 'global priors' is, however, often limited. At the other extreme, a vector of prior probabilities for each individual pixel is pointless, because that would be tantamount to a completed classification. A compromise scale somewhere between the global and individual scales can be derived by first subdividing the image into strata, or segments, according to ancillary context data, and then finding the local prior probability vector for each stratum. For example, Mesev (1998; 2001) used extraneous data from the population census to segment and classify a Landsat Thematic Mapper (TM) image according to contextual and Bayesian rules on housing classes (Table 8.1 and Fig. 8.1).

Prior probabilities for three housing types (high, medium and low density) are entered into the ML decision rule at the stratified enumeration district (local) level. Results in Table 8.1 and Figs. 8.2 and 8.3 are based on comparisons of class area estimates of classifications of the three housing types generated by equal and stratified unequal prior probabilities. Figure 8.3 shows area estimates for most urban land use classes produced from the Bayes' modified-ML classifier to be closer to those derived from the size-ratio transformed census figures. Total absolute error in all settlements is consistently lower under conditions of unequal as opposed to equal prior probabilities. However, in terms of housing, there are considerable variations between types and across the five settlements. No one housing type has consistently lower area estimation error but there is some evidence to suggest that high density housing is under-predicted (i.e. less pixels classified), and conversely low density housing is over-predicted (i.e. more pixels classified). The reason for this may lie in the highly concentrated nature of British housing in central areas of towns.

The spatial extent of individual houses around the central core may sometimes be much smaller than the spatial resolution of the satellite sensor. However, what becomes apparent from these results is that classifications are highly site specific, and they underline the immense problems that arise when sub-residential classifications

Table 8.1 Average prior probabilities for housing density derived from census estimates, and results from maximum likelihood classifications using equal and unequal prior probabilities based on area estimates compared to census numbers

Settlement	Average census prior probabilities			
	Low density	Medium density	High density	Total/absolute
Bristol	0.094	0.297	0.627	1.0
Swindon	0.155	0.266	0.579	1.0
Bath	0.247	0.237	0.516	1.0
Taunton	0.104	0.302	0.594	1.0
	Area estimation (%)			
Bristol equal priors	+12.13	−5.04	−7.42	24.59
Bristol unequal priors	+7.44	−9.89	+2.13	19.46
Swindon equal priors	+11.58	−8.18	−3.94	23.70
Swindon unequal priors	+8.51	+5.56	+4.07	18.14
Bath equal priors	−2.79	+9.94	−8.20	20.93
Bath unequal priors	−1.09	+5.19	−4.11	10.39
Taunton equal priors	+2.35	+4.22	−5.26	11.83
Taunton unequal priors	−0.88	+0.68	+0.31	1.86

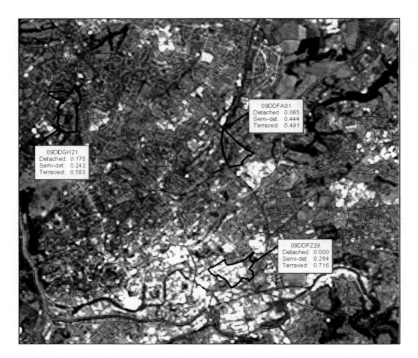

Fig. 8.1 Samples of local prior probabilities displayed alongside a Landsat TM 7 natural color composite for the city of Bristol. Local prior probabilities are generated from housing data recorded by UK Census of Population and are represented by census collection units called enumeration districts (e.g. 09DDFA01, 09DDGH21, and 09DDFZ29)

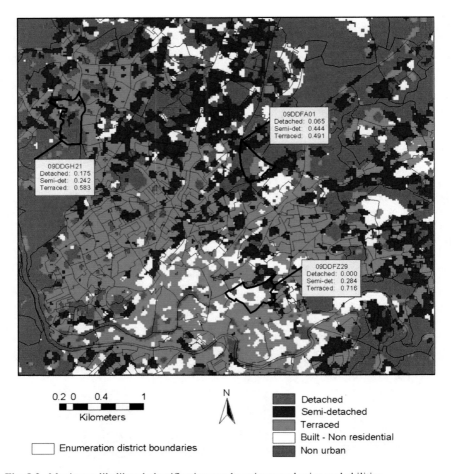

Fig. 8.2 Maximum likelihood classification results using equal prior probabilities

are attempted. For example, consider the case of Bristol in Fig. 8.1 where only the
northwestern part of the city is shown for clarity. A visual comparison of urban land
use coverage between equal priors (Fig. 8.2) and unequal priors (Fig. 8.3) reveals
interesting patterns. More pixels have been classified as detached (single dwelling)
and semi-detached (duplex dwelling) housing under conditions of equal prior probabilities, particularly across the more affluent parts of the west and north east of the
sample area. At the same time, more pixels are classified as terraced (partitioned,
row housing) under unequal prior conditions. In both cases, class estimates from
unequal (stratified and modified) priors are closer to census estimates, and this is
indicative of the flexibility of modified prior probabilities and their ability to incorporate spatial information from beyond the spectral domain. Individual census
tracts can further highlight differences in classification between equal and unequal
priors. The census tracts 09DDGH21 and 09DDFA01 demonstrate how lack of

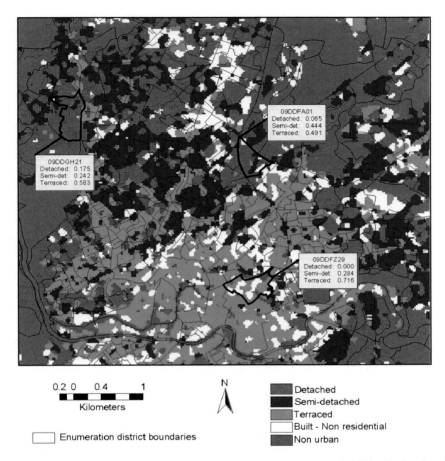

Fig. 8.3 Maximum likelihood classification results using unequal prior probabilities for the city of Bristol. Note the differences in the three housing types between unequal priors in the three highlighted census tracts and results from equal priors shown in Fig. 8.2. In all these cases the equal prior probability classification has over-predicted (or over classified) detached housing (low density) and under predicted semi-detached and terraced (medium and low density). This is most apparent for 09DDFZ29 where no detached housing exists. Under unequal prior conditions, detached and semi-detached are also over-predicted but this time terraced is very close to census estimates

prior knowledge has resulted in the misclassification of detached dwellings. In the case of the census tract (09DDFZ29) within the city center detached housing is again classified using equal priors despite the fact that none are recorded in the census. This zero prior probability is included within the stratified and modified method and results in no pixels classified as detached. What the results reveal is that total absolute error is lower under modified prior probabilities, but that no one housing type has consistently lower area estimation error. However, there is some evidence to suggest that high density housing is underestimated (less pixels classi-fied), and conversely that low density housing is over-estimated (more pixels clas-sified). Further accuracy assessments of other settlements, including one with a detailed accuracy evaluation, can be found in Mesev (1998), with all indicating slight to moderate improvements.

Accuracy from ML per-pixel classifiers is generally increased if the assumption of multi-normality is sustained. However, accuracy is dramatically impaired when spectral distributions representing cover types are far from normal, as in the case of complex heterogeneous urban environments. Moreover, parametric classifiers fre-quent fail to preserve areal estimates; the ability to classify classes with relation to known areal properties. This is particularly noticeable when the distribution of cover types in a scene is highly variable. In attempts to overcome some of the prob-lems, research has led to the development of entirely non-parametric methods using class probabilities from the gray-level frequency histograms to alleviate areal mis-estimation as well as hybrid methods which preserve the benefits of both the para-metric and non-parametric approaches.

8.4   Nearest Neighbor Spatial Pattern Recognition

In contrast to the spectral-based ML decision rule the nearest neighbor spatial pat-tern recognition system is a much more recent development (Mesev 2003, 2005). The system is designed to utilize new disaggregated point-based GIS data for repre-senting individual buildings. Although point based data can be considered the ulti-mate in disaggregation, capable of representing precise locations, in Euclidean geometry terms they are dimensionless, and as such are incapable of measuring the size and shape of objects they represent. Nonetheless, the potential exists for analyz-ing localized two-dimensional spatial patterns of groups of point data for character-izing various morphologies of both residential and commercial developments.

Examples of point-based databases include two in the United Kingdom that were designed to represent the location and type of every postal delivery address. One dataset is known as ADDRESS-POINT™ and is created by the Ordnance Survey of Great Britain, the other as COMPAS™ (COMputerised Point Address Service) in Northern Ireland. Planimetric coordinates of the point data representing postal delivery buildings are claimed to be precise to within 0.1 m (50 m in some rural areas) of the actual location of the building. To achieve this standard of precision both databases were created primarily using the Royal Mail's Postcode Address File (PAF) along with routine ground survey measurements, which are updated every 3 months. Information in both is identical other than an additional "area" attribute in COMPAS™, which indicates the area of the point representing the location of a building. This areal feature is a regular approximation of the actual two-dimensional size but not shape of the building. It is worth noting that postal points are unconstrained, which means that if the area attribute is inappropriately activated it is not inconceivable for build-ings to be represented as overlapping. Nevertheless, the distribution of postal points, including area attributes, are an invaluable source for calculating measures such as density and linearity, both of which can be summarized using standard nearest-neighbor and linear-adjusted nearest-neighbor indices respectively. These indices are then used to identify similar morphological configurations of various types of residential and commercial land uses from second-order, classified satellite sensor imagery. The ultimate objective is to build an automated image pattern recognition system capable of identifying spatial point patterns, representing urban structural morphologies, within image data.

The nearest-neighbor technique is designed to statistically calculate and summarize spatial distributions (Pinder and Witherick 1973). It compares the observed average distance connecting neighboring points (D_{OBS}) and the expected distance among neighbors in a random distribution (D_{RAN}). The statistic is an uncomplicated ratio, where randomness is represented by parity; a clustering tendency has values approaching 0; and perfect uniformity towards a theoretical value of 2.15. The nearest-neighbor statistic R is expressed as,

$$R = D_{OBS} \Big/ D_{RAN} \qquad (8.2)$$

where D_{OBS} is the total measured Euclidean distance between neighboring points divided by the total number of points (N), and D_{RAN} is calculated as,

$$D_{RAN} = \frac{1}{2\sqrt{N/A}} \qquad (8.3)$$

where (N/A) is the density of points within area A. One of the many strengths of the nearest-neighbor statistic is the facility to compare spatial distributions on a continuous scale, especially when area (A) is constant.

Worked examples in Mesev (2005) and Mesev and McKenzie (2005) measured six different residential neighborhoods from the city of Bristol in England (Table 8.2 and Fig. 8.4). Four neighborhoods were successfully identified as having strong

Table 8.2 Density and nearest neighbor statistics with tests for significance of clustering and dispersion

Type	N	R	LN	LR
Residential-1	1479	0.466**	53	0.452
Residential-2	898	0.614**	20	0.503
Residential-3	365	0.942**	11	1.595
Residential-4	906	0.563**	38	0.538
Residential-5	640	0.585**	19	0.589
Residential-6	81	1.564**	14	0.995
Commerical-1	155	0.523**	23	0.509**
Commerical-2	637	1.297**	14	0.903
Commerical-3	18	0.627**	5	0.916
Commerical-4	321	1.289**	10	1.175

N density (area constant); R nearest-neighbor; LN linear density(area constant); LR linear nearest-neighbor

Test for statistical significance in clustering and dispersion using the standard normal deviate are represented by *p values <0.05 and **p values <0.01

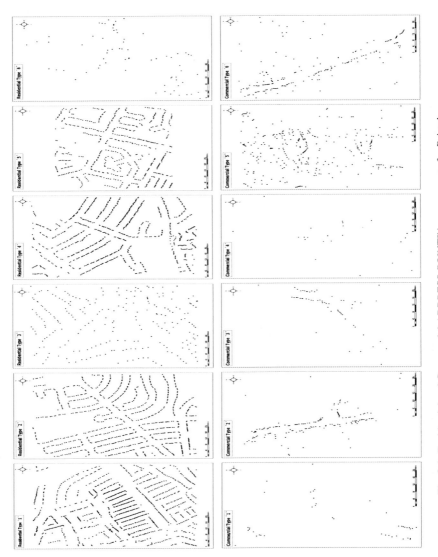

Fig. 8.4 Residential and Commercial ADDRESS-POINT™ arrangements for Bristol

clustering patterns (Residential 1, 2, 4 and 5). As expected, given the compact architecture of terraced (row) housing in the UK, the most clustered neighborhoods are the linear patterns of Residential-1 and Residential-4. However, although their nearest-neighbor values may be very similar, it is plainly apparent that Residential-1 exhibits a far denser concentration of postal points. The same situation applies between Residential-2 (an inner-city local government-owned estate) and Residential-5 (a more affluent 1980s peripheral estate). Conversely, Residential-3 and Residential-4 have very similar densities yet somewhat dissimilar nearest-neighbor values. The remaining pattern, Residential-6, is a highly affluent low-density area of Bristol with large dwellings, and is the only one demonstrating a uniform tendency. Overall, what is clear is that if nearest-neighbor and postal point densities are taken together they are valuable measures for identifying and characterizing different residential types. The same can also be applied to commercial postal points. Again, a strong clustering pattern is indicative of linear developments in Commercial-1, but the city center (Commercial-2) and peripheral estates (Commercial-4) exhibit explicit signs of uniformity. As with the residential values if both nearest-neighbor and density values are taken in combination then unequivocal differences can be revealed. Both Commercial-1 and Commercial-3 (commercial development within residential areas), and Commercial-2 and Commercial-4 have similar nearest-neighbor values but contrasting densities.

The conventional nearest-neighbor statistic is effective for measuring clustering patterns but it lacks the capacity to identify spatial arrangement. A variant of two-dimensional nearest-neighbor analysis is the linear readjustment (LR) devised by Pinder and Witherick (1975). Instead of measuring all observed distances between neighboring points, D_{OBS} is determined from a linear sequence (L) of consecutive points (LN) in all directions (essentially linear subsets), while,

$$LD_{RAN} = 0.5\left(\frac{L}{N-1}\right) \tag{8.4}$$

What constitutes a linear arrangement is of course dependent on scale. Any grouping of buildings regardless of configuration can be considered linear if the scale is fine enough. As a consequence, a minimum of 20 buildings were identified manually but considered representative of a linear formation of both residential and commercial postal points. Although at present this is a subjective process, the intention is to automate future selections and calculate indices that are more representative of each sample type.

Values for both R and LR are documented in Table 8.2. On the whole, they are very similar for both residential and commercial postal point distributions. However, LR values are usually lower for linear patterns of points (Residential-1 through 5, and Commercial-1, 2 and 4) and higher for inherently random or uniform distributions (Residential-6 and Commercial-3). Another difficulty with the linear nearest neighbor is the low numbers of points that are used for calculations. This tends to reduce the significance of LR to levels of randomness demonstrated by Residential-3 and Commercial-2 through 4.

Summaries of residential and commercial point patterns using the nearest-neighbor techniques can now be used to identify similar spatial patterns in classified imagery. In addition, postal points can also be used to identify two types of misclassifications; confusion between built and non-built land covers (especially bare rock or bare soil), and confusion between residential and commercial built land uses. Postal points, on the other hand, represent the entire distribution of individual dwelling units within a city, and as such are the ultimate in disaggregated surfaces. They convey valuable information on local spatial association – density and arrangement – information that is surprisingly overlooked in research on urban image classification, especially given the spatial nature of class distributions and the inherent limitations of spectral data. However, and despite the area attribute, postal points are still models of reality and as such do not precisely delineate the spatial boundaries of buildings. In extreme examples, especially in high density residential areas, the area attribute frequently causes points to overlap. Nonetheless, postal points represent the location of all individual buildings and as such are an invaluable source of information for identifying misclassified pixels that represent not only omitted buildings (errors of omission) but also have also erroneously included the location of buildings (errors of commission). The identification of both types of errors should inevitably lead to a better understanding of the reasons behind misclassified urban pixels.

In an example taken from Belfast (Mesev 2005), the use of postal points has identified a number of misclassified urban pixels from IKONOS imagery. Most of these are at the urban fringe and include the spectral similarities between pixels representing built land cover and those representing bare rock or bare soil. In addition, some misclassifications have been identified between pixels representing residential and commercial land use. In both instances, the identification of misclassified pixels follows the categorization of IKONOS imagery using the ISODATA algorithm available from the ERDAS Imagine™ 8.6 proprietary software. Spatial masks and iterative spectral clustering using panchromatic-sharpening of all four available multispectral IKONOS bands (blue, green, red, and near infrared) then yielded a reasonably accurate classified image of 88% for the built land cover and 72% and 67% for the residential and commercial land uses respectively. Ground data for these accuracy assessments was provided by the interpretation of digital aerial photographs collected in September 2001 by GeoInformation International in Cambridge as part of its Cities Revealed™ series. The aerial photographs were recorded at spatial resolutions of 12.5 and 25 cm, with a 15.25 cm camera focal length, and at a height of 3,200 m. Figure 8.5 illustrates the close correspondence between COMPAS™ points and the 15 cm aerial photograph of Belfast. In addition to identifying misclassified pixels, the spatial distribution of postal points, characterized in terms of density and arrangement, can be used to infer types of residential and commercial developments identifiable in imagery classified as built.

Figure 8.6 illustrates the spatial correspondence between a subset of the classified IKONOS image of Belfast and the location of postal points from the COMPAS™ dataset. Note that the postal points (open squares for residential and shaded squares for commercial) have been scaled in proportion to the area of the

Fig. 8.5   Spatial correspondence between Cities Revealed™ aerial photograph of Belfast (at spatial resolution of 15 cm) and COMPAS™ points (squares proportioned by area attributes)

buildings, a unique feature only available in COMPAS™. White areas of the image have been classified as buildings, light gray as roads, and dark gray as non-built biophysical land covers. Four distinctive residential building types are evident in Figure 8.6 and are labeled as follows.

A. This is an area of very high density residential buildings, showing highly linear arrangement, and is similar to the patterns in Residential-1.
B. In contrast to A, this area has lower density with less of a linear arrangement, and demonstrating a mixture of Residential-3 and Residential-4.
C. Here, medium density and linear arrangements are recognizable and associated with Residential-1 and Residential-2.
D. Lastly, a very low density pattern of residential buildings, indicative of a combination of Residential-3 and Residential-6.

As this subset is a residential suburb of Belfast, only one commercial pattern is observable, at **E**, which is similar to Commercial-3. Note the large area of built pixels to the west of **X**. These correspond to non-residential buildings, in this case, a school. This is because the highly irregular form of non-residential buildings can

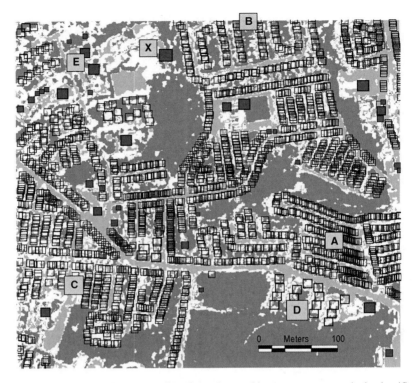

Fig. 8.6 Classified IKONOS image of Belfast subset: white areas represent pixels classified as residential, dark gray areas classified as non-built, and light gray as roads. Residential (*open squares*) and commercial (*shaded squares*) buildings from the COMPAS™ dataset are represented in proportion with area attributes. Note the very high density, linear arrangement of residential buildings at A; lower density, curvi-linear distribution at B; medium density, linear arrangement at C; and very low density, more uniform configuration at D. Only one commercial pattern is observable at E; and a school to the west of X

be seldom represented by convenient square point patterns, regardless of the area attribute. Finally, a quick comparison with Fig. 8.5 reveals many misclassified pixels throughout the subset. Most of these are a result of shadows, a problem rarely confronted in conventional spectral or even spatial classifications.

All building types have been identified visually and are therefore highly subjective. What is now needed is an automated approach that can more precisely and objectively implement the comparison of sample nearest-neighbor indices of both residential and commercial land use with classified imagery. But at the same time be able to also accommodate these same nearest-neighbor indices within an image classification methodology. Research is currently examining the feasibility of an automated image pattern recognition system to facilitate both of these objectives.

8.5   Concluding Thoughts Regarding Classifications

The classification of remotely sensed data is a highly subjective process. Converting radiometric values to user-specified thematic categories requires a level of interpretation that forgoes objective multivariate measurements of reflected and emitted energy for the sake of semantic expediency. The translation is even more tenuous when urban areas are classified from satellite sensor imagery, during both the initial land cover interpretation stage and at the inference of land use from land cover. Spatial variability and spectral heterogeneity, inherent in all images of urban morphologies, severely hamper accurate intraurban classifications, tempered of course by the scale of the image data and the scope of the application. It is therefore unsurprising that a variety of classification and pattern recognition systems have been developed with urban areas particularly in mind. This chapter has reviewed some of these techniques, focusing on the contrast between hard and soft spectral classifications and those oriented on the spatial configuration of pixel groups.

classification of remotely sensed data requires a level of interpretation that forgoes objectivity

Per-pixel spectral techniques, like the Bayes' modification of the ML, allocate one "likely" class to one pixel; whereas soft classifiers, such as mixture models and fuzzy sets, assign proportions of one or more classes per pixel. As pixels representing urban areas are virtually all composed of mixed surfaces, the difference between the two types of classifiers is a matter of scale, generalization, and classification scheme. The development of soft classifiers was widely heralded as a breakthrough in techniques that at last began to mirror the continuum of reality. However, given the severe spatial heterogeneity in the arrangement of urban structures, there is a case to move away from detail (from soft classifiers) in favor of more aggregated or "averaged" pixels (from hard classifiers). In other words, the breakdown of mixed pixels may produce too much information that cannot be easily categorized within a scheme: a situation of "not seeing the wood for the trees". This is particularly evident from new high spatial resolution imagery. Moreover, hard classes, such as the ones generated by the ML illustrated in this chapter, are semantically more communicable where the three urban classes of residential density, high, medium, and low, are directly representative of variations in proportions of buildings to vegetation ratios. For instance, pixels representing mostly, if not all, buildings are classified as high residential density, and pixels representing sizeable proportions of vegetation in conjunction with buildings as low density residential.

The second type of urban classification examined the role of postal point data as major conceptual advances in the de-construction of the geography of urban areas; or at least as an alternative to traditional aggregated census tracts. In disaggregating the geographical distributions of individual households these point data provide unique opportunities to view the urban landscape as a surface of discrete entities rather than the traditional and administratively convenient patchwork of aggregated and uniform surfaces partitioned only by artificial zones. Disaggregated surfaces also offer, for the first time, the possibility of analyzing the spatial configuration and density of individual addresses within neighborhoods, which are typically hidden by zonal representations. The geographic co-ordinates of each point are positioned to correspond with the center of the building, allowing measures relating to the distribution of points indicating density (compactness or sparseness) and arrangement (linearity or randomness). The creation of ADDRESS-POINT™ and COMPAS™ represents an important step forward in the pursuit of 'framework' data that encapsulate the desire for higher quality urban information not just for image pattern recognition and classification improvement but also for all potential urban-based spatial data analyses.

In addendum, it is important to remember that all classification methodologies classify the remotely sensed image not reality. Before over-focusing on algorithm sophistication an appreciation and awareness of what exactly the image represents in reality is of paramount importance. At the same time, effective communication of how reality is represented is critically dependent on the scope and detail of thematic categorization. Finally, it is worth noting that despite the many caveats, compared to conventional mapping, image classification is still a relatively rapid, comprehensive, multidimensional and inexpensive alternative.

> **it is important to remember that all classification methodologies classify the remotely sensed image not reality**

Chapter Summary

This chapter emphasized the main issues prevalent in urban classification; the distinctions between land cover and land use, the difference between hard and soft classification, and the incorporation of ancillary data in spectral and spatial modifications. In all, real world examples were given to demonstrate not only these contrasts but also the emerging importance of large-scale urban remote sensing, particularly, the advent of very high spatial resolution satellite sensor data and the pressing issues of global urbanization and urban sustainability. In addition, differences between spectral based and spatial based classification was introduced, both in relation to the role of ancillary data.

LEARNING ACTIVITIES

List of Software Packages Capable of Spectral Classification

- ENVI (Research Systems Institute). Free evaluation available. 🖳 http://www.rsinc.com/envi/

- ERDAS Imagine (Leica Systems). 🖳 http://www.gis.leica-geosystems.com/products/imagine/
- IDRISI Kilimanjaro (Clark Labs). 🖳 http://www.clarklabs.org/
- eCognition 3.0 (Definiens Imaging GmbH). Free evaluation available. 🖳 http://www.definiens-imaging.com/

Study Questions

- How many practical urban applications would benefit from the classification of satellite sensor data?
- Compare and contrast land cover and land use with respect to various urban applications.
- Discuss the dilemma of, on the one hand, increasing spatial resolution and, on the other, decreasing classification accuracy in the interpretation of urban categories from satellite sensor data.
- Do you think spectral-based classification or spatial-based pattern recognition are more intuitive in how urban areas are structured and how processes are interpreted?
- How processes are interpreted?

LAB EXERCISE

You will need to install ENVI Guided Evaluation software from the following RSI link http://www.rsinc.com/envi/ and download a fully functional version which runs in restricted sessions of 7 min. You will also need to save the image *belfast_subset* from the Book's homepage at springer.com (http://www.springer.com/sgw/cda/frontpage/0,11855,4-40106-22-107950926-0,00.html) to your local hard disc. This is a small subset of a pan-sharpened 4 m multispectral IKONOS image of the city of Belfast in Northern Ireland (alternatively you can follow the same steps using an image subset of your own city).

- *Data import*: Start ENVI Evaluation version. From *File* choose *Open Image File* and select *belfast_subset*. Another window will open, *Available Bands List*. Click on *RGB Color* and make sure that *Layer 4* is loaded in *Red, Layer 3* in *Green* and *Layer 2* in *Blue*. Click on *Load RGB*.
- *Display*: Two other windows of differing resolutions will pop open. You may synchronize these windows to zoom in and out of the image. The subset matches the configuration of built structures as used in Figs. 8.4 and 8.5. Notice the red areas correspond to vegetation, the linear arrangement of the roads is also discernible but the buildings vary in composition and are difficult to delineate.

- *ROI collection*: You will now generate a number of training samples to supervise the classification. These training areas are called regions of interest (ROIs). From *Basic Tools* (on the main horizontal tool bar at the top of your screen choose) choose *Region of Interest* and then *ROI Tool*. The new window will record all of your training areas as you collect them from the image (because of the size of the subset you may want to use the zoom window to navigate). Click on the * next to *Region #1* to highlight it, then begin drawing a polygon by moving the cursor around the area of interest, clicking the left mouse once to draw an arc and clicking the right mouse once to finish the polygon, clicking right again will record the ROI permanently. Collect training areas for four types of residential configurations in Figs. 8.5 and 8.6 as well as for the commercial area labeled as E and the school in the north-west of the subset. Every time a new training area is collected you will need to click on *New Region*. An erroneous ROI can be deleted.
- *Classification*: You are now ready to classify. Again, from the main horizontal tool bar at the top of your screen choose *Classification* and *Supervised* and then *Maximum Likelihood*. Select the *belfast_subset* image. Click *OK*. Another window will list the *Maximum Likelihood Parameters. Select All Items.* Click on *Multiple Values* and then *Assign Multiple Values* the following prior probabilities to each *Region* by highlighting it and typing in the *Edit Selected Item*:

Assign probability threshold	
Region	Probability
A: Very high density, strong linear residential	0.25
B: Low density, weak linear residential	0.21
C: Medium density, strong linear residential	0.31
D: Very low density, weak linear residential	0.14
E: Low density, weak linear commercial	0.07
X: Single commercial	0.02

Click *OK* when done. Do not click *OK* on the main *Maximum Likelihood Parameters* window, yet.

- Experiment: In order to appreciate the power of variable prior probabilities you can *Preview* the classification every time you change the probabilities. Experiment with combinations of prior probabilities to see the effects. When ready, you may generate a full classification by choosing an *Output Result File* name. In the *Available Bands List* select your classified file and *Load Band* in *New Display*. From *Tools, Color Mapping* and *Class Color Mapping* you can change the colors of the classified image. Hopefully, you will have finished within the 7 min!

Acknowledgements ADDRESS-POINT™ and COMPAS™ are Crown Copyright (www.ordsvy. gov.uk and www.osni.gov.uk respectively). Cities Revealed is the copyright of the GeoInformation Group (www.crworld.co.uk). IKONOS image was provided by Space Imaging. The author would like to thank Paul McKenzie for the production of Fig. 8.4.

References

Anderson JR, Hardy EE, Roach JT, Witmer RE (1976) A land use and land cover classification system for use with remote sensor data. LIS Geological Survey Professional Paper 964. USGS, Washington, DC

Baraldi A, Parmiggiani F (1990) Urban area classification by multispectral SPOT images. IEEE Trans Geosci Remote Sens 28:674–680

Barnsley MJ, Steel AM, Barr SL (2003) Determining urban land use through an analysis of the spatial composition of buildings identified in LIDAR and multispectral image data. In: Mesev V (ed) Remotely sensed cities. Taylor & Francis, London, pp 83–108

Batty M, Longley PA (1994) Fractal cities: a geometry of form and function. Academic, London

Berthod M, Kato Z, Yu S, Zerubia L (1996) Bayesian image classification using Markov random fields. Image Vis Comput 14:285–295

Brugioni DA (1983) The Census: it can be done more accurately with space-age technology. Photogramm Eng Remote Sens 49:1337–1339

Carr JR (1999) Classification of digital image texture using variograms. In: Atkinson PM, Tate NJ (eds) Advances in remote sensing and GIS analysis. Wiley, London, pp 135–146

Chen K (2002) An approach to linking remotely sensed data and areal census data. Int J Remote Sens 23:37–48

Corbley KP (1996) One-meter satellites. Geo Inf Syst July:28–42

Couloigner I, Ranchin T (2000) Mapping of urban areas: a multiresolution modelling approach for semi-automatic extraction of streets. Photogramm Eng Remote Sens 66:867–874

Cowan DJ, Jensen JR (1998) Extraction and modeling of urban attributes using remote sensing technology. In: Liverman DM, Moran EF, Rindfuss EE, Stern PC (eds) People and pixels. National Academy Press, Washington, DC, pp 164–188

Dobson JE (1993) Land cover, land use differences distinct. GIS World 6(2):20–22

Donnay JP (1999) Use of remote sensing information in planning. In: Stillwell J, Geertman S, Openshaw S (eds) Geographical information and planning. Springer, Berlin, pp 242–260

Dreyer P (1993) Classification of land cover using optimized neural nets on SPOT data. Photogramm Eng Remote Sens 59:617–621

Fisher PF (1997) The pixel: a snare and a delusion. Int J Remote Sens 18:679–685

Forster BC (1985) An examination of some problems and solutions in monitoring urban areas from satellite platforms. Int J Remote Sens 6:139–151

Geoghegan J, L P, Ogneva-Himmelberger Jr Y, Chowdhury RR, Sanderson S, Turner BL (1998) "Socializing the pixel" and "pixelizing the social" in land use and land cover-change. In: Liverman DM, Moran EF, Rindfuss EE, Stern PC (eds) People and pixels. National Academy Press, Washington, DC, pp 51–69

Grey WMF, Luckman AJ, Holland D (2003) Mapping urban change in the UK using radar interferometry. Remote Sens Environ 87:16–22

Haralick RM, Shanmugam K, Dinstein I (1973) Textural features for image classification. IEEE Trans Syst Man Cybern 3:610–621

Harvey JT (2002) Estimating census district populations from satellite imagery: Some approaches and limitations. Int J Remote Sens 23:2071–2095

Hasse J, Lathrop RG (2003) A housing-level approach to characterizing residential sprawl. Photogramm Eng Remote Sens 69:1021–1030

Heikkonen J, Varfis A (1998) A land cover/land use classification of urban areas: a remote sensing approach. Int J Pattern Recognit Artif Intell 12:475–489

Herold M, Scepan J, Clarke KC (2002) The use of remote sensing and landscape metrics to describe structures and changes in urban land uses. Environ Plann A 34:1443–1458

Ji M, Jensen JR (1999) Effectiveness of subpixel analysis in detecting and quantifying urban imperviousness from Landsat Thematic Mapper imagery. Geocarto Int 14(4):33–41

Lam NSN, Quattrochi DA, Qiu HL, Zhao W (1998) Environmental assessment and monitoring with image characterization and modeling system using multi-scale remote sensing data. Appl Geogr Stud 2:77–93

Mesev V (1997) Remote sensing of urban systems: hierarchical integration with GIS. Comput Environ Urban Syst 21:175–187

Mesev V (1998) The use of census data in urban image classification. Photogramm Eng Remote Sens 64:431–438

Mesev V (2001) Modified maximum likelihood classifications of urban land use: spatial segmentation of prior probabilities. Geocarto Int 16(4):39–46

Mesev V (ed) (2003) Remotely sensed cities. Taylor & Francis, London

Mesev V (2007) Fusion of point-based postal urban data with IKONOS imagery. Int J Inf Fusion 8:157–167.

Mesev V, McKenzie P (2005) Urban neighbourhood patterns: links between high spatial resolution remotely sensed data and point-based GIS data sources. In: Wise S, Craglia M (eds) GIS and evidence based policy making. Taylor & Francis, London

Myint SW (2003) The use of wavelet for feature extraction of cities from satellite sensor images. In: Mesev V (ed) Remotely sensed cities. Taylor & Francis, London, pp 109–134

Ogrosky CE (1975) Population estimates from satellite imagery. Photogramm Eng Remote Sens 41:707–712

Pesaresi M, Bianchin A (2001) Recognizing settlement structure using mathematical morphology and image texture. In: Donnay JP, Barnsley MJ, Longley PA (eds) Remote sensing and urban analysis. Taylor & Francis, London, pp 55–68

Pinder DA, Witherick ME (1973) Nearest neighbour analysis of linear point patterns. Tijdschift voor Economische an Sociale Geographie 64

Pinder DA, Witherick ME (1975) A modification of the nearest-neighbour analysis for use in linear situations. Geography 60:16–23

Rashed T, Weeks JR, Roberts DA, Rogan J, Powell R (2003) Measuring the physical composition of urban morphology using multiple endmember spectral mixture models. Photogramm Eng Remote Sens 69:1011–1020

Ridd M (1995) Exploring a V-I-S (Vegetation-Impervious Surface-soil) model for urban ecosystem analysis through remote sensing: comparative anatomy of cities. Int J Remote Sens 16:2165–2185

Strahler AH (1980) The use of prior probabilities in maximum likelihood classification of remotely sensed data. Remote Sens Environ 10:135–163

Webster CJ (1995) Urban morphological fingerprints. Environ Plann B 22:279–297

Welch R (1982) Spatial resolution requirements for urban studies. Int J Remote Sens 3:139–146

Woodcock CE, Strahler AH (1987) The factor of scale in remote sensing. Remote Sens Environ 21:311–332

Chapter 9
Processing Techniques for Hyperspectral Data

Patrick Hostert

The basic knowledge about the differences between multi- and hyperspectral data is provided and the potential of hyperspectral image analysis is highlighted. Relevant pre-processing steps and different ways to analyze hyperspectral data are presented. The chapter closes with a short outlook on expected developments with relevance for urban applications.

Learning Objectives

Upon completion of this chapter, the student should gain an understanding of:

❶   The principal differences between multi- and hyperspectral data
❷   The need for a dedicated pre-processing of hyperspectral data
❸   Analysis techniques with relevance for urban applications

---

9.1   Introduction: Hyperspectral Data and Urban Remote Sensing

Hyperspectral data, often also referred to as *spectral high resolution data*, *imaging spectrometry data*, or *imaging spectroscopy data*, have been successfully applied in different fields of terrestrial remote sensing since about 20 years (Goetz et al. 1985). However, only recently their value for urban applications has been put forward

P. Hostert (✉)
Geographical Institute, Humboldt-Universitat zu Berlin, Unter den Linden 6,
Berlin 10099, Germany
e-mail: patrick.hostert@geo.hu-berlin.de

T. Rashed and C. Jürgens (eds.), *Remote Sensing of Urban and Suburban Areas*,
Remote Sensing and Digital Image Processing 10,
DOI 10.1007/978-1-4020-4385-7_9, © Springer Science+Business Media B.V. 2010

(e.g. Ben-Dor 2001; Herold et al. 2004). From a remote sensing point-of-view, the urban setting differs from natural or semi-natural environments due to a few distinct characteristics:

- Object heterogeneity or texture: Many urban features exhibit sharp borderlines, while their inner-object variance may vary substantially. A large parking lot with cars may appear extremely heterogeneous, while the neighboring industrial complex is represented by a few homogeneous roof constructions.
- Landscape heterogeneity and object size: Object size and heterogeneity are often interlinked. It is also sometimes difficult to specify average object sizes for a complex environment such as the city. However, the size of most objects (houses, cars, street width) may be regarded as relatively small (Small 2003), compared to other situations (agricultural fields, forest plots, open water surfaces). The amount of mixed pixels resulting from this circumstance varies, depending on the pixel size, but is usually much higher than in most other cases.
- Combination of natural and anthropogenic materials: Urban surfaces include a great variety of spectrally distinct surfaces. Urban areas may well include large areas consisting of natural materials (vegetation, soils, water), as "urban" is not necessarily defined through the built environment. Theoretically, mixtures of all natural and anthropogenic materials may occur.
- Geometric complexity: The application of airborne sensors and the associated wide field-of-view angles (compare 9.2) results in extreme differences in object illumination. A sensor records the shaded backside of built-up areas with scan angles opposite to the sun azimuth, scan angles parallel to the sun azimuth lead to a view on illuminated facades. The strength of this effect varies with sun elevation/azimuth, object geometry/spectral behavior, and flight direction, i.e. it is a spectrally varying function depending on sun-object-sensor geometry.

Details and explanations on the spectral and geometric behavior of urban surfaces are given elsewhere in this volume. However, even from this short introduction it becomes apparent that the analysis of such an environment provides an enormous challenge for remote sensing based data analysis, monitoring approaches, and thematic assessments. One of the options to tackle object and landscape heterogeneity is to employ high spectral resolution remote sensing data.

There is no precise definition of which number of bands separates multispectral from hyperspectral data. One may, for example, agree that sensors allowing for a detailed analysis of absorption features in their spectral range fall in the category of hyperspectral data. To date, there is no operational hyperspectral satellite sensor offering an adequate geometric resolution for urban applications. With the advent of the Airborne Visible / Infrared Imaging Spectrometer (AVIRIS) in 1987, the first airborne hyperspectral imager with 224 contiguous spectral bands between 400 and 2,500 nm was available for a wide range of applications. Sensors like the Digital Airborne Imaging Spectrometer (DAIS 7915) featuring

Hyperspectral data differ from multispectral data in the number of bands and band widths. Spectral signatures from hyperspectral data appear contiguous

hyperspectral thermal infrared capabilities offer additional prospects for urban analysis; however, thermal infrared devices for narrow band sensors offer a critical signal-to-noise ratio and a stable calibration appears difficult. For this chapter, examples from field or laboratory measurements, sampled with an ASD FieldSpec Pro II spectroradiometer (Hostert and Damm 2003), and a subset of HyMap data acquired in July 2003 over Berlin, Germany, (DLR 2003) are given.

9.2 Pre-processing

An adequate pre-processing of hyperspectral data is a mandatory prerequisite to extract useful information from hyperspectral data, regardless of working in urban or other environments. However, the analysis of urban properties must be regarded among the most demanding applications in terms of hyperspectral image pre-processing. This applies on one hand to the requirements for a precise co-registration with other raster or vector data sets (van der Linden and Hostert 2009). On the other hand, the

Hyperspectral data require a dedicated pre-processing. This is particularly true in the case of urban environments

spectral variability and complex illumination geometry ask for a precise definition of radiometric correction processes. It is therefore not surprising that pre-processing of hyperspectral data consists of a not to be underestimated series of processing steps, in terms of complexity as well as in terms of the amount of effort and time.

From the end-user point of view, pre-processing of remote sensing data can be divided into preliminary quality assessment, correction of bidirectional effects, geometric correction, and radiometric correction. A screening for spatial, spectral or radiometric errors should be performed to detect problematic regions of an image. Usually, bands with particularly low signal-to-noise-ratio (SNR) are discarded. Such a screening may also include further steps such as cloud and cloud shadow mapping or the definition of areas with uncharacteristic directional reflectance behavior (e.g. regions of specular reflectance in the case of water targets).

Most airborne scanners are characterized by a wide field-of-view (FOV) resulting in different directional reflectance behavior of similar targets depending on sun-surface-sensor geometry. As hyperspectral data are almost exclusively acquired with airborne sensors today, correcting for wavelength dependent bidirectional effects is obligatory for most analyses (Schiefer et al. 2006). This can be achieved by calculating and individually applying a view angle dependent and band-wise polynomial function in across-track direction, also referred to as "across-track illumination correction". Mean column-wise reflectance values are calculated for each spectral band and differences interpreted as the scan-angle dependent variations in reflectance. It is obvious that such a simplistic approach does not account for land cover dependant differences in bidirectional behavior. As the urban environment is spatially extremely heterogeneous, a pre-classification in dominating land cover classes allows for a class-wise calculation and correction of directional properties.

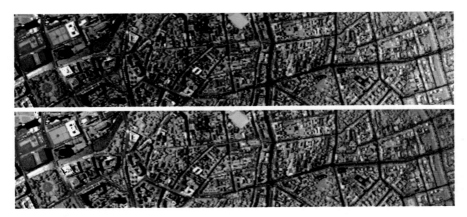

Fig. 9.1 Comparison of an urban subset before (*top*) and after across-track illumination correction (*bottom*). R-G-B: band 29–band 80–band 15 (equivalent to a Landsat-TM false color composite with R-G-B: band 4–band 5–band 3; *white arrow*: flight direction; *yellow arrow*: North)

A view angle dependent correction results in comparable radiances of similar urban surfaces in across-track direction (Fig. 9.1).

9.2.1 Radiometric Correction

Usually, hyperspectral data will be distributed as scaled radiance values (e.g. in $\mu W * cm^{-2} * nm^{-1} * sr^{-1}$) and calibration is carried out by the data provider. However, to compare hyperspectral imagery with field-based measurements and to open up the pathway towards quantitative analysis, radiance (variable with illumination) has to be converted to reflectance (invariable for comparable surfaces). This process is termed "radiometric correction". Various methods of empirical and parametric radiometric pre-processing methods can be distinguished. A simple and useful approach is the empirical line correction method, relating spectral ground measurements with radiance values of the respective targets in the imagery. The urban environment offers abundant invariant and well identifiable targets, which may serve as input from the image. Applying the resulting band-wise transfer functions leads to values close to reflectance. However, due to the linear approach non-linear radiometric distortions will usually not be adequately corrected. Disturbance patterns that vary over the scene – especially the highly variable water vapor content – can also not be tackled. Nevertheless, for many cases empirical line corrected data may serve as a valid input for further processing steps.

If a more precise correction of radiometric properties is required, parametric approaches need to be implemented. Atmospheric properties are measured, modeled or estimated to pixel-wise invert the respective disturbance processes and result in reflectance values. Non-linear effects like the influence of second-order radiometric disturbances from the target environment can be incorporated via a window-based determination of scattering processes. Aerosol scattering in the shorter wavelength regions is corrected by applying pre-defined aerosol models and distributions along with appropriate aerosol scattering functions. The most problematic factor is the water vapor content that varies over short distances. As hyperspectral data sets are spectrally quasi-continual measurements including water vapor absorption bands, it is possible to determine the water vapor quantities by analyzing the absorption bands at wavelengths of 940 and 1,140 nm, which correlate well with water vapor quantities. A pixel-wise water vapor estimate from the image itself can hence be included in the correction process (Gao and Goetz 1990).

Finally a correction of topography effects is necessary to precisely account for illumination dependent differences. Direct and diffuse illumination along with shading effects largely varies the target reflectance properties. In an urban environment, the influence of topography and the influence of the built environment are to be distinguished. The first can be included via aspect, slope, shading, and visible sky view properties extracted from a digital elevation model (DEM). However, large scale geometric properties, such as building height or roof angles, are only provided by precise digital object models (DOM). Such models are available from high resolution stereo data, light detecting and ranging (LIDAR), or interferometric synthetic aperture radar (IFSAR). Though, state-of-the-art techniques do not yet allow for a geometric co-registration of these models and hyperspectral data in the cm-range, which would be necessary to apply the appropriate calculations. Nevertheless, a parametric radiometric pre-processing relying on an adequate parameterization of atmospheric parameters and including a DEM is the most accurate way of radiometrically correcting hyperspectral imagery (Fig. 9.2).

9.2.2   Geometric Correction

The geometric correction of airborne hyperspectral scanner data is similar to the geometric correction of multispectral scanner data apart from the amount of spectral

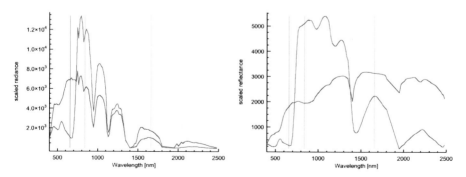

Fig. 9.2 Spectral comparison of paving stones and photosynthetic active vegetation from HyMap imagery before and after parametric radiometric correction

bands to be rectified. Considering urban environments, the precise co-registration with cadastral data or similarly high resolution geometries is particularly demanding. The advantage may be that precise reference data often exist for urban environments, which is not necessarily the case for other settings.

In an ideal case, airborne data are provided as an image cube accompanied by an auxiliary data stream of differential global positioning system positions (DGPS) and inertial navigation system data (INS). The first provides sub-meter accurate position data of the sensor during image acquisition (x-, y-, and z-coordinates), the latter information on roll, pitch, and yaw movements of the platform (κ-, φ-, and ω-angles). Assuming a correct synchronization between scan lines and auxiliary data, it is possible to calculate the acquisition geometry for every pixel. A DEM has to be included to correct for terrain induced distortions (Schläpfer and Richter 2002).

It will usually be necessary to incorporate ground control points (GCPs) in this processing scenario to correct for inaccuracies in the measurements itself and for potential erroneous synchronization between data and auxiliary data. This is a rather straightforward task in the case of urban environments, as either ground-based DGPS measurements, orthophotos, or accurate vector data are available or may be retrieved (in the case of DGPS measurements) for many urban areas. The diversity and crispness of urban features supports the identification of accurate GCPs. Additionally, accurate ground truth allows for a high-quality assessment of geometrically corrected data sets.

9.3   Spectral Libraries

One of the most advantageous conceptual frameworks in hyperspectral remote sensing is based on the opportunity to relate field- or laboratory based spectrometric measurements with imaging spectrometry data from airborne or spaceborne sensors. The spectral behavior of distinct objects on the Earth's surface is determined by their physical and chemical properties. While a few working groups have started to collect such spectra, the available databases are far from exhaustive (ASTER 1998; Ben-Dor 2001; Heiden et al. 2001; Hostert and Damm 2003). Recently, a structured approach to acquiring a more complete urban spectral library and to analyze material separability has been exemplified for the Santa Barbara region by Herold et al. (2004) and is illustrated in this textbook.

Spectral libraries contain reference spectra for subsequent analysis

Measurements of the respective components under controlled conditions in the laboratory or under real-world conditions in the field can hence be related to the surface's physical or chemical properties (quantitative approaches); alternatively, such measurements may serve as well-defined samples to identify similar components (qualitative approaches) in imaging spectrometry data. The ability to relate radiometrically corrected hyperspectral data from diverse sensors with ground-based spectroradiometric data can be regarded as a spectral upscaling.

Field or laboratory measurements are performed with spectrally very high resolution instruments. Spectra, or so called *spectral endmembers*, are usually normalized to reflectance values and stored in a spectral database, along with an appropriate set of meta-data. Spectral data may be combined with coordinate information in a geo-database to provide an urban spectral cadastre. While such data sets are abundant for many natural environments, there is still the need for more extensive urban spectral libraries that allow selecting a great range of very high resolution urban spectra from pre-defined sources. Once collected, very high resolution spectral references may be resampled to the spectral resolution of imaging spectrometers based on their band dependent sensitivity functions (Fig. 9.3).

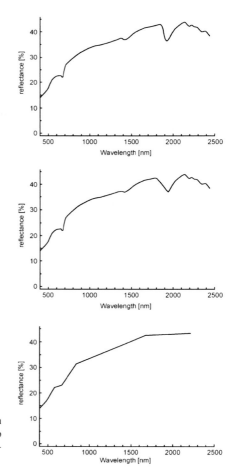

Fig. 9.3 Cobblestone pavement spectra from laboratory measurements (*top*), resampled to HyMap (*centre*) and Landsat TM spectral resolution (*bottom*)

9.4   Analysis Techniques

High resolution spectral data differ from multispectral data in their ability to detect subtle differences in surface components. While other sensor concepts focus on the utilization of different wavelength regions or fundamentally different acquisition techniques (e.g. radar sensors or sounding sensors), high resolution spectral data work in the same wavelength domains as most multispectral devices, but in very narrow spectral windows per band. As a consequence, the high number of bands not only offers different analysis options, but actually *requires* different analysis techniques. While conventional classification approaches may be utilized, comparable to those employed for multispectral data analysis, the full potential of such data is made accessible when more sophisticated or adapted methods are utilized. In the following a focus is put on data optimization, classification/ material detection, and spectral mixture analysis.

Qualitative and quantitative analysis techniques may be employed with hyperspectral data. The large number of bands may require a data optimization to retrieve optimum results

9.4.1   Data Optimization

The high number of spectral bands can be regarded as an advantage and a problem at the same time. A high spectral autocorrelation between neighboring wavelengths leads to redundant information. Considering that hyperspectral data sets may easily grow to GByte sizes, processing performance will unnecessarily suffer, depending on hard- and software capabilities. While such problems will be overcome with more powerful tools, the ability to derive useful information from such data sets may also be impeded by redundant information. Data transformations are therefore a standard pre-processing option in cases when the original spectral information is not inevitably needed (e.g. for optimized classification).

The *Minimum* (or Maximum) *Noise Fraction* (MNF) is widely used to optimize hyperspectral data analysis. Comparable to a principal component analysis, an MNF transformation sorts the bands of a data set regarding variance explanation. It then decorrelates the noise content in the data and orthogonalizes feature space (Green et al. 1988). The resulting MNF bands with low noise components may then be analyzed during further processing steps (Fig. 9.4).

Alternatively, the first bands that are considered to be noise-free may be extracted and inverted again to yield noise-free reflectance data. It has to be remarked that such a procedure has always to be considered in the light of the analysis goal. Depending on the original feature space and the thematic question at hand, important information may be found in less important MNF bands and a careful screening of individual bands is necessary before either spectrally subsetting or inverting subsetted data. In any case, a transformation of spectral library information is also mandatory when using transformed data along with ground-based spectrometry.

Fig. 9.4 R-G-B 1-3-5 of an MNF transformation (same subset as Fig. 9.1)

9.4.2 Classification and Material Detection

In principal, the same fundamentals apply to the classification of multispectral and hyperspectral data sets. Well known supervised and unsupervised classification techniques will hence not be considered here. More recent developments, such as the use of image segmentation and object oriented analysis techniques, are also applicable to spectral high resolution data and described elsewhere in this volume. In this chapter, a focus is put on those methods that are more often used with hyperspectral data or that appear particularly advantageous when applied with hyperspectral data.

There are numerous techniques focusing on either the ability to detect absorption features in surface materials from imaging spectrometer data or on the extended feature space of hyperspectral imagery as a whole (or MNF-transformed input). Absorption based detection of single materials originates from geological applications, but is also useful in urban environments, where diverse and spectrally distinct materials occur. This capability of spectrometric data is generally enhanced by normalizing spectra via a so-called *convex-hull transformation*. A mathematically derived curve is fitted to envelop the original spectrum (hull), utilizing local spectral maxima to connect the hull segments, while leaving absorption features as spectral gaps below the hull. Dividing the original spectrum by the hull values results in a baseline along 1 (or 100% of the hull) and relative absorption features with depths between 0 and 1 (Fig. 9.5). These features are quantifiable in a sense that for example the absorption depth or the full width at half maximum (FWHM) of the absorption feature can be measured regardless of potential albedo differences in the individual materials.

It is important to choose the appropriate analysis technique depending on the questions to be answered. This may include feature based methods or illumination insensitive techniques

It is then possible to compare transformed spectra from imaging spectrometry data with equally processed spectra from a spectral library. This may be done by calculating the band-wise residuals between image and reference spectrum and cumulating these in a root mean squared error (RMSE). A perfect match (which is a rather theoretical assumption) should yield in zero residuals and would indicate

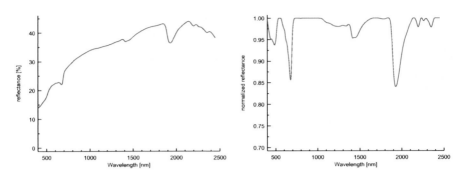

Fig. 9.5 Original reflectance spectrum from a cobblestone pavement and continuum removed spectrum

image areas that are 100% pure concerning the respective material. Usually, even pure materials will not perfectly fit library spectra due to diverse error components (measurement setup, SNR, directional reflectance differences, calibration, atmospheric correction, etc.). It is very likely that the majority of image pixels will rather be mixed than pure in urban environments. Absorption features will therefore be masked or enhanced by other material characteristics on one hand and new absorption features may appear on the other hand. There is in any case the need to account for such effects beyond the RMSE as a global measure of spectral fit. Individual absorption feature depth or FWHM comparisons between image and reference spectra may hence serve as a measure of material abundance in mixed pixels.

As view angle dependent effects are critical in urban environments and illumination geometry is complex, it might be advantageous to employ methods that are fairly insensitive to illumination effects. Spectral angle mapping (SAM) is such a technique. Differently from other classification techniques, SAM compares reference signatures with individual pixels not by their statistical representation in feature space per se, but by their angular differences in feature space position. Considering multidimensional feature space as axes starting from a zero-reflectance point, reference signatures and pixels are aligned along these axes and the multidimensional angle between all references and the respective target pixel are calculated. This angle is independent from changes in pixel albedo, as all pixels of the same spectral character exhibiting for example illumination differences will align along the same vector starting from the zero reflectance point. As the vector direction does not change the angle between a reference target and a pixel vector is fixed either.

9.4.3 Analysis Focusing on Mixed Pixels

Ridd (1995) has proposed a conceptual framework to analyze urban remote sensing data based on the major urban surface components vegetation, impervious surfaces and soil. This model has became a kind of standard concept for many remote sensing based analysis approaches focusing on the urban environment. Authors like Phinn et al. (2002) have shown that such an approach can be successfully transferred to an analysis at subpixel level. Hyperspectral data are well suited for applying such methods, as their spectral information content allows the discrimination of diverse materials in a pixel. This paragraph provides an overview on the analysis of urban areas with methods capable to quantify material components at a sub-pixel level, commonly referred to as *spectral mixture analysis* (SMA) or *spectral unmixing*.

Spectral mixture analysis is a well adapted concept to work with spectral high resolution data

A straightforward unmixing procedure is a linear spectral unmixing approach. In a simplifying approach, pixel reflectance is supposed to depend on a linear combination of a limited set of pure urban surface components (or endmembers) and can hence be decomposed by calculating the respective fractional component abundances.

For statistical reasons, the maximum number of possible endmembers depends on the dimensionality of the data set and the spectral contrast of the individual endmembers. It will usually be close to the independent feature space bands (represented, for example, by an MNF-transformation). Potentially, this leads to uncertainties in the unmixing process and to unexplained surface components not represented by a limited number of endmembers. The unexplained components are accounted for by band-wise residuals; residuals may be summarized as root mean squared error (RMSE). The RMSE is particularly relevant in highly diverse urban areas to keep track of uncertainties in the data analysis. A second indicator is that the resulting fraction image for every endmembers contains positive values only. Every unmixing model should sum to unity, which is mathematically also possible with endmember abundances below zero or above 100%. Largely positive fractions indicate the validity of the unmixing, as the mathematical solution represents physically meaningful results (Fig. 9.6).

Assuming that suitable spectra are available in a spectral library, one way to overcome this limitation is to employ multiple endmember models, i.e. to use individual endmember combinations depending on the respective components present in every individual pixel. It may, for example, be adequate to model a pixel in a homogeneous industrial area with two endmembers only, e.g. concrete and asphalt. Heterogeneous urban areas such as many residential quarters may result in pixels

Fig. 9.6 Results from a linear spectral unmixing with five endmembers: Vegetation, soil, concrete, asphalt, clay shingle. Here: R-G-B vegetation-concrete-clay shingle (water along the left-image border is masked)

containing much more surface features such as grass, asphalt, concrete, roof shingles, and colored metal surfaces from cars. It is then possible to either leave it to the software to define appropriate endmember models for each pixel or to define possible combinations in advance and only chose among those.

9.5   Future Developments

Imaging spectroscopy is at present a tool largely driven by technological improvements. In the near future, advanced spectrometers will emerge that will open up the road to new analysis tools and new ways to employ them. Such sensors will enhance our ability to differentiate materials or to model quantitative indicators from primary parameters like surface reflectance. One of these near future developments is the Airborne Reflective and Emissive imaging Spectrometer (ARES) with 155 spectral bands including the thermal infrared and an excellent SNR (Wilson and Cocks 2003). Also, spaceborne high resolution spectrometers with satisfactory SNR will become available in a few years, such as the Environmental Mapping and Analysis Program (EnMAP, Buckingham and Staenz 2008).

Moreover, the combination of hyperspectral with other remote sensing data and enhanced analysis techniques offers a high potential of further improvements in data analysis. Sensor integration may include data fusion concepts between very high geometric resolution and hyperspectral data (Lehmann et al. 1998). Such sensor combinations are particularly valuable for urban applications as an improved geometric resolution will result in less mixed pixel surfaces. From a processing point-of-view, combined analysis schemes such as the integration of supervised classification and spectral unmixing (Segl et al. 2000) or the use of machine learning classifiers (van der Linden et al. 2007) offer new opportunities, especially in the heterogeneous urban environment. Finally, it has to be remarked that quantitative analyses and modeling approaches will become more relevant in the future. While there are examples of quantitative models of soil or vegetation properties (e.g. Schlerf et al. 2005) such approaches have not yet been implemented for urban applications.

Chapter Summary

Hyperspectral remote sensing data differ from multispectral data in the number of spectral bands and hence in the analysis options associated with such data. These extended analysis opportunities are on one hand particularly useful in a heterogeneous urban environment. On the other hand, this heterogeneity results in demanding pre-processing schemes and accuracy level that need to be achieved. Radiometric pre-processing focuses on illumination and atmospheric corrections. As all hyperspectral data used for urban applications are acquired by airborne sensors nowadays, the geometric pre-processing including DGPS and INS information is demanding.

The huge number of bands and information redundancy might also require data optimization, such as MNF transformations. Relevant analysis approaches include material detection techniques, spectral angle mapping, or spectral mixture analysis. Hyperspectral data will further gain in importance in the future with the advent of new sensors and dedicated analysis options.

LEARNING ACTIVITIES

- Download the trial version of *The Environment for Visualizing Images (ENVI)*: http://www.ittvis.com/Academic/Students/ENVIStudentEdition.aspx. Use the online help system to learn more about the different options to analyze hyperspectral data.
- Read Goetz et al. from the list of references to get an insight on imaging spectroscopy.
- Read Section 4 (Urban and Land Use Applications) and Section 13 (Collecting Data at the Surface – Ground Truth; The "Multi" Concept; Hyperspectral Imaging Spectroscopy) in the NASA Remote Sensing Tutorial: http://rst.gsfc.nasa.gov/Front/tofc.html.

Exercises

- Discuss in how far multispectral and hyperspectral data differ. Why are these differences particularly useful in analyzing urban remote sensing data?
- Why is it mandatory to perform a radiometric pre-processing of hyperspectral data to compare spectra with those from a spectral library?
- Access the ASTER spectral library (see list of references). Compare the available list of urban materials with the prevailing materials of an urban area you are familiar with. Evaluate the completeness of the database.
- Use the trial version of ENVI to load the John Hopkins University spectral library (choose: Spectral – Spectral library viewer – Open Spec Lib; change to the folder jhu_lib and chose manmade2.sli; confirm your choice with ok). Click the "red smooth-faced brick" (a plot window will open to visualize the spectrum) and then "Reddish asphalt roofing shingle". Why would Spectral Angle Mapping not necessarily be the best way to separate these materials?
- A modeling approach for simulating the urban heat island requires an estimate of the amount of different roofing materials. Which analysis methods could be chosen to extract such information from hyperspectral data?

- Fig. 9.3 shows a cobblestone pavement in different spectral resolutions. Use this figure to explain why hyperspectral resolution is needed to employ feature based analysis techniques.
- Comparing Figs. 9.4 and 9.1, it is obvious that the 3-band representation in Fig. 9.4 contains more differentiated information. What does this mean for an analysis process based on original and MNF-transformed data? Why is it not appropriate to employ an MNF-transformation when considering feature based analysis concepts?

References

ASD – Analytical Spectral Devices (2004) Spectroradiometers – FieldSpec3. http://www.asdi. com/products-fs3.asp. Accessed 11 Mar 2009

ASTER (1998) ASTER spectral library. http://speclib.jpl.nasa.gov/search-1/manmade. Accessed 11 Mar 2009

Ben-Dor E (2001) Imaging spectrometry for urban applications. Kluwer, Dordrecht/Boston/ London, pp 243–281

Buckingham R, Staenz K (2008) Review of current and planned civilian space hyperspectral sensors for EO. Can J Remote Sens 34:187–197

Gao BC, Goetz A (1990) Column atmospheric water vapour and vegetation liquid water retrievals from airborne imaging spectrometer data. J Geophys Res 95:3549–3564

Goetz AFH, Vane G, Solomon JE, Rock BN (1985) Imaging spectrometry for earth remote sensing science. Science 228:1147–1153

Green AA, Berman M, Switzer P, Craig MD (1988) A transformation for ordering multispectral data in terms of image quality with implications for noise removal. IEEE Trans Geosci Remote Sens 26(1):65–74

Heiden U, Rößner S, Segl K (2001) Potential of hyperspectral HyMap data for material oriented identification of urban surfaces. In: Proceedings of remote sensing of urban areas, Regensburg, Germany, pp 69–77

Herold M, Roberts DA, Gardner ME, Dennison PE (2004) Spectrometry for urban area remote sensing – Development and analysis of a spectral library from 350 to 2400 nm. Remote Sens Environ 91:304–319

Hill J, Mehl W (2003) Geo- und radiometrische Aufbereitung multi- und hyperspektraler Daten zur Erzeugung langjähriger kalibrierter Zeitreihen. Photogrammetrie-Fernerkundung-Geoinformation 2003(1):7–14

Hostert P, Damm A (2003) Sensitivity analysis of multi-source spectra from an urban environment. 3rd EARSeL Workshop on Imaging Spectroscopy, Herrsching, Germany, pp 215–219

Lehmann F, Bucher T, Hese S, Hoffmann A, Mayer S, Oschütz F, Zhang Y (1998) Data fusion of HyMap hyperspectral with HRSC-A multispectral stereo and DTM data. 1st EARSeL Workshop Imaging Spectroscopy, Zürich, Switzerland, pp 105–117

Phinn S, Stanford M, Scarth P, Murray AT, Shyy T (2002) Monitoring the composition and form of urban environments based on the vegetation – impervious surface – soil (VIS) model by sub-pixel analysis techniques. Int J Remote Sens 23(20):4131–4153

DLR – Deutsches Zentrum für Luft- und Raumfahrt/German Aerospace Centre (2003) The Quicklooks of HyEurope 2003 http://www.op.dlr.de/dais/hyeurope2003/hyeurope2003_ql.html. Accessed 11 Mar 2009

Richter R, Stanford D (2002) Geo-atmospheric processing of airborne imaging spectrometry data. Part 2: Atmospheric/topographic correction. Int J Remote Sens 23(13):2631–2649

Ridd MK (1995) Exploring a V–I–S (vegetation–impervious surface–soil) model for urban ecosystem analysis through remote sensing: comparative anatomy for cities. Int J Remote Sens 16:2165–2185

Schiefer S, Hostert P, Damm A (2006) Correcting brightness gradients in hyperspectral data from urban areas. Remote Sens Environ 101:25–37

Schläpfer D, Richter R, Damm A (2002) Geo-atmospheric processing of airborne imaging spectrometry data. Part 1: Parametric orthorectification. Int J Remote Sens 23(13):2609–2630

Schlerf M, Atzberger CG, Hill J (2005) Remote sensing of forest biophysical variables using HyMap imaging spectrometer data. Remote Sens Environ 95(2):177–194

Segl K, Rößner S, Heiden U (2000) Differentiation of urban surfaces based on hyperspectral image data and a multi-technique approach. In: Proceedings of the IEEE IGARSS 2000, Honolulu, pp 1600–1602

Small C (2003) High spatial resolution spectral mixture analysis of urban reflectance. Remote Sens Environ 88:170–186

van der Linden S, Janz A, Waske B, Eiden M, Hostert P (2007) Classifying segmented hyperspectral data from a heterogeneous urban environment using support vector machines. J Appl Remote Sens 1. doi:10.1117/1.2813466

van der Linden S, Hostert P (2009) The influence or urban surface structures on the accuracy of impervious area maps from airborne hyperspectral data. Remote Sens Environ 113:2298–2305

Wilson IJ, Cocks TD (2003) Development of the Airborne Reflective/Emissive Spectrometer (ARES) – a progress report. In: Proceedings of the 3rd EARSeL Workshop Imaging Spectroscopy, Herrsching, Germany, pp 50–55

Chapter 10
Segmentation and Object-Based Image Analysis

Elisabeth Schöpfer, Stefan Lang, and Josef Strobl

This chapter focuses on segmentation of remotely sensed image data and object-based image analysis. It discusses the differences between pixel-based and object-based image analysis; the potential of the object-based approach; and, the application of eCognition software for performing image segmentation and classification at different levels of detail.

Learning Objectives

Upon completion of this chapter, you should be able to:

❶ Explain the principles of image segmentation
❷ Differentiate between pixel-based and object-based image analysis
❸ Speculate on object-based classification

10.1 Introduction

The number of operational high-resolution satellite and air-borne sensors has increased significantly during the last years. The obtained images provide more and more detail about land and water surfaces and hence permit us to observe additional

---

E. Schöpfer (✉)
German Aerospace Center (DLR), German Remote Sensing
Data Center (DFD) Oberpfaffenhofen, 82234 Wessling, Germany
e-mail: elisabeth.schoepfer@dlr.de

S. Lang
Centre for Geoinformatics, University of Salzburg, Schillerstraße 30, Salzburg 5020, Austria
e-mail: stefan.lang@sbg.ac.at

J. Strobl
Austrian Academy of Sciences, Geographic Information Science, Schillerstraße 30,
Salzburg 5020, Austria
e-mail: josef.strobl@oeaw.ac.at

T. Rashed and C. Jürgens (eds.), *Remote Sensing of Urban and Suburban Areas*,
Remote Sensing and Digital Image Processing 10,
DOI 10.1007/978-1-4020-4385-7_10, © Springer Science+Business Media B.V. 2010

facets in both anthropogenic and natural settings. Accordingly, we are able to detect more categories of geographical features and the principle set of classes addressed is being extended. With refined image resolution there likewise is an increasing demand for adequate image analysis and interpretation techniques. A significant amount of structure-related information is revealed by high-resolution imagery (see a related review in Chapter 7). This additional information can be used to describe classes with properties beyond their spectral reflectance. By complementing the spectral behavior of geographical features with their spatial characteristics we are bridging remote sensing techniques and GIS functionality. The term 'semantic class definition' has been suggested to signify structural and geometrical properties of image features. This term helps us to conceptually distinguish between 'target classes' and 'ancillary classes'. The former addresses the final set of classes the user wants to report as a result from image analysis. The latter comprises all intermediate categories used during the processing of an image, which are used to describe the internal structure of the target classes. This kind of semantic class definition of geographical features requires (1) representations of the scene over several scales and (2) rules to integrate heuristics about these representations. A toolset supporting this approach is implemented in the image analysis software eCognition from Definiens GmbH, Munich (eCognition User Guide 2004). This chapter explains the rationale for object-based image analysis, presents and discusses image segmentation techniques, and finally demonstrates examples from urban scene classifications.

> **a definition of semantic classes combines the spectral behavior of geographical features with their spatial characteristics**

10.2   Image Segmentation

10.2.1   *Limitations of Pixel-Based Classification Techniques*

Most of the methods for image processing developed since the early 1970s until present are based on classifications of individual pixels clustered in a multi-dimensional feature space (Chapter 8). Although a range of sophisticated and now well established techniques have been developed, the current demand from the remote sensing community is not fully met due to different characteristics of high resolution imagery. These new sensors significantly increase the within-class spectral variability and, at the same time, decrease the potential accuracy of a purely pixel-based approach to classification (Tadesse et al. 2003). Consequently, traditional image processing algorithms are being complemented and sometimes replaced by novel classification methods. The currently evident trend towards image segmentation suggests exploring 'segments' as spatially contiguous and spectrally homogeneous groups of

> **image segmentation suggests exploring spatially homogeneous groups of pixels rather than single pixels**

pixels (rather than dealing with individual pixels), supporting geometrical, topological and/or textural properties (Antunes et al. 2003). The need for image segmentation is especially important in high-frequency images of urban and sub-urban settings since we are dealing with very significant scene complexity and detailed structures. Moreover, many urban features are primarily defined by their spatial inter-relationships instead of spectral characteristics.

Traditional image analysis uses individual pixels as the basic unit in classification of remotely sensed imagery. Pixels carry an integrated spectral signal and are characterized by their ground sampling cell size. Essentially, the set of pixels making up an image defines its entire amount of information. Nevertheless, working from individual pixels is a limiting strategy, since pixels are not treated as building blocks within their spatial context, but rather as independent samples (which they in fact are not!). The alternative approach is to consider image objects made of homogeneous clusters of adjacent pixels with meaningful geometric and other spatial properties. These clusters promise a much richer and more powerful working environment throughout the classification process (Blaschke and Strobl 2001). Spatial relationships describing hierarchical ('vertical') or lateral pixel neighborhoods can be fully considered during classification. Other strategies are implemented through the 'Expert Classifier' by ERDAS Imagine®/Leica-Geosystems or the 'Feature Analyst' from Visual Learning Systems (Erdas Imagine® 2004 and Visual Learning Systems 2004). The former provides a rule-based approach to characterize pixels through the integration of evidence and previous knowledge. The latter extracts features by inspecting certain spatial arrangements of pixels representing the target features (Lang et al. 2003).

10.2.2   Reasons for Object-Based Classification

As pointed out above, shortcomings of pixel-based classification have stimulated new classification concepts to be investigated. With pixels grouped into homogeneous regions (image objects) by segmentation, these aggregates may reflect semantic objects of interest from the real world in a more appropriate way than purely spectrally defined agglomerations of single pixels. Classification starting from image objects rather than individual pixels can utilize spatial and geometrical properties as well as relationships among objects. Blaschke and Strobl (2001) make a case for the use of segmentation algorithms to delineate objects based on contextual information in an image.

Using textural information and additional characteristics like the size or shape of the objects the per-pixel classification process is complemented by a new image processing technique. In particular, information from high resolution images is being aggregated at multiple levels of detail, resulting in hierarchical sets of homogenous regions according to the given application semantic (Fig. 10.1). Different scales of an image lead to vertically and horizontally interconnected objects that could potentially enlarge the set of target classes.

Overall, object-based classification techniques are a new and innovative approach, especially for applications handling human-made features. In the realm

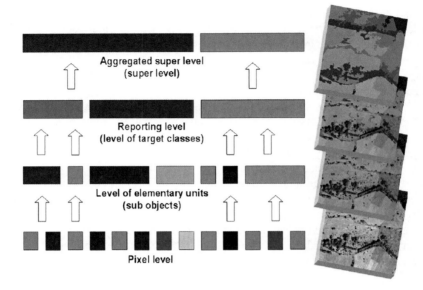

Fig. 10.1 Multi-scale object representation of an image scene. Note that the reporting level does not necessarily have to be the highest aggregated level

of urban remote sensing there are several applications demonstrating that object-based approaches are superior to per-pixel analysis especially when urban land cover classes, as airport, roads, etc., are to be separated (e.g. Darwish et al. 2003; Hofmann 2001). In comparison to most natural environments the built environment is characterized by sharp, discrete boundaries and a high-frequency change of different surfaces with similar reflectance properties. Anthropogenic features can be reproduced rather unambiguously in an object-based environment through an iterative segmentation process. Shadowed areas for example can be characterized by specific relationships to neighboring objects and the shape of an object may help in discerning between a roof and a road of similar reflectance. In summary, object-based image analysis combines GIS functionality and remote sensing techniques by working with polygonal, homogeneous clusters instead of single pixels.

image segmentation suits the urban environments due to the nature of its anthropogenic features that are characterized by sharp, discrete boundaries and a high-frequency change of different surfaces with similar reflectance properties

10.2.3 Principles and Realization

A crucial step for object-based image analysis is image segmentation. Image segmentation aims at delineating spectrally homogenous regions within an image to generate

Area-Based vs. Edge-based Algorithms

The numerous algorithms for image segmentation can be separated into two basic groups of algorithms: 'area-based' and 'edge-based' algorithms. Edge detection makes sense if an image scene is dominated by linear elements such as geological faults. As we are interested in larger and contiguous geographical features being better represented by area objects, this chapter subsequently focuses on area-based segmentation producing polygonal image objects. Three principal approaches exist for area-based segmentation: (1) histogram thresholding, (2) region-based techniques (including region-growing and split-and-merge), and (3) 'blobs' or scale space analysis (Pinz 1994). Histogram or global threshold techniques seek out 'image events' (i.e. significant changes in contrast), but are only applicable on very high contrast images. These techniques have lost more and more significance with recently available multi-spectral high resolution images of high variance but lower contrast. 'Blob' detection techniques do not primarily seek the delineation of a consistent set of objects but rather they are focused on the dynamic nature of scale-dependent objects (Hay et al. 2003). This is achieved by a stepwise smoothing of the image, usually through Gaussian filtering. Finally, region-based algorithms aggregate single pixels into homogenous groups. A crucial component of a region-based segmentation is the criterion of homogeneity used in the grouping of pixels. Usually homogeneity is considered in terms of spectral or textural similarity, but it can also be defined by geometric properties of newly created objects.

a set of exhaustive 'image objects'. Adjacent pixels with similar reflectance values are grouped into 'segments' by constraints of size and shape. The entire image information is aggregated onto a higher level of detail. Koch et al. (2003) point out that any kind of segmentation involves a certain level of generalization. Segmentation can be performed over several nested levels, i.e. lower level image objects are grouped again and again helping to represent image information on different levels of homogeneity and thus scales. This very point arguably makes image objects correspond to our conceptual understanding of the hierarchical human cognition process of scenes, thus better matching human perception of the real world.

Image segmentation seeks to delineate spectrally homogenous regions known as 'image objects'

The Fractal Net Evolution Approach (FNEA) as documented by Baatz and Schäpe (1999) serves as a framework for operational image segmentation implemented in eCognition software from Definiens AG. The core of the eCognition software is built around the extraction of homogeneous image objects with a specified level of resolution. This multi-resolution segmentation technique fits into the category of bottom-up region-growing techniques. Image objects are generated according to

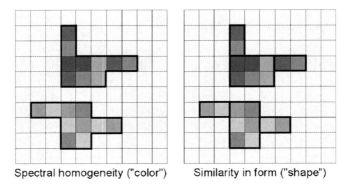

Spectral homogeneity ("color")   Similarity in form ("shape")

Fig. 10.2 Two dimensions of 'homogeneity' as implemented in eCognition

defined criteria of homogeneity through a combination of parameters: scale (i.e. average size of objects), color (i.e. mean spectral value) and shape (i.e. geometric form of the objects) (Fig. 10.2).

While the scale parameter controls the average size of the generated objects, the color and shape parameters reflect the respective criteria of homogeneity being applied. The software allows absolute and relative weighting of these parameters, and in addition it splits the shape parameter into outline smoothness and shape compactness. It helps to extract different object shapes, each representative of different categories in the respective level. Based on these per-level parameters, a hierarchical network of image objects can be built, which allows for the simultaneous representation of image information at different resolutions ('scales').

10.3   Object-Based Classification

10.3.1   Object Features

After performing multi-scale segmentation the original image is represented by image objects on different levels of detail, organized into a hierarchical network. The feature space for the description of image objects is characterized by a vast set of available measures. Image objects are described by spectral, textural and contextual measures that are stored in an object feature database. Selected features are then used to subsequently classify objects.

image objects are described by spectral, textural and contextual properties

10.3.2   Classification

Objects generated through segmentation can be classified using two different classification methods: (1) classification based on samples and (2) classification based on the integration of prior external knowledge stored in rule bases.

The high number of potential characteristics of objects leads to a high-dimensional feature space. For image classification, representative objects for each class are first identified. This process can be compared to the delineation of training areas in a supervised classification. These sample objects should be well separated in feature space by using the most representative and clearly distinguished features. Algorithms for the optimization of the feature space can support this process. That means that the entire set of possible feature space dimensions is reduced to a smaller set of features optimizing separation of image objects. After selecting the samples and their corresponding feature values, a standard classification algorithm (such as box classifier, nearest neighbor, maximum likelihood) is applied to all image objects.

Multi-scale segmentation creates a network of image objects connected by vertical and horizontal relationships. Binnig et al. (2002) explain the way in which this network reflects a hierarchical knowledge structure as a "self-organizing, semantic, self-similar network." By operating with the relations among networked objects and integrating prior knowledge of the target classes, analysts can classify objects through a set of rules. Those reflect both the spectral response and the contextual information. Integrating knowledge provides a way to overcome the spectral similarity of different geographical features (e.g. to separate rural grassland from urban green space using geometrical properties). Rules are established to characterize the lateral relationships between objects from specific classes (e.g. "relative border length with neighboring class XY"). Moreover, some object-based classification approaches use fuzzy logic to soften crisp distinctions and characterize the degree of membership of an object to a certain class. One of the strengths of fuzzy techniques comes from the way they can be used to describe real-world phenomena with some degree of uncertainty.

two groups of classification methods are used to delineate objects from a segmented image: sample-based methods and rule-based methods

Furthermore it is possible to support any classification with ancillary thematic data like GIS layers or digital elevation models. Ancillary thematic data can be used for image segmentation and for class descriptions as additional information in the object feature database (Fig. 10.3).

10.3.3 Post Processing

An object-based classification approach offers the possibility of the spectral, as well as the semantic grouping of objects. Contiguous objects from the same class can be merged into one polygon. Objects belonging to the same class on a higher semantic level can also be merged (e.g. a football field and the surrounding area can be merged into the target class 'urban'). Objects and classification results can be stored in either raster or vector formats.

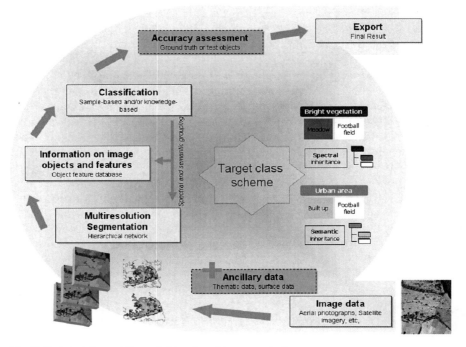

Fig. 10.3 Principle workflow of object-based image analysis

10.3.4   Accuracy Assessment

Quantitative site-specific accuracy assessment (Congalton and Green 1998) is an important means to evaluate the quality of classification results. Error matrices and corresponding assessment values (error of commission, error of omission, kappa statistic) are frequently used for evaluating pixel-based classifications. The thematic assessment can be performed by generating random points within objects and checking the classification labels against a ground truth layer. Alternatively a specific set of randomly selected objects can be chosen as test areas (similar to training areas) and then inspected as reference information. Geometrical fitting is however by far harder to evaluate.

a challenging issue in assessing the accuracy of object-based classification is the geometrical fitting of objects

While the classified image objects can be visually checked against manual delineation of interpreted imagery (Koch et al. 2003), the geometrical congruence is assessed by comparing existing vector data with overlay techniques.

10.4   Case Studies

Remote sensing in urban areas covers a huge spectrum of applications and scientific research, where object-based analysis can be very supportive. In this chapter two case studies are briefly presented, one based on the detection of informal settlements, the other on textural analysis for mapping urban structure.

10.4.1   Detecting Urban Features from IKONOS Data Using Methods of Object Oriented Image Analysis (Hofmann 2001)

Most urban image classifications target formal settlements like big cities, which are already studied in detail and mapped extensively. The detection of informal settlements is a challenging task due to their microstructure and irregularity in object shapes. Based on spectral and spatial resolution two different levels of detail can be examined: (1) the detection of single shacks or (2) the location of entire informal settlements and their boundaries.

The study area, a part of Cape Town (South Africa), is characterized by several different forms of settlement areas. The first step was an image enhancement by applying a principal-component pan-sharpening method. Next, multiple hierarchical image object levels were created in eCognition. The smallest level of image segments reveals single houses, while the top-most level represents entire settlement areas or parts thereof. After conducting a nearest-neighbor classification the result was improved by using inheritance mechanisms and form criteria. Textural information described by reflectance and shape of lower-level objects helps to identify and classify different types of settlement areas.

Hofmann argues that the object-based approach is well suited to detecting the complex structures of informal settlements. Especially when working with enhanced high resolution satellite imagery, image segmentation identifies 'real-world' objects that show a typical texture according to the different types of settlements.

10.4.2   Analysis of Urban Structure and Development. Applying Procedures for Automatic Mapping of Large-Area Data (De Kok et al. 2003)

The analysis of urban structure starts with the central role of texture analysis for city footprint extraction. In general, texture can be expressed by calculating Grey Level Co-occurrence Matrices (GLCM). A bottleneck in textural applications is the relationship between the object of interest and a fixed filter area size used for the moving window processing. Contrary segment-size is not restricted to a certain width. In the object-oriented approach segments represent objects of interest in different area sizes. Therefore the segmentation parameter with various manipulations is crucial for successful texture analysis and the following classification. The resulting mask of the city footprint is used to differentiate between sealed and unsealed areas within the city border.

Chapter Summary

The main idea of object-based image analysis is to work on homogenous image objects rather than on single pixels. Using spectral and spatial information the pixels are merged into homogenous groups (segments, image objects). After segmentation the features of generated objects are used for image interpretation and subsequent classification. Thus the user's knowledge can be integrated into image processing and the potential set of target classes can be extended. There are several segmentation techniques available, with this chapter focusing on the algorithms implemented in eCognition. These allow image segmentation into an arbitrary set of hierarchical scale levels, grouping pixels according to a scale parameter and two aspects of homogeneity: color and shape/form. Object-based classification utilizes external knowledge by means of a rule base. Intuitive knowledge can be formalized and made operational through fuzzy membership functions. Two case studies demonstrate practical applications of object-based image analysis in an urban context.

LEARNING ACTIVITIES

Internet Resources for Segmentation Approaches and Software Packages

- Definiens Professional/eCognition: Software based on a region growing algorithm.
  🖳 http://www.ecognition.com
- Feature Analyst (Visual Learning Systems): Extension for ESRI's ArcView and ArcGIS and ERDAS' IMAGINE to extract object-specific geographic features.
  🖳 http://www.featureanalyst.com/
- PARBAT:
  🖳 http://parbat.lucieer.net

- ImageJ & watershed JAVA plugin:
  🖥 http://rsb.info.nih.gov/ij/
- InfoPACK (InfoSAR Ltd.) and CAESAR 3.1 (N.A. Software Ltd.): Both simulated annealing.
  🖥 http://www.infosar.co.uk/; http://www.nasoftware.co.uk/
- SPRING / Freeware:
  🖥 http://www.dpi.inpe.br/spring/english/
- RHSEG (NASA`s Goddard Space Flight Center): linux–based segmentation algorithm with labeling tool.
  🖥 http://opensource.gsfc.nasa.gov/projects/HSEG/index.php
- Berkeley Image Segmentation – 30 days trial software.
  🖥 http://berkenviro.com/berkeleyimgseg/

Study Questions

- How would you use altitude information to classify settlements in mountainous areas?
- Which scale parameters would you use to classify different categories of roads (highway, etc.)?
- Which semantic description (rule) would you use to tell city green from urban grassland?

References

Antunes AF, Lingnau C, Da Silva JC (2003) Object oriented analysis and semantic network for high resolution image classification. In: Proceedings of Anais XI SBSR Conference, Belo Horizonte, Brazil, 05–10 Apr 2003, INPE, pp 273–279

Baatz M, Schäpe A (1999) Object-oriented and multi-scale image analysis in semantic networks. In: Proceedings of the 2nd international symposium on operationalization of remote sensing, Enschede, ITC, 16–20 Aug 1999

Binnig G, Baatz M, Klenk J, Schmidt G (2002) Will machines start to think like humans? Europhy News 33(2). Online http://www.europhysicsnews.com/full/14/article2/article2.html. Accessed Sept 2005

Blaschke T, Strobl J (2001) What`s wrong with pixels? Some recent developments interfacing remote sensing and GIS. GIS Zeitschrift für Geoinformationssysteme 6:12–17

Congalton RG, Green K (1998) Assessing accuracy remotely sensed data principles practices. Mapping science series. CRC Press, London

Darwish A, Leukert K, Reinhardt W (2003) Urban land-cover classification: an object based perspective. In: Proceedings of the 2nd GRSS/ISPRS Joint Workshop on Data Fusion and remote sensing over urban areas. URBAN 2003, Berlin, pp 277–282

De Kok R, Wever T, Flockelmann R (2003) Analysis of urban structure and development applying procedures for automatic mapping of large area data. In: Juergens, C (ed) Remote sensing of urban areas. In: Proceedings of the 4th international symposium held in Regensburg/Germany, 27–29 June 2003. (The International Archives of Photogrammetry, Remote Sensing and Spatial Information Sciences, vol XXXIV–7/W9) (CD-ROM): 41–46

eCognition User Guide (2004) Munich, Germany. http://www.definiens-imaging.com. Accessed 20 Feb 2009

Erdas Imagine®/Leica-Geosystems, Image Segmentation (2004) http://www2.erdas.com/support-site/downloads/tools/descriptions/tool_descriptions.html#image_seg. Accessed 20 Feb 2009

Hay GJ, Blaschke T, Marceau DJ, Bouchard A (2003) A comparison of three image–object methods for the multiscale analysis of landscape structure. Photogramm Remote Sens 57:327–345

Hofmann P (2001) Detecting informal settlements from IKONOS image data using methods of object oriented image analysis – an example from Cape Town (South Africa). In: Juergens C (ed) Remote sensing of urban areas – Fernerkundung in urbanen Räumen. In: Proceedings (Abstracts and Full papers on Supplement CD-ROM) of the 2nd international symposium held in Regensburg/Germany, 22–23 June 2001-Regensburger Geographische Schriften 35, Regensburg

Koch B, Jochum M, Ivits E, Dees M (2003) Pixelbasierte Klassifizierung im Vergleich und zur Ergänzung zum objektbasierten Verfahren. Photogrammetrie Fernerkundung Geoinformation 3(2003):195–204

Lang S, Schöpfer E, Blaschke T (2003) Object-specifc change detection based on assisted feature extraction: a case study of an expanding suburban area. In: Juergens C (ed) Remote sensing of urban areas. In: Proceedings of the 4th international symposium held in Regensburg/Germany, 27–29 June 2003. (The International Archives of Photogrammetry, Remote Sensing and Spatial Information Sciences, vol XXXIV-7/W9) (CD-ROM): 93–98

Neubert M, Meinel G (2003) Evaluation of segmentation programs for high resolution remote sensing applications. In: Schroeder M, Jacobsen K, Heipke C (eds) In: Proceedings of the Joint ISPRS/EARSeL Workshop "High Resolution Mapping from Space 2003", Hannover, Germany, 6–8 Oct 2003 (published on CD only)

Pinz A (1994) Bildverstehen. Springer, Vienna (in German)

Tadesse W, Coleman TL, Tsegaye TD (2003) Improvement of land use and land cover classification of an urban area using image segmentation from Landsat ETM+ data. In: Proceedings of the 30th international symposium on remote sensing of the environment, 10–14 Nov 2003, Honolulu, Hawaii

Visual Learning Systems, Feature Analyst (2004) http://www.featureanalyst.com/. Accessed 20 Feb 2009

Chapter 11
Data Fusion in Remote Sensing of Urban and Suburban Areas

Thierry Ranchin and Lucien Wald

This chapter focuses on techniques related to image fusion and data fusion from different sources with varying spatial and spectral resolutions, and presents and discusses some of the technical issues that influence data fusion in the urban context.

Learning Objectives

Upon completion of this chapter, you should be able to:

❶ Understand the process of data fusion in remote sensing of urban areas
❷ Explain the scope of and relations between multi-scale methods and wavelet transforms for data fusion
❸ Understand and speak about applications of data fusion in urban areas

---

11.1 Introduction

Data fusion is becoming of paramount importance in the Earth observation science. Data fusion is an approach oriented to information extraction adopted in several domains. It is based on the synergetic exploitation of data originating from different sources. It aims at producing a better result than that obtained by a separate exploitation of the same sources. The utilization of satellite images and more generally of observations of the Earth and our environment is presently one of the most productive applications in data fusion (Wald 2002). Observation of the Earth is performed by means of satellites, planes, ships, and ground-based instrument, and results in a great

---

T. Ranchin (✉) and L. Wald
Center for Energy and Processes, MINES ParisTech, rue Claude Daunesse,
Sophia Antipolis Cedex 06904, France
e-mail: thierry.ranchin@mines-paristech.fr; lucien.wald@mines-paristech.fr

T. Rashed and C. Jürgens (eds.), *Remote Sensing of Urban and Suburban Areas*,
Remote Sensing and Digital Image Processing 10,
DOI 10.1007/978-1-4020-4385-7_11, © Springer Science+Business Media B.V. 2010

variety of measurements, partly redundant and partly complementary. These measurements may be punctual or time-integrated, bi-dimensional or instantaneous (images), vertical profiles with time-integration or not, as well as three-dimensional information (e.g. oceanic/atmospheric profiler/sounder at ground level, satellite-borne, ship-borne). With the large amount of archives and numerical models representing the geophysical/biological processes in mind, it is obvious that the quantity of information available to describe and model the Earth and our environment rapidly increasing. Data fusion as a subject is becoming increasingly relevant because it efficiently helps scientists to extract precise and relevant knowledge from available information.

By itself, the process of data fusion is not new in environmental studies. Meteorologists, for example, have been using it in weather prediction for decades. In remote sensing, many of the classification procedures widely used in the field are obviously relevant to data fusion. Data fusion allows the combination of the measurements produced by classification, as well as the monitoring of the quality of information generated by these measurements.

Data fusion: a formal framework in which are expressed means and tools for the alliance of data originating from different sources. It aims at obtaining information of greater quality; the exact definition of 'greater quality' will depend upon the application

The European Association of Remote Sensing Laboratories (EARSeL) created a "Special Interest Group for Data Fusion" in 1996 (Wald 2000). This group has been contributing to a better understanding and use of data fusion in the field of Earth observation through organizing regular meetings for its members to discuss and tackle the fundamentals of data fusion in remote sensing. A series of bi-annual international conferences called "Fusion of Earth Data: Merging point measurements, raster maps and remotely sensed images" was launched in 1996 with the aim of exploring this field of research and helping the scientific community to fully understand the benefits of data fusion in the Earth observation domain (Ranchin and Wald 1996a, 1998, 2000a). A set of terms of reference emerged from this work, including the definition of data fusion as "a formal framework in which are expressed means and tools for the alliance of data originating from different sources. It aims at obtaining information of greater quality; the exact definition of 'greater quality' will depend upon the application" (Wald 1999). "Quality" is used in the definition as a generic word to imply when the user is more satisfied by the results obtained through a fusion process than without it.

Many techniques for data fusion already exist. The focus of this chapter is on the fusion of images. The general approach in the fusion of images is to create a new set of images, I, usually of reduced dimensions, from the original sets of images, as the following equation indicated:

$$I = f(A, B, C, D, \ldots) \tag{11.1}$$

where A, B, C, D, … are the original sets of images and eventually characteristics derived.

These sets may or may not be commensurate, and could originate from various modalities (e.g. panchromatic, microwave, hyperspectral) taken at different points of times and with different times of integration. They may also have different spatial resolutions. All images with an individual set, however, must be geometrically aligned and have the same spatial resolution, or pixel size (Wald 2002). Here the term image comprises any information that is presented in a raster, or gridded, format in two dimensions. The grid cell is called a pixel.

Methods and objectives of image fusion vary by application (Wald 2002). A classical example is the classification process in environmental applications (see Chapter 8). Several images of commensurate or non-commensurate measurements, and possibly of other information, are used as an input to a classifier. In the case of a supervised classification, a fusion algorithm is included in the classification to produce an image of taxons and possibly another image of the related accuracy (or plausibility, or probabilities, etc.). In the case of an unsupervised procedure, the state vectors of the pixels are grouped based on the similarity of certain properties. The unsupervised classification is usually an iterative fusion process with successive refinements until a threshold is met. In either type of classification, the original dimension of the information is reduced and as such, the semantic level of the fused product is typically higher than that of the original set of images.

Other approaches in image fusion exist. Some include the extraction of features from each input image and then the fusion of features. For examples, road maps can be produced by fusing several sets of images, where the final product is a GIS layer of roads. Other approaches utilize visual analysis and interpretation in the fusion with the aim of creating a set of images of reduced dimension, which contains all the information of interest that are present in the original sets of images. Fusion may also be performed to create new sets of images in various modalities with a better spatial resolution.

This chapter focuses on few methods applied to images and imaging sensors but also to gridded data and punctual measurements. These include: (1) encrustation of images within another image; (2) synthesis of images based on the best spatial resolution available in the original sets of images; and (3) fusion of images, gridded data and punctual measurements. Each method is described in general below and illustrated by an example. These methods call upon advanced mathematical tools that are presented in the following section.

Different forms of Data Fusion

Data fusion may be sub-divided into many domains. For example, the military community uses the term "positional fusion" to denote aspects relevant to the assessment of the state vector or "identity fusion" when establishing the identity of the entities is at stake. If observations are provided by sensors and only by sensors, one will use the term "sensor fusion". "Image fusion" is a sub-class of sensor fusion; here the observations are images. If the support of the information is always a pixel, one may speak of "pixel fusion". "Evidential fusion" means that the algorithms behind call upon the evidence theory. Other terms commonly used are "measurement fusion", "signal fusion", "features fusion", and "decision fusion".

11.2   Advanced Mathematical Tools for Data Fusion

Wavelet transform (WT) analysis appeared in the early 1980s at the junction of signal processing, applied mathematics, and quantum mechanics to provide an efficient mathematical framework for the study of non-stationary signals. Combined with the multi-resolution analysis (MRA), introduced by Mallat (1989), the WT was applied to the analysis and processing of images. In the field of Earth observations, the first applications of these two mathematical tools (i.e. MRA and WT) appeared in the early 1990s and focused on the analysis of remotely sensed images in different fields of investigation, including geology, urban areas, and oceanography, as well as on the processing of SAR imagery and data fusion for the improvement of spatial resolution of images. Currently, the use of WT and MRA is quite common for modeling and fusing data in the field of remote sensing.

11.2.1   Wavelet Transform (WT)

> wavelet transform produces a time–frequency representation of signals and images obtained from decomposition over a base generated by dilations and translations of a single function called the mother wavelet

The main property of the WT is to adapt the analysis window to the phenomenon under study based on local information. The WT leads to a time–frequency representation. In the case of images, it leads to a scale-space representation. As in Fourier transform, WT is equivalent to a decomposition of the signal on the basis of elementary functions: the wavelets. Wavelets are generated by dilation and translation of a single function called the mother wavelet:

$$\psi_{a,b} = |a|^{\frac{-1}{2}} \, \psi\left(\frac{x-b}{a}\right) \tag{11.2}$$

where a and b are reals and $a \neq 0$. a is called the dilation step and b the translation step.

Many mother wavelets exist. They are all oscillating functions, which are well localized both in time and frequency. All the wavelets have common properties such as regularity, oscillation and localization. They also need to satisfy an admissibility condition (see Meyer (1990) or Daubechies (1992) for more details about the properties of the wavelets).

Despite the common properties, each wavelet brings a single decomposition of the image signal related to the used mother wavelet. In the one dimension case, the continuous WT of a function $f(x)$ is:

$$WT_f(a,b) = \langle f, \psi_{a,b} \rangle = \frac{1}{\sqrt{|a|}} \int_{-\infty}^{+\infty} f(x) \overline{\psi(\frac{x-b}{a})} dx \tag{11.3}$$

where $\Psi_{a,b}$ is defined as in Eq. 11.2 and $\overline{\psi(\dfrac{x-b}{a})}$ is the complex conjugated of Ψ.

$WT_f(a,b)$ represents the information content of $f(x)$ at scale a and location b. For fixed a and b, $WT_f(a,b)$ is called the wavelet coefficient.

The application of the WT for each scale and each location of a signal provides a local representation of this signal. The process can be reversed and the original signal reconstructed exactly (without any loss) from the wavelet coefficients according to the following equation:

$$f(x) = \frac{1}{C_\psi} \int\limits_{-\infty}^{+\infty} \int\limits_{-\infty}^{+\infty} WT_f(a,b)\psi_{a,b}(x)\frac{da\,db}{a^2} \qquad (11.4)$$

where C_ψ is the admissibility condition of the mother wavelet. This equation can be interpreted in two ways:

- $f(x)$ can be reconstructed exactly if one knows its wavelet transform.
- $f(x)$ can also be thought of as a superimposition of wavelets.

These two points of view lead to different applications of the WT: signal processing and signal analysis.

11.2.2   Multi-resolution Analysis (MRA)

The concept of MRA introduced by Mallat (1989) is derived from the Laplacian pyramids (Burt and Adelson 1983). In this approach, the size of a pixel is defined as a resolution of reference and is used as a basis to measure local variations in the image. Note that the resolution is related to the inverse of the scale used by cartographers.

the larger the resolution of an image, the smaller the size of visible objects

Hence, the larger the resolution of an image, the smaller the size (or the characteristic length or characteristic scale) of visible objects on the image scene.

Figure 11.1 shows a description of MRA and more generally of pyramidal algorithms. MRA computes successive, coarser and coarser approximations of the same original image. The base of the pyramid corresponds to the original image. Climbing the pyramid, the different steps represent the successive approximations of the image. The theoretical limit of these algorithms is the top of the pyramid, which corresponds to a unique pixel. The difference of information existing between two successive approximations of the same image is described by the wavelet coefficients.

The application of the Mallat's algorithm to images is well known for the wavelet community. Another algorithm of interest has been proposed by Dutilleux (1989), the so-called "à trous" algorithm. In this algorithm, only a scale function is used. The approximation of the original image is obtained by filtering the original image, the

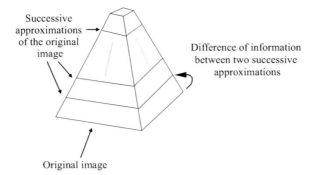

Successive
approximations
of the original
image

Difference of information
between two successive
approximations

Original image

Fig. 11.1 Representation of the successive approximations of an image based on a multi-resolution algorithm

wavelet coefficient image is obtained by subtracting the approximation to the original image in a pixel-by-pixel basis. This algorithm provides at each step one context and one non-directional wavelet coefficient image. To perform a dilation of the scale function, one adds zero between each coefficient of the filter. Hence, no sub-sampling of the image is performed and all images will have the same size. The reconstruction is done by summation of the last context and the wavelet coefficient images computed.

> **multi-resolution analysis offers a hierarchical modeling, also called a scale-by-scale representation of signals and images**

These two mathematical tools: the wavelet transform and the multi-resolution analysis enable an image to be decomposed into the structures of different sizes that participate to its information content. Some examples of using these tools in remote sensing applications can be found in Ranchin (1997).

11.3 Encrustation of Higher Spatial Resolution Quickbird Image in Low Resolution Image

This section discusses with an example the problem of merging partially overlapping images (including raster maps) with different spatial resolutions. Specifically we discuss the problem of the encrustation of a higher spatial resolution "imagette" (image of small size) into an image of lower resolution with the attenuation of the discontinuities created at the edges of the imagette by the difference in resolutions.

> **encrustation is a means to benefit from images with high spatial resolution in a localized area within an image of lower resolution**

It is assumed that both the image and imagette offer the same level of "radiometric" accuracy, i.e. none of these images is superior to the other with respect to radiometry.

In that case, the encrustation of the higher resolution imagette into the wider, lower resolution image is the best way to merge the data and obtain a final image offering the highest spatial accuracy. This will also be true if the imagette exhibits higher level of "radiometric" accuracy than the image. For the purpose of the following discussion, spatial resolution refers to the smallest area on the ground that an imaging system, such as a satellite sensor, can distinguish.

Firstly, we illustrate the benefit of the encrustation of higher resolution data into lower resolution image. Then, we propose a method for the attenuation of the edges, which can help identify the limits between both resolutions and better simulate the actual data at the highest resolution. A Quickbird image is utilized to provide a didactic example of the problems and solutions associated with the encrustation process, as well as a test case to assess quantitatively the qualities of the image resulting from the merging. Other case studies and examples can be found in Ranchin (1997).

11.3.1 Benefits of Encrustation

The benefit of encrustation is illustrated by an examination of the loss (respectively gain) of information when degrading (respectively upgrading) the resolution of an image. A Quickbird image of the city of Strasbourg, located in the Northeast of France, is used in this example. Its spatial resolution is 0.7 m. It is called the original or reference image.

The smoothing of this image provides an imagette whose pixel size is still 0.7 m, but its effective resolution is 5.6 m, hereafter referred to as a "5.6 m smoothed image". Figure 11.2 displays an encrusted image resulting from the encrustation of

Fig. 11.2 A lower resolution image with a higher-resolution encrusted imagette

a high resolution imagette extracted from the original image at 0.7 m into the 5.6 m smoothed image. Thus apart the encrusted part, this image is similar to the smoothed one.

The examination of the high and low resolution portions in Fig. 11.2 and of their differences clearly demonstrates the large amount of information which is gained when the spatial resolution increases from 5.6 to 0.7 m. Contours are blurred in the low resolution part. Buildings and streets are clearly visible in the original image while even large streets are hardly visible in the 5.6 m resolution image. The loss of information between the reference image and the 5.6 m smoothed image is quantified by computing the difference between both images for each pixel of 0.7 m. The differences are synthesized by a few parameters: bias, difference of variances, and a correlation coefficient. There is no difference in average values; that is, the bias is equal to zero. The difference in variances expresses the difference in information and is equal to 47% of the mean value. It means that 47% of the information content is lost when decreasing the resolution from 0.7 to 5.6 m. In other words, the small size structures lost in the range between 0.7 and 5.6 m account for approximately half of the information contained in the original image. The correlation coefficient between both images is only 0.85. This example clearly demonstrates the interest to have images with the best available spatial resolution and therefore the need for efficient encrustation procedures.

11.3.2   *Methods*

Encrustation is accomplished using the following approach. First, a high resolution imagette and a low resolution image of a larger geographical coverage are re-mapped onto each other or onto a common geographical reference. The resulting re-mapped image and imagette now have the same pixel size, but different effective spatial resolutions. The encrusted image is then computed from the values of the re-mapped imagette, when available. Otherwise it can be computed from the values of the re-mapped image.

As Fig. 11.2 shows above, the encrustation creates a clearly defined border which is created at the periphery of the imagette due to the dramatic change of resolution. This border is often disturbing in photo-interpretation and image analysis. Also the discontinuities at the periphery (abrupt changes in derivatives) can prevent digital processing methods such as edge detection and pattern recognition from performing well. There is, therefore, a need for improved methods for encrustation.

One common solution is to apply a filter to smoothen the inner periphery (or even the inner and outer peripheries) of the imagette as shown in Fig. 11.3 below. This filtering may efficiently attenuate the border but has a drawback since the high resolution content of the periphery of the imagette is replaced by information of lower accuracy. Likewise, if the smoothing is applied to the outer periphery, it would result in a degradation of the accuracy for this area.

To solve this problem, we propose a method for encrustation which attenuates the border but still preserves the information content of the whole imagette.

Fig. 11.3 Low resolution image with a high-resolution encrusted imagette with smoothed inner periphery

This method provides a better simulation of the actual information at the outer periphery and hence increases the accuracy of this area. This method mostly makes use of the wavelet transform and multiresolution analysis. The paragraph below summarizes the overall approach in a simplified manner without detailing the mathematical basis of the proposed encrustation method.

The border is due to the abrupt transition from one resolution to another at the periphery of the imagette. In the inner periphery, structures of any size are present, including those of smaller size (high frequencies in signal theory). In the outer periphery, only structures of larger size are present. The lower resolution means an absence of structures of small size which can only be observed at higher resolutions. The principle of our approach is that attenuation of the border discontinuities can be made if some of smaller size structures can be injected in the outer periphery. This will result in an increase of the effective resolution, thus creating a smooth transition in resolution, and hence attenuating the border effects. Multiresolution analysis allows extraction of small size structures from the imagette. Here only the periphery is of interest. Wavelet transformation models change in information for each pixel between two consecutive resolutions by means of the so-called wavelet coefficients. The larger these coefficients in absolute values are, the more visible the corresponding structures will be. For higher resolutions, i.e. smaller structures sizes, the wavelet coefficients are large at the inner periphery and null or very weak at the outer periphery. Increasing the resolution of the latter is equivalent to an injection of structures of small size, which is our objective. The wavelet coefficients in the outer periphery are replaced by a linear combination of the wavelet coefficients in the inner periphery. The parameters of the combination are a function of the distance to the edge. Finally, an inverse wavelet transform is applied to produce the encrusted image with smooth edges.

11.3.3 Illustration Example

In order to illustrate the application of the proposed method, it is applied to a test case and compared to standard procedure (raw encrustation). In the test case, an imagette is extracted from the reference image and is encrusted into a 5.6 m smoothed image. Two encrusted images are made: one based on the raw procedure (previously shown in Fig. 11.2), and one based on our proposed smooth encrustation procedure (shown in Fig. 11.4 below). An urban area has been selected for this test case because this is certainly the most difficult type of landscape to process from a numerical point of view, and therefore it helps point out the drawbacks and qualities of algorithms. In addition, urban areas entail high variability of information induced by the diversity of the features sizes.

In the raw encrusted image (Fig. 11.2), pixels have either an effective resolution of 5.6 m (outside the imagette area) and 0.7 m (inside the imagette area). Large size features such as streets can be followed across the image whatever the resolution.

Fig. 11.4 Low resolution image with a high-resolution encrusted imagette with improved outer periphery

However the change of resolution is well shown and would clearly impact photo-interpretation. Figure 11.4 shows the result of our smoothed encrustation procedure in which the visual differences are small. However they still exist. Close examination shows that the outer periphery exhibits a larger variability in Fig. 11.4 than in Fig. 11.2. This higher variability makes a smoother transition from the high resolution part to the lower resolution portion of the scene. The shapes in the periphery are also more visible. For example, note the white shape at the upper left corner of the imagette: its contours are less blurred in Fig. 11.4 than in Fig. 11.2. The same is true for several features along the upper periphery of the imagette. Though the differences are faint, it visually confirms the better results achieved by our proposed encrustation method. The border has been smoothed outside the imagette area without the alteration of the features larger than 5.6 m.

The disturbing effect of the discontinuity along the periphery of the imagette can be quantified by smoothing the two encrusted images in order to have a uniform effective resolution of 5.6 m across the image scene including the higher resolution imagette. These degraded images are compared to the 5.6 m smoothed image. In the inner and outer periphery, the difference in variance between the encrusted images and the 5.6 smoothed image account for about 22% of the variance of the latter image. On a pixel-by-pixel basis, 50% of the pixels have a similar value in the smooth encrusted and the 5.6 m smoothed images, plus or minus 5 digital numbers (images are coded in 16 bits). The better results demonstrate quantitatively the attenuation of the edge effects by the proposed method.

A second quantified test deals with the accuracy of the information of small sizes injected in the outer periphery in order to decrease discontinuity. The results indicate that our proposed encrusted image better represents the 0.7 m reference image than the raw encrustation method does. In order to assess the quality of the simulation of the actual information in the reference image, structures of any size down to 0.7 m were examined. The differences between the reference and the encrusted images were computed for the outer periphery. As expected, the new approach provides better results than the raw encrustation, i.e. closer to the ideal values. For example, the difference of variances decreases from 63% in the raw encrustation approach down to 59% for our smooth encrustation method (ideally, this difference should be null), with a correlation coefficient equals 0.87. This demonstrates that the new approach provides a better simulation of the actual information than the other method does.

11.4   The ARSIS Concept for Fusion of Images in Urban Areas

11.4.1   Introduction

Several studies have documented the utility of merging broadband higher spatial resolution images with imagery that are low in spatial resolution imagery but high in spectral resolution. Many methods have been developed for that purpose, which typically result in multispectral images having the highest spatial resolution available within the data set. They are applied on data sets comprising multispectral images B_{kl} at a low spatial resolution l and images A_h at a higher spatial resolution h but with a different spectral content. Examples of such data sets are SPOT-XS (3 bands, 20 m) and SPOT-P (panchromatic, 10 m) images, SPOT-4 (3 bands at 20 m) and the band XS2 (10 m), IRS-1C (3 LISS bands, 23.2 m) and a panchromatic band (5.8 m), or IKONOS (4 multispectral bands at 4 m) and a panchromatic band (1 m).

In this section, we discussed some of the methods which claim to provide a synthetic image close to what would be observed by a similar multispectral sensor with a better spatial resolution (Ranchin and Wald 2000b). Those methods which only provide a better visual representation of the set of images (Carper et al. 1990) are not addressed here.

Wald (2002) groups the synthetic merging methods under three categories:

- Projection and substitution methods: These project original data sets into another space, substitute one vector by the higher resolution image, and then inverse projection into the original space. Examples of these methods is IHS (Intensity, Hue, and Saturation) method (Carper et al. 1990).
- Relative spectral contribution methods: These include the Brovey transform (Pohl and Van Genderen 1998) which can be applied to any set of images, and the CNES P+XS method (Anonymous 1986) developed for SPOT imagery. It should be noted that the Brovey transform does not represent well synthetic merging methods because of its poor principles in construction (Wald 2002). Nevertheless, it is often used.
- Methods relevant to the ARSIS concept: In these methods a scale-by-scale description of the information content of both images is generated to facilitate a high-quality transformation of the multispectral content (Ranchin and Wald 2000b; Ranchin et al. 2003). The HPF (High Pass Filtering) method is an example of these methods but does not usually provide quality transformation of the multispectral content.

11.4.2   Synthetic Merging Methods

Let's denote the acquired images of lowest spatial resolution by B_l, and the images of highest spatial resolution by A_h. The subscripts l and h denote the spatial resolution of images B or A, i.e. lower and higher resolutions respectively.

The fusion methods aim at constructing synthetic images $B*_h$, which are close to what would be observed by a similar multispectral sensor with a better spatial resolution. The methods perform a high-quality transformation of the multispectral content of B_l, when increasing the spatial resolution from l to h. The general problem is relevant to the fusion of representations and is the creation of a new representation $B*$ from the original representations A and B (Wald 2002):

$$B* = f(A, B) \tag{11.5}$$

It may be seen as the inference of the information that is missing in the images B_{kl} for the construction of the synthesized images B^*_{kh}.

The ARSIS concept is based on the assumption that the missing information is linked to the high frequencies of the representations A and B. It searches for a relationship between the high frequencies in B and A and models this relationship. A method belonging to the ARSIS concept performs typically the following operations: (i) the extraction of a set of information from A, (ii) the inference of the information that is missing in the images B_{kl} using information extracted from A, and (iii) the construction of the synthesized images B^*_{kh}. Current methods perform a scale-by-scale description of the information content of both images. High frequency information that is missing is synthesized so that low spatial resolution images are transformed into high spatial resolution, high spectral content images. Ranchin and Wald (2000b) show that many schemes can be accommodated within the ARSIS concept.

The images A and B do not need to be commensurate. Some studies have been published where images acquired in thermal infrared bands have been synthesized with images acquired in the visible range to create an image of satisfactory quality and better spatial resolution (Kishore Das et al. 2001; Liu and Moore 1998; Nishii et al. 1996; Wald and Baleynaud 1999).

It is difficult to sketch the general scheme for the application of the ARSIS concept. In the HPF and other methods (Liu and Moore 1998; Cornet et al. 2001; Diemer and Hill 2000; Pradines 1986; Price 1999), the modeling of the missing information from the image A to the image B is performed on moving windows of these images themselves. It is possible to focus more on the modeling of the missing high frequencies, expressed by Fourier coefficients, wavelet coefficients or other appropriate spatial transformations (Wald 2002).

Figure 11.5 presents the general scheme that applies in the case of a multi-scale model (Ranchin and Wald 2000b), which is used the following section to describe the ARSIS concept. Input to the fusion process includes the images A at high spatial resolution (A_h, resolution n°1) and the spectral images B at low spatial resolution (B_{kl}, resolution n°2).

Three models appear in this scheme. A Multi-scale Model (MSM) performs a hierarchical description of the information content relative to spatial structures of an image. An example of such a model is the combination of the wavelet transform and multiresolution analysis (Ranchin 1997). Ranchin and Wald (2000b) provide practical details for the implementation of the algorithm of Mallat combined with a Daubechies wavelet. When applied to an image, the MSM provides one or more images at higher frequencies (detail images), and one image of approximation at lower frequencies. For instance, imagine an Ikonos image with 1 m resolution. The first iteration of the MSM gives one image of the structures comprised between, say, 1 and 2 m (details image) and one image of the structures larger than 2 m (approximation image). The spatial variability within an image can thus be modeled and the model can be inverted (MSM^{-1}) to perform a synthesis of the high-frequency information to retrieve the original image at 1 m.

The Inter-Band Structure Model (IBSM) deals with the transformation of spatial structures with changes in spectral bands. It models the relationships between the

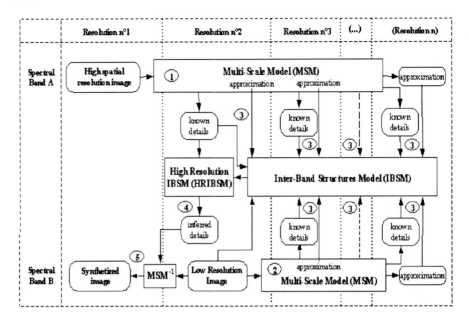

Fig. 11.5 General scheme for the application of the ARSIS concept using a multiscale model (MSM) and its inverse (MSM-1)

details or approximation observed in the representation *A* and those observed in the representation *B*. The IBSM may relate approximations and/or details for one or more resolutions and one or more spectral bands. As an example, the Model 2 relates the details observed at resolution n°3 in the image *A* and the image B_k using a linear relationship (Ranchin and Wald 2000b).

The High Resolution Inter-Band Structure Model (HRIBSM) performs the transformation of the IBSM with change in resolution. This operation is not straightforward as the former ones. Many studies have demonstrated the influence of the spatial resolution on the quantification of parameters extracted from satellite imagery (Lillesand and Kiefer 1994; Woodcock and Strahler 1987). To our knowledge, there is no published fusion method that has paid particular attention to this point and the HRIBSM is often set identical to the IBSM. Ranchin et al. (1994, 2003) performed a multiscale synthesis of the parameters of their IBSM from resolution n°3 to resolution n°2.

To summarize, the operations are typically performed as follows. First, the MSM is used to compute the details and the approximations of image *A* (Step 1 in Fig. 11.5). The same operation is applied to image *B* (Step 2). The analysis is performed for several resolutions, up to *n* in Fig. 11.5 – that is $(n-1)$ iterations for the analysis of the image *A* and $(n-2)$ iterations for that of B_{kl}. These analyses provide one approximation image and several images of details for *A* and *B*. The known details at each resolution are used to adjust the parameters of the IBSM (Step 3). The HRIBSM (at resolution n°2 in Fig. 11.5) is derived from this model, which converts the known details of image A into the inferred details of image B_k (inferred details, Step 4). Finally, MSM⁻¹ from resolution n°2 to resolution n°1 performs the synthesis of the image B^*_{kh} (Step 5).

11.4.3 Illustration Example

Ranchin et al. (2003) propose a set of examples of different implementations of the ARSIS concept. In the proposed example, the selected MSM is implemented through the Undecimated Wavelet Transform (UWT) (Dutilleux 1989), the IBSM is implemented through the Model 2 (Ranchin and Wald 2000b; Ranchin et al. 2003) and the HRIBSM is the implemented through the Identity model. The three models combined are referred to as the UWT-M2 method.

An Ikonos dataset has been used to test the UWT-M2 method. This Ikonos dataset is composed of a panchromatic image Pan at a spatial resolution of 1 m and of 4 multispectral MS images at a spatial resolution of 4 m. The spectral bands of the images are detailed in Table 11.1. The geographical area is the city of Hasselt in Belgium. The images were acquired the 28th of April 2000 at 10:39 UTC. The original data are delivered with a dynamic range of 11 bits (gray values). Table 11.1 reports the mean value and standard deviation of each band in gray values. It also provides the correlation coefficient between each band and the panchromatic image resampled at 4 m. The application of the methods leads to the synthesis of MS images at the spatial resolution of 1 m with the same dynamic range in 11 bits

Figure 11.6 (left) shows an excerpt of the Pan image at 1 m. On the left, there is a river, crossed by two bridges, with small boats and several barges. Along the riverbank is a main street. Several cars are visible. This area is mainly an industrial district with large buildings, surrounded by numerous trees. Note those features at the center of the image that appear in white in the upper left corner. On the top right is a stadium. Its lawn is partly degraded. Figure 11.6 (right) presents an excerpt of the same area in the near infrared (NIR) band with a spatial resolution of 4 m.

Figure 11.7 shows an excerpt of the NIR image synthesized at 1 m. The difference between this image and the original image at 4 m is striking. More details are visible and the visual interpretation of the scene is easier and much more accurate. Even the boats can be distinguished. One may note that the details (high frequencies)

Table 11.1 Spectral bands, mean value and standard deviation of the panchromatic (PAN) and multispectral images (in gray values). Correlation coefficient between the original spectral bands and the Pan image resampled at 4 m

	Pan	Blue	Green	Red	Near infrared
Spectral band	450–900	450–520	520–600	630–690	760–900
Mean value	352	328	339	249	467
Standard deviation	55	26	39	48	124
Correlation coefficient	1.00	0.271	0.409	0.323	0.875

Fig. 11.6   An excerpt of the Pan image at 1 m (*left*) and NIR image at 4 m (*right*)

Fig. 11.7   An excerpt of the NIR image synthesized at 1 m

are very similar to those in the Pan image. However, a close examination reveals differences: the gray values and local contrasts are not the same as exemplified in the upper left corner of the image or the above-mentioned large buildings. The gray values in the synthesized image are very close to those in the original NIR image.

In summary, the ARSIS concept provides an excellent framework for the development of accurate methods that can be tailored to specific user needs and can be assembled within a toolbox. It is also an open framework with many areas for the development of different cases of applications and approaches for implementation. Different methods can be developed based on this concept, depending upon the multiscale description and synthesis model MSM, the IBSM relating the content of both representations A and B, and the HRIBSM transforming the parameters of the IBSM when increasing the spatial resolution (Wald 2002; Ranchin and Wald 2000b; Ranchin et al. 2003).

The methods belonging to the ARSIS concept are constructed in such a way that once the synthesized images B^*_h are degraded to a lower resolution l, they reproduce the original spectral content of B_l. These methods perform a high-quality transformation of the multispectral content when increasing the spatial resolution. The synthesized images can therefore be used for purposes other than visual interpretation. In the case of the SPOT imagery, it was demonstrated that the accuracy and the quality of the road network automatically extracted was increased by the use of images synthesized by the ARSIS concept (Ranchin and Wald 1997). For urban mapping, the benefits of using ARSIS synthesized images were enhanced by Ranchin and Wald (1996b), Raptis et al. (1998), Terretaz (1997), and Couloigner et al. (1998). The benefits of ARSIS fused products for the analysis and mapping of the city were demonstrated over the city of Marseille (France) through the use of color images having a higher spatial resolution clearly permits a more accurate interpretation of the features in the city (Wald and Ranchin 2001). Furthermore, such fused products may be the object of further image processing techniques, such as contrast enhancement, edge detection or classification, without the creation of visible artifacts. It illustrates the capability of the synthetic images to support further image processing dealing with the high frequencies (Wald and Ranchin 2001). The ARSIS concept was also employed to increase the quality in mapping air quality in city (Wald and Baleynaud 1999).

11.5   Fusion of Images, Databases and Punctual Measurements for Air Quality

11.5.1   Introduction

Presently, most large cities in Europe have an air quality surveillance program. Such a network is composed of a few static measuring stations, which perform a continuous surveillance of air pollution at station locations. In France, pollution data are collected in near-real time and used to compute an air quality index, the ATMO index (Garcia and Colosio 2001). This index aims at informing local authorities, as well as the public, about air quality in the city. In instances of high levels of pollution, public authorities are able to implement restrictive measures on car traffic and on activities of some polluting companies.

However, the actual exposure of persons to ambient pollution cannot be estimated with the present networks. The costs of a measuring station and its maintenance limit the generation of index values to specific points of towns. Given the few measuring stations composing a standard air quality network, an accurate knowledge of the spatial distribution of the atmospheric pollutants over a city is presently impossible. Several tools based on numerical modeling of the airflow and chemical transportation, and conversion provide maps of pollutant concentrations. However, these are produced at a regional scale with a grid cell of 1–10 km, which is insufficient.

A methodology based on a multisource approach for mapping pollutants concentrations over a city has been proposed to overcome this problem (Ung et al. 2002a, b). The notion of a "virtual station" is defined, and sources related to air pollution and urban shapes and morphology are exploited to virtually increase the number of measuring stations, thus increasing the quality of mapping by interpolation techniques. The approach was applied to the city of Strasbourg (France) and a ground truth campaign achieved in June 2003 confirms the validity of the proposed approach.

11.5.2   Methods

The proposed methodology used in the above-mentioned exercise has four steps: (1) the creation of identity cards of each actual measuring station, (2) the evaluation of the sites of the city in order to create pseudo-stations, (3) the selection of virtual stations in this set of pseudo-stations, and (4) the creation of a map.

The main idea is to study and evaluate the urban environment as factors influencing the air pollution in the city. Urban space is a complex domain composed of built-up areas, roads and streets, bare soil, residential areas, industrial areas, wooden and parks areas. Atmospheric pollution does not behave the same in each of these areas (Derbez et al. 2001). Several studies demonstrate the heterogeneity of air pollution and the influence of the building positions and heights and the street orientations according to the dominant wind situation. Indeed, dramatic differences in pollutant concentrations have been observed for two adjacent streets (Derbez et al. 2001; Croxford and Penn 1998; Croxford et al. 1996; Scaperdas and Colvile 1999). Hence, a characterization of the city's morphology is necessary in order to model its influence on the behavior of air pollutants. This can be obtained by jointly analyzing and processing images and databases, and the organization of urban features. It allows the establishment of the so-called "identity card" of each place of the city.

This ID card comprises a set of attributes, such as:

- The geographical position of the area
- Its land use

- Its proximity to emission sources
- Morphology of the buildings around the area, impinging on air flow
- Climatic and meteorological conditions of the area under study

Each element of the town, including the measuring stations, is identified by an ID card. Then, the relationships existing between the ID card and the pollutant concentration at each measuring station can be studied. The ID card is built from three sources of data: the measuring stations, the geographical database and remotely sensed data. Combined exploitation of this dataset allows the study of the morphological configuration of the city and the characterization of the measuring stations (Ung et al. 2002a, b; Weber et al. 2002).

The sitting of measuring stations is done according to objectives of air pollution control, their neighborhoods, population density, and sources of air pollution. Due to the cost of a measuring station, the agencies in charge of air quality control have a restricted number of stations. The ID cards can be used to detect areas of the city similar to those surrounding the measuring stations. These areas are called the pseudo-stations; their attributes are similar to those of the measuring stations. The similitude is defined from the ID cards. At this stage, a hypothesis can be suggested regarding the possibility to model the air pollution concentrations for these areas through a combination of satellite images according to a relationship between these images and the measuring stations. If this hypothesis is valid, then it is possible to obtain a densification of the measuring network with virtual stations. The estimation of air pollutant concentrations is based on a law linking satellite measures with atmospheric transmittance and then to air pollutant concentration measured by actual stations (Wald and Baleynaud 1999; Retalis et al. 1999; Sifakis et al. 1992, 1998; Finzi and Lechi 1991; Basly 2000). The establishment of such a law is only possible for a restricted set of pseudo-stations; a virtual station is such a pseudo-station.

Once the virtual stations are generated, a surface interpolation can be applied based on both the actual and virtual stations. This produces a surface map of the concentration for each pollutant. Actually, the fusion process is fully achieved by imposing some constraints on the interpolation process. The result should reproduce what is obtained by the numerical models of airflow at a resolution of 10 km and, of course, it should reproduce what has been measured at the actual stations and assessed at the virtual stations. There are several techniques for fusing gridded data and point measurements with constraints at both ends of the multiresolution pyramid. Here, below we show an example in which we adopted that of Beyer et al. (1997) and Lefèvre et al. (2004).

11.5.3   Illustrated Example

The area of interest is the urban community of Strasbourg, in the Northeast of France, close to Germany. The data available are concentrations of pollutants provided by the measuring network, a set of satellite images (Landsat and SPOT) and a geographical database of the French Geographical Institute (IGN), the so-called BD TOPO© (Weber et al. 2001).

The agency controlling air pollution, ASPA (Association pour la Surveillance et l'étude de la Pollution atmosphérique en Alsace), is in charge of 32 measuring stations within the Alsace region with a subset of 14 dedicated to the Strasbourg area. The pollutants measured are SO_2, NO_x, CO, CO_2, O_3, PM_{10} and $PM_{2.5}$. All measures are available every 15 min and allow continuous study and surveillance of local air pollutants concentration. The set of satellite images is composed of eight Landsat scenes and a SPOT scene acquired at different seasons between 1998 and 2001. The geographical database BD TOPO© is georeferenced and contains the land use, streets network, railway network, built-up areas, building heights, hydrological information, topography, and administrative limits

From this set of data, ID cards of the urban area are built for each available satellite image. Apart from the spectral signatures, these ID cards include morphological descriptors such as spots measurements descriptors, urban spot descriptor, geometric descriptors, volumes descriptors (Weber et al. 2002; Basly 2000). A part of the ID card is stable regardless of the acquisition date of satellite images, an other part is dependent on the images itself. For a given data, a classification of the ID cards is performed and similitude with IDs of the actual stations determines the pseudo-stations (Fig. 11.8). Some artifacts appear because of incomplete ID cards and should be removed.

An example of pseudo-station relating to the actual station, the STG Centre 2 station, is presented in Fig. 11.9. Figure 11.9a is a superimposition of the

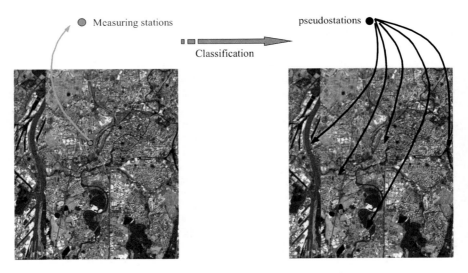

Fig. 11.8 Location of pseudo-stations after classification of the ID cards

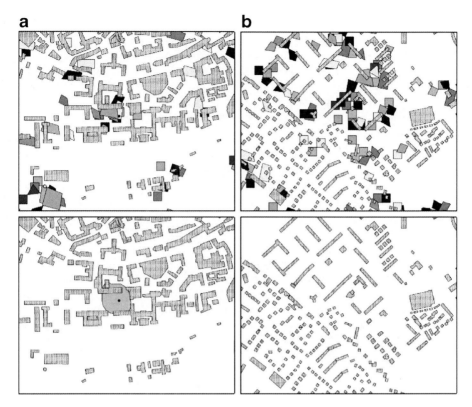

Fig. 11.9   Determination of pseudo-station identical to the measuring station STG Centre 2

BD TOPO© and of the results of the individual classifications achieved for each date. The measuring station is the spot in the middle of the figure. The gray tones denote the number of times a particular site appears as a pseudo-station in the ensemble of classifications. Figure 11.9b is the same but for another place in the city with no actual measuring station. Figure 11.9c, d represent the intersection of all the classifications taking into account a circle of influence. In Fig. 11.9c, the intersection area is the area surrounding the measuring station, showing the normal behavior of the process. Figure 11.9d presents a pseudo-station, i.e. an area that has the same ID card than the measuring station.

From the analysis of the full set of data, 28 pseudo-stations were determined. A measuring campaign in June 2003, has confirmed the behavior of some of the pseudo-stations with respect to pollutant concentration. Mobile means for air pollution measurement have been installed at several of these 28 pseudo-stations.

A comparative study has been conducted allowing the verification of several hypotheses (Weber et al. 2001):

- A strong correlation between the mobile means of measurement and the measuring station has been established. A law was established allowing the use the pseudo-stations as virtual stations. The behavior of the virtual stations was similar to that of the measuring stations.
- The influence of urban morphology on the spatial repartition of air pollutants has been confirmed.

Using these virtual stations, it is now possible to make the network denser. Two maps of PM_{10} have been computed to show the benefits of the virtual stations (Fig. 11.10). In both cases, the interpolator is a thin-plate operator. This interpolator should not be used to preserve the information at both ends of the multiscale representation (10 km and 10 m) but is used here in a didactic purpose. On the left the map is obtained with three measuring stations (black dots in Fig. 11.10 left). The background of these images is the TM4 channel of Landsat, for a visualization of the network of streets, the highways, and the Rhine River. Due to a few number of measurement points, this map is rather homogeneous and is not representative of the complexity of the air pollution. The right part of Fig. 11.10 is a similar map, but obtained using 301 pseudo-stations determined with less restrictive classification rules. The relationship between the concentrations measured at the actual stations and the virtual stations was established, thus providing an assessment of the concentration at each virtual station. Then, the same interpolation technique was applied and a map is obtained (Fig. 11.10 right). Though the error in the pollutant concentration can be high, the spatial repartition of the pollutants seems close to reality and in any case, is much better than what can be obtained presently.

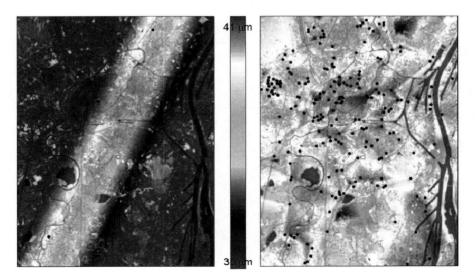

Fig. 11.10 Map of the concentration in PM10 obtained from interpolation of the measuring stations (*left*) and of the measuring and virtual stations (*right*)

Chapter Summary

The main objective of data fusion is use a set of datasets to obtain information of greater quality than what could be obtained by each single data considered separately. It is a formal framework in which are expressed means and tools for the alliance of data originating from different sources. It aims at obtaining information of greater quality; the exact definition of 'greater quality' will depend upon the application. Several fusion cases studies were discussed in this chapter to illustrate the potential of data fusion techniques. The increasing complexity of the examples is designed to gradually help students understand data fusion. The diversity of data fusion is so important that the few examples provided cannot fully describe its complexity. This field is still a strong and active research in urban remote sensing and the other civilian domains.

LEARNING ACTIVITIES

Data and Image Fusion, and Software

- For a better understanding of what data fusion is and what it does.
  🖥 http://www.data-fusion.org
- The Online Resource for Research in Image Fusion.
  🖥 http://www.imagefusion.org
- The IEEE Geoscience and Remote Sensing Society Data Fusion Committee (DFC).
  🖥 http://www.grss-ieee.org/community/technical-committees/data-fusion/
- The International Society for Information Fusion.
  🖥 http://www.inforfusion.org
- Free trial version of ENVI software from Research Systems Inc limited to 7 min of use. Contains a set of sharpening algorithms.
  🖥 http://www.rsinc.com/download/index.asp
- The wavelet digest with access to information, preprints, softwares, etc.
  🖥 http://www.wavelet.org

Study Questions

- What are the different image fusion algorithms? Discuss their advantages and disadvantages.
- How do you quantitatively evaluate the quality of an image fusion product?

Acknowledgements These works have been partly supported by the French program Action Concertée Incitative Ville of the French Ministry of Research, the French Programme National "Télédétection Spatiale", and the Canadian GEOIDE program. The authors also thank the GIM Company for providing Ikonos data.

References

Anonymous (1986) Guide des utilisateurs de données SPOT, 3 tomes, Editeurs CNES et SPOT-Image. Toulouse, France

Basly L (2000) Télédétection pour la qualité de l'air en milieu urbain. Thèse de Doctorat (Ph.D. Thesis). Sciences de technologies de l'information et de la communication, University of Nice-Sophia Antipolis, France, p 186

Beyer HG, Czeplak G, Terzenbach U, Wald L (1997) Assessment of the method used to construct clearness index maps for the new European solar radiation atlas (ESRA). Solar Energy 61(6):389–397

Burt PJ, Adelson EH (1983) The Laplacian pyramid as a compact image code. IEEE Trans Commun 31(4):532–540

Carper WJ, Lillesand TM, Kiefer RW (1990) The use of Intensity Hue Saturation transformations for merging SPOT panchromatic and multispectral image data. Photogramm Eng Remote Sens 56(4):459–467

Cornet Y, de Béthune S, Binard M, Muller F, Legros G, Nadasdi I (2001) RS data fusion by local mean and variance matching algorithms: their respective efficiency in a complex urban context. In: Proceedings of the IEEE/ISPRS joint Workshop on Remote Sensing and Data Fusion over Urban Areas, Roma, Nov 8–9, 2001, pp 105–111

Couloigner I, Ranchin T, Valtonen VP, Wald L (1998) Benefit of the future SPOT 5 and of data fusion to urban mapping. Int J Remote Sens 19(8):1519–1532

Croxford B, Penn A (1998) Sitting consideration for urban pollution monitors. Atmos Environ 32(6):1049–1057

Croxford B, Penn A, Hillier B (1996) Spatial distribution of urban pollution: civilizing urban traffic. Sci Total Environ 189(190):3–9

Daubechies I (1992) Ten lectures on wavelets. CBMS-NSF regional conference series in applied mathematics 61, SIAM, Philadelphia

Derbez M, Mosqueron L, Nedellec V (2001) Quelles sont les expositions humaines à la pollution atmosphérique? Rapport de synthèse de PRIMEQUAL-PREDIT 1995–2000, La Documentation Francaise, Paris

Diemer C, Hill J (2000) Local correlation approach for the fusion of remote sensing data with different spatial resolutions. In: Ranchin T and Wald L (eds) Proceedings of the third conference on fusion of earth data, SEE/URISCA, Nice, France, pp 91–98

Dutilleux P (1989) An implementation of the "algorithme a trous" to compute the Wavelet Transform. In: Combes JM, Grossman A, Tchamitchian Ph (eds) Wavelets: time–frequency methods and phase space. Springer, Berlin, pp 298–304

Finzi G, Lechi GM (1991) Landsat images of urban air pollution in stable meteorological conditions. Il Nuovo Cimento 14C(5):433–443

Garcia J, Colosio J (2001) Les indices de qualité de l'air: élaboration, usages et comparaisons internationals. Collection – Sciences de la terre et de l'environnement. Presses de l'École des Mines, Paris, France

Kishore Das D, Gopal Rao K, Prakash A (2001) Improvement of effective spatial resolution of thermal infrared data for urban landuse classification. In: Proceedings of the IEEE/ISPRS Joint Workshop on Remote Sensing and Data Fusion over Urban Areas, Roma, Nov 8–9, 2001, pp 332–336

Lefèvre M, Albuisson M, Ranchin T, Wald L, Remund J (2004) Fusing ground measurements and satellite-derived products for the construction of climatological maps in atmosphere optics. In:

Goossens R (ed) Proceedings of the 23rd EARSeL Annual Symposium, Milpress, Rotterdam, The Netherlands, pp 85–91

Lillesand TM, Kiefer RW (1994) Remote sensing and image interpretation, 3rd edn. Wiley, New York

Liu JG, Moore JM (1998) Pixel block intensity modulation: adding spatial detail to TM band 6 thermal imagery. Int J Remote Sens 19(13):2477–2491

Mallat SG (1989) A theory for multiresolution signal decomposition: the wavelet representation. IEEE Trans Pattern Anal Mach Intell 11(7):674–693

Meyer Y (1990) Ondelettes et opérateurs 1: Ondelettes. Hermann, Paris, France

Nishii R, Kusanobu S, Tanaka S (1996) Enhancement of low spatial resolution image based on high resolution bands. IEEE Trans Geosci Remote Sens 34(5):1151–1158

Pohl C, van Genderen JL (1998) Multisensor image fusion in remote sensing: concepts, methods and applications. Int J Remote Sens 19(5):823–854

Pradines D (1986) Improving SPOT image size and multispectral resolution. Proceedings of SPIE: Earth Remote Sensing using the Landsat Thematic Mapper and SPOT Systems 660:78–102

Price JC (1999) Combining multispectral data of differing spatial resolution. IEEE Trans Geosci Remote Sens 37(3):1199–1203

Ranchin T. (1997) Wavelets, remote sensing and environmental modelling. In Proceedings of the 15th IMACS world congress on scientific computation, modelling and applied mathematics, August 24–29, 1997, Berlin, Germany, vol 6: applications on modelling and simulation, pp 27–34

Ranchin T, Wald L (Eds.) (1996a) In: Proceedings of the international conference on fusion of earth data, Cannes, France, 6–8 Feb 1996

Ranchin T, Wald L (1996b) Benefits of fusion of high spatial and spectral resolutions images for urban mapping. In: Proceedings of the 26th international symposium on remote sensing of environment and the 18th annual symposium of the Canadian remote sensing society, pp 262–265

Ranchin T, Wald L (1997) Fusion d'images HRV de SPOT panchromatique et multibande à l'aide de la méthode ARSIS : apports à la cartographie urbaine. In Actes des 6èmes journées scientifiques du Réseau Télédétection de l'AUPELF-UREF "Télédétection des milieux urbains et périurbains", Ed. AUPELF-UREF, pp. 283–290

Ranchin T, Wald L (Eds.) (1998) In: Proceedings of the 2nd international conference on fusion of earth data, Sophia Antipolis, France, 28–30 Jan 1998

Ranchin T, Wald L (eds) (2000a) In: Proceedings of the 2nd international conference on fusion of earth data, Sophia Antipolis, France, January 26–28, 2000

Ranchin T, Wald L (2000b) Fusion of high spatial and spectral resolution images: the ARSIS concept and its implementation. Photogramm Eng Remote Sens 66(1):49–61

Ranchin T, Wald L, Mangolini M (1994) Efficient data fusion using wavelet transforms: the case of SPOT satellite images. In: Proceedings of SPIE 1993 international symposium on optics, imaging and instrumentation, vol 2034. Wavelet Applications in Signal and Image Processing, pp 171–178

Ranchin T, Aiazzi B, Alparone L, Baronti S, Wald L (2003) Image fusion. The ARSIS concept and some successful implementation schemes. ISPRS J Photogramm Remote Sens 58(1–2):4–18

Raptis VS, Vaughan RA, Ranchin T, Wald L (1998) Assessment of different data fusion methods for the classification of an urban environment. In: Ranchin T and Wald L (eds) Proceedings of the 2nd conference on Fusion of Earth Data, SEE/URISCA, Nice, France, pp 167–182

Retalis A, Cartalis C, Athanassiou E (1999) Assessment of the distribution of aerosols in the area of Athens with the use of Landsat Thematic Mapper data. Int J Remote Sens 20(5):939–945

Scaperdas A, Colvile RN (1999) Assessing the representativeness of monitoring data from an urban intersection site in central London. Atmos Environ 33:661–674

Sifakis NI, Bildgen P, Gilg JP (1992) Utilisation du canal 6 (thermique) de Thematic Mapper pour la localisation de nuages de pollution atmosphérique. Application à la région d'Athènes (Grèce). Pollution Atmosphérique 34(133):96–107

Sifakis N, Soulakellis NA, Paronis DK (1998) Quantitative mapping of air pollution density using Earth observations: a new processing method and application to an urban area. Int J Remote Sens 19(17):3289–3300

Terretaz P (1997) Comparison of different methods to merge SPOT P and XS data: evaluation in an urban area. In: Gudmansen P (ed) Proceedings of the 17th Symposium of EARSeL, A.A. Balkema Rotterdam, Brookfield, pp 435–445

Ung A, Wald L, Ranchin T, Weber C, Hirsch J, Perron G, Kleinpeter J (2002a) Satellite data for the air pollution mapping over a city – the use of virtual station. In: Begni G (ed) Proceedings of the 21th EARSeL Symposium. , A. A. Balkema Rotterdam, Brookfield, pp 147–151

Ung A, Wald L, Ranchin T, Weber C, Hirsch J, Perron G, Kleinpeter J (2002b) Cartographie de la pollution de l'air: une nouvelle approche basée sur la télédétection et les bases de données géographiques. Application à la ville de Strasbourg. PhotoInterprétation 3/4:53–64

Wald L (1999) Some terms of reference in data fusion. IEEE Trans Geosci Remote Sens 37(3):1190–1193

Wald L (2000) The presents achievements of the EARSeL – SIG 'Data Fusion'. In: Buchroithner M (ed) Proceedings of EARSeL Symposium 2000 in Dresden, Germany, A.A. Balkema, Rotterdam, Brookfield, pp 263–266

Wald L (2002) Data fusion. Definitions and architectures – fusion of images of different spatial resolutions. Presses de l'Ecole, Ecole des Mines de Paris, Paris, France, 200 p

Wald L, Baleynaud J-M (1999) Observing air quality over the city of Nantes by means of Landsat thermal infrared data. Int J Remote Sens 20(5):947–959

Wald L, Ranchin T (2001) Data fusion for a better knowledge of urban areas. In: Proceedings of the IEEE/ISPRS Joint Workshop on remote sensing and data fusion over urban areas, Roma, Nov 8–9, 2001, pp 127–132

Weber C, Hirsch J, Perron G, Kleinpeter J, Ranchin T, Ung A, Wald L (2001) Urban morphology, remote sensing and pollutants distribution: an application to the city of Strasbourg, France. In: Proceedings of the International Union of Air Pollution Prevention and Environmental Protection Associations (IUAPPA) Symposium and Korean Society for Atmospheric Environment Symposium, 12th World Clean Air & Environment Congress, Greening the New Millennium (CD-ROM), Abstract: vol 1, p 274

Weber C, Hirsch J, Ranchin T, Wald L, Ung A, Perron G, Kleinpeter J (2002) Urban morphology and atmospheric pollutants distribution. In: Proceedings of the 23rd international symposium of the urban data management society (CD-ROM), pp 47–60

Woodcock CE, Strahler AH (1987) The factor of scale in remote sensing. Remote Sens Environ 21(3):311–332

Chapter 12
Characterization and Monitoring of Urban/ Peri-urban Ecological Function and Landscape Structure Using Satellite Data

William L. Stefanov and Maik Netzband

This chapter utilizes a case study from Phoenix, Arizona to examine the relationships between ecological variables and landscape structure in cities. The relationships are assessed using ASTER and MODIS data; and through the techniques of expert system land cover classification and grid-based landscape metric analysis.

Learning Objectives

Upon completion of this chapter, you should be able to:

❶ Describe the unique characteristics of ASTER and MODIS and how they differ from other satellite observation data
❷ Explain the role of expert classification systems and how they can be used in urban land cover classification
❸ Speculate on the use of landscape metrics with remotely sensed data and their application to urban ecosystems, in terms of both structure and function

---

W.L. Stefanov (✉)
Image Science & Analysis Laboratory/ESCG, Code KX, NASA Johnson Space Center, Houston, TX 77058, USA
e-mail: william.l.stefanov@nasa.gov

M. Netzband
Geography Department, Ruhr-University, Bochum, Universitätsstraße 150, 44801 Bochum, Germany
e-mail: maik.netzband@rub.de

T. Rashed and C. Jürgens (eds.), *Remote Sensing of Urban and Suburban Areas*, Remote Sensing and Digital Image Processing 10, DOI 10.1007/978-1-4020-4385-7_12, © Springer Science+Business Media B.V. 2010

12.1   Introduction

12.1.1   Why Study Cities?

Urbanization is a significant, and perhaps the most visible, anthropogenic force on earth – affecting its surface, atmosphere, and seas; its biodiversity and its people. Reliable baseline data on the state of many urban area's ecosystems and biodiversity is lacking, and our progress in obtaining these data is moving slower than our ability to alter the environment. Characterization and monitoring of urban center land cover/land use change is only of limited use in understanding the development pathways of cities and their resilience to outside stressors (Longley 2002). Geological, ecological, climatic and social/political data are also necessary to describe the developmental history of a given urban center and to understand its ecological functioning (Grimm et al. 2000). The data available from the NASA Earth Observing System (EOS) satellite-based instruments presents an opportunity to collect this information relevant to urban (areas of high population concentration with high building density and infrastructure) and peri-urban (adjacent agricultural and undisturbed regions with low population concentration) environments at a variety of spatial, temporal and spectral scales. EOS sensors offer two advantages essential for characterization and monitoring of urban/peri-urban regions: (1) they can supply a large volume of surficial multispectral data at relatively low or no cost, and (2) data for the same region can be repeatedly acquired over relatively short periods (days to weeks).

urban centers are the logical starting point for study of the effects of humans on ecosystems and climate

12.1.2   Remote Sensing and Urban Analysis

There is a long legacy of urban and peri-urban analysis using automated, passive satellite-based sensors, however much of this work has focused on delineation of urban vs. nonurban land cover at coarse to moderate spatial resolutions (Donnay et al. 2001; Longley 2002; Mesev 2003). Extensive use has been made of the Landsat Multispectral Scanner (MSS), Thematic Mapper (TM), and Enhanced Thematic Mapper Plus (ETM+) sensors to characterize urban extent and materials (Buyantuyev et al. 2007; Forster 1980; Jackson et al. 1980; Jensen 1981; Haack 1983; Haack et al. 1987; Seto et al. 2007; Stefanov et al. 2001b, 2003) and to conduct basic comparisons between urban centers (Ridd 1995; Ridd and Liu 1998; Chapter 6 in this book). As presented by Fugate et al. in Chapter 7 of this book, these sensors provide coarse to moderately high spatial resolution (80–15 m/pixel in the visible and near-infrared wavelengths); fairly low spectral resolution (four to seven bands in the visible through shortwave infrared and 1–2 thermal infrared bands); and excellent temporal resolution (typically 14–16 day repeat cycle from 1972 to present). Other satellite-based sensors with greatly improved spatial resolution (15 m/pixel to less than 1 m/pixel) have been developed primarily by the commercial sector and include the Système Probatoire d' Observation de la Terre, or SPOT (Martin et al. 1988), IKONOS (Dial et al. 2003), and Quickbird (Sawaya et al. 2003). These high-resolution systems enable highly detailed land cover/land use and ecological characterization of urban and suburban regions (Weber 1994; Greenhill et al. 2003; Sawaya et al. 2003; Small 2003, 2007; Weber and Puissant 2003). Data from these commercial systems are typically limited in both spatial and temporal coverage, and spectral coverage is generally limited to the visible and near infrared wavelengths (Jensen 2000). Active remote sensing of urban areas using radar and lidar technologies is also becoming more common, and provides new tools for use in urban ecology (Dell'Acqua and Gamba 2001; Gamba et al. 2006)

The ready availability of both commercial and governmental satellite data has led to several comparative studies of urban centers under national and multinational auspices. For example, several such programs have focused on European cities (Eurostat 1995; Churchill and Hubbard 1994; Weber 2001). The project MOLAND was initiated in 1998 (under the name of Murbandy – Monitoring Urban Dynamics) with the objective to monitor the development of urban areas and identify trends at the European scale (Lavalle et al. 2001). The United States' Defense Meteorological Satellite Program Operational Linescan Systems (DMSP OLS) nighttime imagery of global light distributions has been used to estimate urban to suburban population densities and urban extents throughout the world (Sutton 2003; Elvidge et al. 2003; Imhoff et al. 1997). Astronaut photography has also been used to track urban growth in several US cities (Robinson et al. 2000). The 100 Cites Project (formerly known as the Urban Environmental Monitoring Project) based at Arizona State University (Stefanov et al. 2001a; Netzband and Stefanov 2003; Ramsey 2003; Netzband et al. 2007; Stefanov et al. 2007; Wentz et al. 2009) seeks to foster collaborative use of remotely sensed data – primarily EOS datasets – and analysis techniques to advance understanding of urban development trajectories around the world.

In this chapter we will focus on the complementary use of two EOS- based sensors for urban ecological analysis to assess the question "does urban landscape structure influence biophysical parameters at the 1 km scale?" We use the Advanced Spaceborne Thermal Emission and Reflection Radiometer (ASTER) on board the Terra satellite, and the Moderate Resolution Imaging Spectroradiometer (MODIS) sensors on board both the Terra and Aqua satellites to answer this question. The ASTER instrument was built by the Japanese Ministry of International Trade and Industry, and acquires surface data in the visible to near-infrared (three bands at 15 m/pixel), shortwave infrared (six bands at 30 m/pixel; data from these bands acquired after April 2008 are not usable due to anomalously high detector temperatures), and thermal-infrared (five bands at 90 m/pixel) wavelength regions of the electromagnetic spectrum. An additional panchromatic band is included to allow for the generation of high-resolution (30 m postings) digital elevation models from ASTER scenes (Abrams 2000). Each ASTER scene captures a 60 × 60 km area. The expanded wavelength range, spectral resolution, and increased spatial resolution of ASTER allow for increased characterization and investigation of urban/peri-urban land cover and biophysical parameters (biomass, albedo, spatial metrics, and surface temperature/emissivity) relative to the Landsat sensors (Ramsey et al. 1999; Stefanov et al. 2001a, 2007; Zhu and Blumberg 2002; Ramsey 2003; Netzband and Stefanov 2003; Schöpfer and Moeller 2006; Stefanov and Netzband 2005; Wentz et al. 2008).

> in addition to traditional platform- and radiometrically-corrected data, the EOS science teams provide a variety of science data products including albedo, vegetation indices, and surface temperature

The MODIS sensors are similar to ASTER in that they obtain spectral information in the visible through mid-infrared wavelengths over 36 bands with a swath width of 2,300 km. However, the spatial resolution of MODIS data is significantly lower and ranges from 250 m/pixel (two visible bands), 500 m/pixel (five visible to shortwave infrared bands), and 1,000 m/pixel (29 visible, near infrared, shortwave infrared, and mid-infrared bands; Parkinson and Greenstone 2000). As both the Terra and Aqua satellites are equipped with MODIS sensors, repeat coverage over any given area of the Earth is acquired every 1–2 days. This makes MODIS data especially attractive for fine-scale temporal monitoring of regional land surface processes associated with urban centers (Schaaf et al.

2002; Schneider et al. 2003). In addition, the MODIS science team produces data products useful for characterization and monitoring of regional-scale biophysical and climatic variables in urban/peri-urban areas. The MODIS data products used in the current study are listed in Table 12.1, and they have been validated for scientific use by the MODIS Science Team.

Surface temperatures from MODIS are calculated using paired day/night measurements in seven thermal infrared bands and atmospheric profile data also derived from MODIS (Wan and Li 1997). Day and night surface temperature data are important for understanding of urban climatology and urban heat island effects (Brazel et al. 2000; Stefanov and Brazel 2007; Voogt and Oke 2003). The acquisition times for temperature data are 11:00/22:54 (day/night). The normalized difference vegetation index (NDVI) produced every 16 days provides a gross measure of relative photosynthetically active vegetation abundance per pixel (Botkin et al. 1984; Tucker 1979). The Leaf Area Index (LAI) and fPAR (fraction of photosynthetically active radiation) are both calculated on an 8-day basis with the highest value during that period recorded as the pixel value (Running et al. 2000). These parameters are measurements of the capacity of the biophysical landscape to utilize incoming radiation for photosynthesis (LAI) and the amount of radiation so utilized (fPAR; Smith 1980). Albedo is the ratio of radiation reflected from the ground

Table 12.1 MODIS biophysical datasets used for comparison with Phoenix landscape metrics

Data product	Time period	Days	Resolution (m)	Data description (units)
MOD11A1	19-September-00	1	1,000	Land surface temperature (K)
MOD13A2	13-September-00 to 28-September-00	16	1,000	Normalized difference vegetation index (unitless)
MOD15A2	13-September-00 to 21-September-00	8	1,000	Leaf area index (%) and fraction of photosynthetically active radiation (m^2/m^2)
MOD43B3	13-September-00 to 28-September-00	16	1,000	Albedo (unitless)

surface to the incident radiation, and can be considered a gross measure of landscape "brightness" (Sabins 1997). The MODIS 16-day albedo product contains several different measurements; we use the broadband visible through shortwave infrared (0.3–5.0 μm) reflectance albedo value as this represents the full wavelength range of surface reflectance (Schaaf et al. 2002).

Landscape Metrics

A promising way to study urban landscapes is the spatially focused approach of patch dynamics (Zipperer et al. 2000). The urban landscape is a mosaic of biotic and abiotic patches (or land cover types) within a matrix of human-induced settlements, technical infrastructure and other human-caused landscape modifications. Spatial heterogeneity within an urban landscape has both natural and human sources. This spatial heterogeneity exerts environmental pressures, both on the "natural" areas in cities (forests, large parks and wetlands) and beyond the boundaries of cities, by reducing the size and contiguity of ecosystem patch types (Smith 1980). The impacts from urbanization around and within cities affect areas of high economic, recreational and ecological value such as agricultural and forest areas. The mechanisms of these impacts include increasing run-off, deforestation, soil erosion, habitat fragmentation and change in biodiversity (Haff 2002). In contrast, the creation of new green spaces also occurs in some urban areas; this enables recreation and water infiltration (drinking water reservoirs) possibilities.

Within the last 10 years landscape metrics have been implemented on remote sensing data for different mapping scales to emphasize the spatial content and patch distribution of classified remote sensing data (Wu et al. 2000). The field of quantitative landscape ecology has been the primarily developer of landscape metrics or indices. Since these metrics vary in time and space they provide useful tools for monitoring a particular landscape by providing various measures of the distribution and shape of ecological patches on the landscape. Numerous metric algorithms have been developed to quantify the spatial contiguity and shape of patches (i.e. area of contiguous pixels, perimeter vs. area, etc.; McGarigal and Marks 1995). These algorithms can be applied to both pixels and vector polygon data.

As many urban monitoring programs are searching for useful indicators of landscape change, landscape metrics deliver a starting point for the comparison of urban areas that is not dependant on their physical and cultural setting. Several recent studies (Alberti and Waddell 2000; Barnsley and Barr 2000; Herold et al. 2002, 2005; Huang et al. 2007; Narumalani et al. 2004; Rainis 2003; Whitford et al. 2001; Yu and Ng 2007) demonstrate the usefulness of the application of landscape metrics for the assessment and evaluation of urban structure, planning and ecology. The challenge remains to find an adequate way to standardize recently developed approaches in order to ensure the comparability of results on a regional, national or even worldwide scale.

12.2   Analytical Methods and Techniques

12.2.1   Study Area

We illustrate the use of ASTER and MODIS data for urban ecological analysis using the Phoenix, Arizona metropolitan area. Phoenix was selected because it is one of the fastest-growing conurbations in the United States, and is the focus of the Central Arizona-Phoenix Long-Term Ecological Research Project (CAP LTER; Grimm et al. 2000). This project has been the locus of significant remote sensing investigation and characterization of the Phoenix urban/peri-urban area (Stefanov 2002; Jenerette et al. 2007; Stefanov et al. 2007) combined with ground truthing and allied studies (Hope et al. 2003; Wentz et al. 2006).

The greater Phoenix metropolitan area is situated on an alluvial plain formed by the Salt River and alluvial fans derived from the surrounding mountain ranges (Nations and Stump 1996), at an elevation of 305 m in an arid environment that averages less than 20 cm of annual precipitation. Mean monthly temperatures in the region range from 12°C in January to 34°C in July. The area contains 300,000 ha of highly productive farmland, and 3.2 million people (2000 US Census data) are concentrated in an expanding metropolitan area. The growth of industry related to World War II, the introduction of air conditioning, the rise of automobile traffic, expanding tourism, and a growth-minded citizenry propelled Phoenix into the largest population center of the American Southwest, converting it to an industrial, commercial, and administrative hub and the fastest-growing metropolitan area in the United States (Gammage 1999; Kupel 2003).

12.2.2   ASTER and MODIS Image Processing

The work presented here uses ASTER data for Phoenix, AZ acquired on September 19, 2000. Scenes were obtained as "Level 2" atmospherically corrected data products. Atmospheric correction for the visible, shortwave infrared, and thermal infrared Level 2 products is accomplished using a radiative transfer model and atmospheric parameters derived from the National Centers for Environmental Prediction (NCEP) data (Abrams 2000). Two contiguous scenes were required to provide coverage for the eastern Phoenix metropolitan area. The visible to near infrared (VNIR) bands of each scene were georeferenced to the Universal Transverse Mercator (UTM) coordinate system using nearest-neighbor resampling and the pixel location grid incorporated into ASTER data. The two scenes were then mosaiced to form a single contiguous dataset using a commercial image-processing software package. No manipulation of the individual scene histograms was performed as each was already calibrated to reflectance.

Appropriate subsets of the MODIS datasets listed in Table 12.1 were extracted to match the ASTER data coverage for Phoenix. Each of these data products are atmospherically corrected using a similar approach to that described above for ASTER data. Data for September 19, 2000, or for date ranges including this day (Table 12.1), were downloaded for comparison with spatial metrics calculated from the ASTER classified data. The MODIS subsets were georeferenced to the UTM coordinate system using nearest-neighbor resampling. The data number (DN) values for each dataset were then converted to the correct units using appropriate scaling factors in a commercial image-processing software environment.

12.2.3   Land Cover Classification of ASTER Data

The 15 m/pixel VNIR ASTER data (bands 1 – visible green, 2 – visible red, and 3 – near infrared) were used as the initial base data for land cover classification. These bands were chosen as they provide the highest spatial resolution. The six 30 m/pixel shortwave infrared bands were also assessed for their usefulness in land cover classification, but no significant improvement in discrimination of urban classes was observed. This observation is in agreement with the results of Zhu and Blumberg (2002), who found that the VNIR ASTER bands performed better than the shortwave bands in classification of the Beer Sheva, Israel urban area. The classification approach presented here is broadly similar to work previously per-formed for the Phoenix metropolitan area using Landsat TM data by Stefanov et al. (2001b), and was also used in an associated multiscalar investigation of ASTER and MODIS data for urban landscape structure characterization (Stefanov and Netzband 2005).

An initial minimum distance to means (MDM; Jensen 1996) supervised classi-fication was performed on the ASTER mosaic using 16 classes: Desert Soil, Low Vegetation; Desert Soil, Vegetated; Bedrock; Fluvial Sediments; Bare Soil; Fallow Agricultural Soil; Water; Canopied Vegetation; Grass; Riparian Vegetation; Active Agricultural Vegetation; Mesic Built Materials; Xeric Built Materials; White Rooftops; Blue Rooftops; and Asphalt. The term "mesic" refers to land cover types with significant vegetation in the form of grass, shrubs, and canopied woody plants. Xeric land cover types are typified by little to no grass or shrub cover and open-canopy plant types with significant bare rock and soil (i.e. similar to equatorial deserts). The MDM classification was run using each training sub-area as a separate class, followed by aggregation of the results into the original 16 classes.

The ASTER mosaic was also used to calculate a Normalized Difference Vegetation Index (NDVI: Botkin et al. 1984; Tucker 1979). This index highlights actively photosynthesizing vegetation by comparing reflectance values in the visible red (low for vegetation) and near infrared (high for vegetation) bands. The index is calculated using:

$$NDVI = \frac{(Band3 - Band2)}{(Band3 + Band2)} \tag{12.1}$$

where Band 3 is near infrared and Band 2 is visible red reflectance. The index returns pixel values ranging from −1 (no vegetation; low reflectance in both bands 2 and 3) to 1 (pixel dominated by actively photosynthesizing vegetation). Spatial variance texture was also calculated from the VNIR mosaic. This operation highlights large changes in brightness value (or reflectance) between adjacent pixels and has been shown to correlate well with urban versus non-urban land cover types (Irons and Petersen 1981; Gong and Howarth 1990; Stuckens et al. 2000). Variance texture is calculated using:

$$V = \sum \frac{(x_{ij} - M)^2}{n - 1} \qquad (12.2)$$

where x_{ij} is the reflectance value of pixel (i,j); n equals the number of pixels in a moving window; and M is the mean value of the moving window (Leica Geosystems 2003) as defined by:

$$M = \sum \frac{x_{ij}}{n} \qquad (12.3)$$

Spatial variance texture was calculated for all three VNIR bands using both a 3 × 3 and 5 × 5 pixel moving window. This was done to capture fine-scale spatial texture in urbanized regions as well as coarser-scale texture in undeveloped regions. The NDVI and variance texture raster data were then each separated into low, medium, and high data values using an unsupervised ISODATA algorithm. This approach takes advantage of the inherent statistical clustering within each NDVI and texture dataset, and provides a simple means of objective thresholding of the data.

Qualitative assessment of the MDM classification results indicated that significant misclassification was present both within and between the various soil, vegetation, and built classes. We then constructed an expert classification system similar to that used by Stefanov et al. (2001b, 2003) to perform post-classification recoding of the MDM classification result.

An expert classification system applies a sequence of decision rules to a set of georeferenced datasets using Boolean logic (Vogelmann et al. 1998; Stefanov and Netzband 2005; Stefanov et al. 2001b, 2003; Stuckens et al. 2000). This approach allows for the introduction of a priori knowledge into the classification data space and can significantly reduce errors of omission and commission. Figure 12.1 presents a schematic example where the dashed rectangle indicates the hypothesized pixel classification ("Soil and Bedrock"), hexagons are alternative decision pathways, and solid rectangles indicate the variables being tested. If any one of the decision pathways ("Path") is satisfied by the variables, the pixel will receive the hypothesized classification value. There is no limitation on the number of variables or decision pathways that can be combined within an expert system framework. Most image processing software packages now include tools for constructing expert system or decision tree classification frameworks.

The datasets combined in the expert system framework include the initial MDM land cover classification, unsupervised classifications of the NDVI and spatial

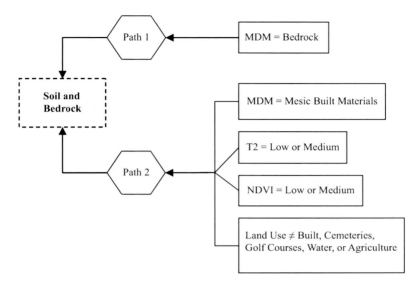

Fig. 12.1 Schematic example diagram of expert system model described in text. Variable definitions are: MDM – initial minimum distance to means classification result; T2 – ISODATA classification for variance texture calculated from ASTER Band 2; NDVI – ISODATA classification for vegetation index value; Land Use – categories extracted from land use vector dataset (figure published in Stefanov and Netzband 2005, copyright Elsevier)

variance texture data, and a land use vector polygon dataset. The land use data were acquired from the Maricopa Association of Governments (MAG; Maricopa Association of Governments 2000) and are contemporaneous with both the ASTER and MODIS data. The land use data are constructed from a combination of survey questionnaires, site visits, and aerial photograph data. This dataset contains 46 separate land use categories which were aggregated to seven for use in the expert system model: Open Residential, Built, Cemeteries, Open Space, Golf Courses, Water, and Agriculture. Incorporation of land use polygon data provides additional discriminatory power for spectrally similar pixels such as asphalt and bedrock. For example, a pixel classified as Asphalt located within an Open Space polygon would be reclassified as Soil and Bedrock. A series of decision rules were then constructed to recode misclassified pixels in the MDM classification product. The MDM classes White Rooftops and Blue Rooftops were also recoded into one class, Reflective Built Surfaces, within the expert system model. The expert classification model was run using the area of overlap of the MDM classification and the MAG land use dataset only (Fig. 12.2).

the expert system approach facilitates the integration of different data sources or "knowledge" to increase classification accuracy

The 15-class output of the expert classification model was then further aggregated to 11 classes prior to accuracy assessment. Classes were aggregated if they were functionally similar landscape elements (i.e. Canopied and Riparian Vegetation) to

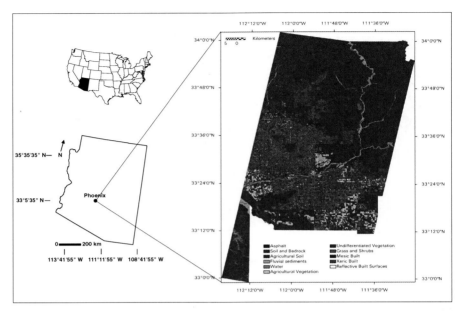

Fig. 12.2 Location of the Phoenix, AZ study region and recoded expert land cover classification results

minimize potential confusion in the reference dataset used for accuracy assessment. The reference dataset was constructed from 3 m/pixel digital aerial orthophotos for the Phoenix metropolitan region collected in 1999. The 2000 ASTER VNIR data were also examined during the assessment process to minimize reference dataset error due to temporal change in classes such as Agricultural Vegetation. Assessment points were selected using a stratified random approach to ensure that each class had at least 50 associated reference points. Points were selected using a 3 × 3 moving window and a 9-pixel majority rule; in other words all 9 pixels in the moving window were required to be the class of interest for the central point to be selected. Each accuracy assessment point was then examined to ensure that it did not fall within the associated class of interest's training regions; any point that did was removed from the reference dataset. If the number of reference points for a given class fell below 40, new points were selected to achieve this minimum number. Table 12.2 presents producer, user, and conditional Kappa accuracy assessment values.

12.2.4 Landscape Metric Calculation and Grid Construction

The land cover classification data were further aggregated to six superclasses (Asphalt, Soil and Bedrock, Agriculture, Undifferentiated Vegetation, Built, and Water) to reduce the computational load for landscape metric calculation. The Water class was not included in further analysis due to its minimal representation in the land

Table 12.2 Accuracy assessment results (table published in Stefanov and Netzband 2005, copyright Elsevier)

Class name	Reference totals	Classified totals	No. correct	Accuracy (%) Producer	User	k
Asphalt	50	50	45	90.00	90.00	0.890
Soil/bedrock	94	77	68	72.34	88.31	0.860
Agricultural soil	50	50	44	88.00	88.00	0.869
Fluvial sediments	43	50	42	97.67	84.00	0.827
Water	43	43	42	97.67	97.67	0.975
Agricultural vegetation	52	50	44	84.62	88.00	0.868
Undifferentiated vegetation	45	50	41	91.11	82.00	0.805
Grass and shrubs	61	69	56	91.80	81.16	0.789
Mesic built	45	45	38	84.44	84.44	0.831
Xeric built	51	50	49	96.08	98.00	0.978
Reflective built surfaces	44	44	40	90.91	90.91	0.902
Total	578	578	509			
Overall classification accuracy = 88.06%						0.868

cover data. A grid with 1 km^2 elements was created from the reprojected MODIS data pixels to allow for direct comparison with the aggregated land cover classification. The full extent of the Phoenix urban area is not captured by the ASTER data, but we selected the location and extent of the analysis grid to capture a representative portion of both the urban and peri-urban regions of Phoenix (Fig. 12.3). Four metrics were selected for analysis: Class Area, Mean Patch Size, Edge Density and Interspersion/Juxtaposition Index. These metrics were computed using the FRAGSTATS software package (McGarigal and Marks 1995). This suite of metrics was selected as best capturing the key spatial aspects of urban landscape structure that can be easily interpreted in terms of biophysical variables obtained from 1 km MODIS data.

Class Area (CA), as an essential descriptive statistical number, equals the area (m^2) of the given land cover type divided by 10,000 (to convert to hectares). Mean Patch Size (MPS), as a measure for the grain of the investigated landscape, indicates the mean land cover pixel size in hectares and is a function of the total area of the landscape and the number of land cover types. Smaller values indicate a higher fragmentation of the landscape. Edge Density (ED) for the evaluation of edge complexity and the density of a given landcover class equals the sum of the lengths of all edge segments involving the corresponding land cover type divided by the total landscape area (converted to hectares). The Interspersion/Juxtaposition Index (IJI), as an indicator for the extent of interspersion of landscape patches, describes the observed interspersion over the maximum possible interspersion for a given number of patch types within the landscape. IJI approaches 0 when the corresponding land cover type is adjacent to only 1 other land cover type. IJI equals 100 when the corresponding land cover type is equally adjacent to all other land cover types (i.e., maximally interspersed and juxtaposed to other land cover types) within the landscape.

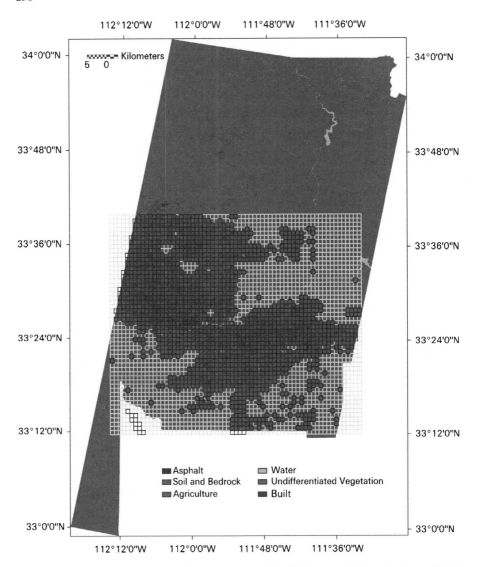

Fig. 12.3 One kilometer grid overlain on land cover classification recoded to six classes. Black grid cells are urban, all others are nonurban (figure published in Stefanov and Netzband 2005, copyright Elsevier)

The vector grid was merged with the raster land cover data within a Geographic Information System (GIS) environment. Each pixel in the land cover classification dataset was assigned a unique identifier (Element1_Class1, Element1_Class2, etc.) within the GIS environment for input into the FRAGSTATS software. The metric results were then averaged over each 1 km² grid element for comparison to the MODIS data using linear regression analysis. The large areal extent of the grid made it necessary to conduct the analysis in parts since the FRAGSTATS software has a limitation of approximately 4,000–5,000 input classes per calculation run. Applying a uniform grid to the land cover data guarantees a uniform external limitation on different test areas throughout the dataset (Netzband and Kirstein 2001). It also facilitates raster-based intra-regional characterizations of features. Gradients between the city center and the outskirts, or through other divisions of urban agglomerations, can be analyzed and quantified in a differentiated manner. Furthermore, different patterns of residential or green areas can be reconstructed, if they are larger than the pixel size of the remotely sensed data (minimum 15 m/pixel for ASTER).

gridding of the land cover and landscape metric data provides a uniform framework for regression analysis, and allows for rigorous comparison with results from other urban centers

12.3   Results and Discussion

The general urban structure of Phoenix is the product of only 150 years of development. The majority of the current built-up area was constructed after 1940, and currently defines a northwest–southeast trending agglomeration set along a regular north–south and east–west transportation grid (Fig. 12.2). This development pattern is the result of physiographic constraints (mountains), availability of water delivery infrastructure, the dominance of individual automobile transport rather than mass transit, and political boundaries related to federal and Native American lands (GP2100 2003). The Phoenix metropolitan area has a fairly well defined urban core defined by the skyscrapers of downtown Phoenix proper. This is surrounded by a heterogeneous mixture of residential, commercial/industrial, and parkland areas of varying age and spatial extent comprising the numerous surrounding municipalities.

Older residential developments within the urban agglomeration tend to be more open and occupy larger plots of land. More expensive and exclusive developments located in desert areas and next to scenic landforms also tend to be less dense. A burst of construction of dense multi-unit residential developments that began in the 1990s defines the location of the current urban fringe together with a fragmenting halo of agricultural fields and farms undergoing land use conversion to residential/commercial uses. Most of the socially and economically weak segments of society in Phoenix are concentrated well within the urbanized region rather than along the outskirts (Gammage 1999).

The majority of variance in the metric results is expressed in the urban grid cells (Fig. 12.4), as the nonurban grid cells include relatively little variation in land cover types. Representative results for the CA, ED, and IJI metrics for the Built aggregate land cover class are presented as greyscale raster images in Fig. 12.5. An example of the Mean Patch Size results is not included as there is little to no variation in this metric at the 1 km analysis scale (leading to monotonic raster results). Raster images were generated for each aggregate land cover class (data not shown) with the exception of the Water class.

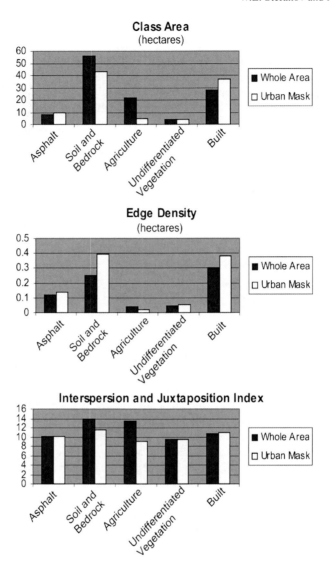

Fig. 12.4 Class Area (CA), Edge Density (ED), and Interspersion and Juxtaposition Index (IJI) for the aggregated land cover classes (whole investigated area and for urban masked grid cells)

The Water class was not included as it had minimal representation in the grid analysis area. The Asphalt class exhibits relatively low IJI values in areas associated with urban mountain parks and along the urban/peri-urban fringe area. Highest values are associated with older mesic residential areas, while significantly variable IJI is obtained for the remainder of the urbanized region. Older residential areas in the Phoenix region are generally comprised of lots with high amounts of vegetation (grass, trees) in addition to built materials and asphalt roadways (Hope et al. 2003).

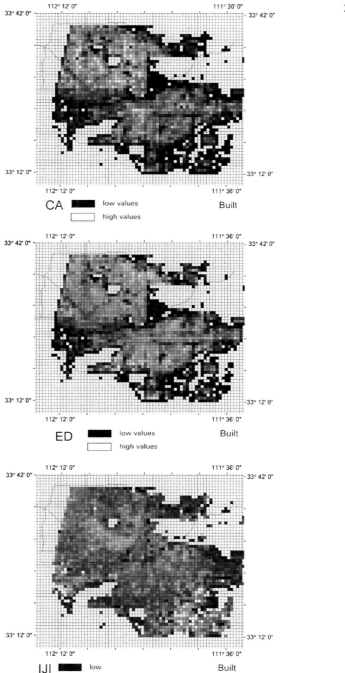

Fig. 12.5 Class Area (CA), Edge Density (ED), and Interspersion/Juxtaposition Index (IJI) for the Built aggregate land cover class. Grid corresponds to that depicted in Fig. 12.3. Background vector data are major highways

The metric results for the Soil and Bedrock class suggest that while the majority of surficial soil and rock material in the urban area is not highly intermingled with other land cover types (relatively low IJI), these areas are irregular in shape (relatively high ED). The boundaries between urban and peri-urban regions are not as sharply defined for the Soil and Bedrock class as for the other classes. This is to be expected as the majority of the peri-urban region is comprised of desert soils with low vegetation cover or exposed bedrock (GP2100 2003). The Agriculture class has minor representation in the urban region, however it exhibits low to moderate ED in keeping with the generally rectilinear plan of agricultural fields in this region. The Undifferentiated Vegetation class shows highest ED values along the Salt River bed and in areas of mesic residential land cover/land use. While the Salt River bed is generally dry, there is enough water released in association with an upstream impoundment of water (Tempe Town Lake) to support a sparse riparian vegetation community (Musacchio 2003). The IJI values for the Undifferentiated Vegetation class are similar to those for Asphalt; relatively high values are recorded in older mesic regions, while variable results are associated with the remainder of the urban area. Both ED and IJI values for the Built class are generally high throughout the urban region. This suggests that this class is evenly distributed, highly intermingled with other land cover types throughout the urbanized area, and has generally irregular boundaries with other classes at the 1 km scale of analysis. The land cover classification presented in Fig. 12.2 illustrates the degree of mixing present between soil, built, and vegetation land cover types in the Phoenix region.

We hypothesized that the landscape structure of the Phoenix urban/peri-urban area correlated with a variety of biophysical parameters at the 1 km scale of a MODIS pixel. The results presented in Table 12.3 suggest that this hypothesis is not generally correct. Table 12.4 presents descriptive statistics for the MODIS datasets. Examination of the results for both the urban and nonurban grid cells indicates generally low variance in the MODIS data; standard deviations are typically an order of magnitude or more about the means. This suggests that at the 1 km/pixel scale of the MODIS data, the Phoenix metropolitan area appears fairly uniform. The exceptions to this general conclusion are the fPAR and LAI datasets. The observed high variance of these datasets is however the result of small sample size rather than actual variation across the metropolitan region (discussed below).

Nevertheless, there are indications of some weak correlations between MODIS variables and the landscape metric results. We therefore limit the following discussion of results to positive and negative correlations greater than 0.2 and −0.2 respectively. Albedo measurements show a positive correlation with CA for the aggregate Agriculture class in the nonurban area. This is a result of using the full wavelength range albedo (0.3–5.0 μm) which includes the near-infrared reflectance peak from plants. The Agriculture class also contains both vegetated and fallow agricultural fields with high soil brightness. Finally, the Agriculture class is concentrated primarily in the nonurban grid cells. Relatively strong negative correlation with albedo is observed for both urban Asphalt and Built classes in the CA, ED, and IJI metrics. Small (2003) demonstrates that the characteristic spatial scale of albedo variations with land cover for 14 urban centers (including Phoenix) is 10–20 m using IKONOS data.

Table 12.3 Land cover and MODIS variable correlations for urban (U) and non-urban (NU) grid cells

	Asphalt		Soil and bedrock		Agriculture		Undifferentiated vegetation		Built	
	U	NU	U	NU	U	NU	U	NU	U	NU
Class area										
Albedo	−0.231	−0.076	0.155	−0.07	0.189	0.221	−0.008	−0.018	−0.188	0.061
fPAR	−0.002	0.156	−0.334	−0.107	0.085	0.024	0.210	−0.027	0.231	0.038
LAI	0.030	0.176	−0.088	−0.031	0.029	−0.095	0.031	0.000	0.037	0.035
Day surface temperature	0.149	0.060	0.258	0.109	−0.083	−0.011	−0.271	−0.028	−0.190	0.007
Night surface temperature	0.034	0.196	0.097	0.293	−0.232	−0.289	−0.112	0.047	0.113	−0.004
NDVI	−0.048	−0.100	−0.289	−0.367	0.076	0.390	0.182	0.039	0.218	0.024
Mean patch size										
Albedo	0.029	0.047	0.085	−0.131	0.180	0.196	0.079	−0.020	−0.066	0.014
fPAR	−0.066	0.094	−0.140	−0.185	0.038	−0.014	−0.023	−0.012	0.187	0.058
LAI	0.041	0.108	−0.047	−0.128	−0.012	−0.063	−0.008	0.000	0.109	0.021
Day surface temperature	0.088	0.016	0.131	0.079	0.023	0.017	−0.021	0.000	−0.194	−0.013
Night surface temperature	0.015	0.073	0.094	0.222	−0.199	−0.102	−0.074	0.005	0.155	−0.046
NDVI	−0.091	−0.048	−0.129	−0.298	0.047	0.194	−0.038	−0.029	0.177	0.168
Edge density										
Albedo	−0.298	−0.114	−0.174	0.109	0.146	0.183	−0.033	−0.019	−0.165	0.052
fPAR	0.031	0.122	0.049	0.090	0.078	−0.009	0.275	−0.022	0.159	0.007
LAI	0.018	0.142	−0.018	0.098	0.002	−0.090	0.043	0.005	−0.010	0.014
Day surface temperature	0.157	0.070	−0.039	0.030	−0.167	−0.025	−0.341	−0.029	−0.160	0.023
Night surface temperature	0.019	0.216	0.023	−0.005	−0.124	−0.257	−0.134	0.041	0.055	0.022
NDVI	−0.014	−0.107	0.052	−0.010	0.068	0.245	0.248	0.066	0.167	−0.018
Interspersion/Juxtaposition										
Albedo	−0.144	−0.027	0.056	−0.066	0.099	0.094	−0.049	0.099	−0.213	−0.044
fPAR	0.246	0.120	−0.059	−0.022	0.193	0.066	0.191	−0.011	0.353	0.147
LAI	0.048	0.096	0.004	0.017	0.047	−0.043	0.058	0.061	0.102	0.073
Day surface temperature	−0.279	−0.070	0.104	0.042	−0.164	−0.040	−0.191	0.026	−0.281	−0.070
Night surface temperature	−0.062	0.009	−0.091	0.177	−0.225	−0.293	−0.137	−0.054	−0.067	−0.156
NDVI	0.230	0.105	−0.055	−0.187	0.186	0.378	0.167	0.184	0.307	0.335

Table 12.4 Statistics for MODIS data for urban (U) and nonurban (NU) grid cells

Dataset (units)	Minimum (U)	Minimum (NU)	Maximum (U)	Maximum (NU)	Mean (U)	Mean (NU)	±1 σ (U)	±1 σ (NU)
Albedo (unitless)	0.10	0.05	0.30	0.31	0.19	0.20	0.03	0.04
fPAR (%)	0.10	0.08	0.81	0.82	0.28	0.26	0.10	0.11
LAI (m^2/m^2)	0.10	0.10	3.30	4.30	0.38	0.37	0.22	0.29
Day surface temperature (K)	24.50	24.50	51.00	52.00	47.26	47.74	1.66	2.96
Night surface temperature (K)	0.00	0.00	30.00	30.00	27.55	26.82	2.08	4.48
NDVI (unitless)	0.04	0.07	0.72	0.74	0.24	0.23	0.08	0.11

This characteristic scale is an order of magnitude below our scale of investigation, however the weak negative correlations we obtain suggest that some effect of the distribution of the Asphalt and Built classes is still discernable in the 1 km scale reflectance measured by MODIS. Figure 12.4 illustrates the overall distribution and high degree of mixing of the Built class with other land cover types in the urban region; the metric results for the Asphalt class are similar. The results of Small (2003) also suggest that the 15 m spatial scale of ASTER is adequate to capture land cover variations (and by extension, landscape structure) of importance to urban ecological and climatological studies.

The results for fPAR indicate that this variable is weakly correlated with urban CA for both the Undifferentiated Vegetation and Built classes. Weak correlation is also obtained for urban ED with the Undifferentiated Vegetation class, and for urban IJI with the Built class. These results suggest a connection between fPAR and older mesic residential neighborhoods which tend to have larger lot sizes and higher amounts of trees, grass, and shrubs (GP2100 2003). No correlations are observed for LAI with any metrics or classes which most probably reflect the low proportion of broad-leafed plant species in the Phoenix area (Whitford 2002; Hope et al. 2003). Somewhat surprisingly there is also a weak correlation observed for fPAR with the urban Asphalt class for IJI. This may be a result of a high degree of association between asphalt roadways and residential areas. It should be noted that the fPAR and LAI values associated with "urban" pixels, as classified using the MODIS Land Cover data product, are not modeled and therefore invalid (Knyazikhin et al. 1999). A total of 466 out of 1,599 urban grid cells (29%) are invalid. By contrast only 32 out of 1,781 nonurban grid cells (2%) are invalid. This aspect of the fPAR and LAI datasets is unfortunate, as it limits detailed study of these variables in urban core areas using these MODIS data products.

Daytime surface temperature shows a weak positive correlation with urban CA for the Soil and Bedrock class. This may result from the presence of several urban mountain parks (Piestewa Peak/Dreamy Draw, Papago, South Mountain, and Camelback Mountain) with relatively little soil or vegetation cover. Relatively strong negative correlations are obtained with both CA and ED for the urban Undifferentiated Vegetation class. This reflects the relatively high extent of vegetated surfaces in the Phoenix metropolitan region (as parks, golf courses, and mesic residential developments) and suggests that distribution of this land cover type may influence surficial thermal properties even at the relatively coarse 1 km scale. Relatively strong negative correlations of daytime surface temperature with urban IJI for both Asphalt and Built classes are also observed. This could reflect the high degree of association of these two classes with vegetated regions (hence cooler surface temperatures).

Relatively strong positive correlations of night surface temperature with nonurban CA and MPS are noted for the Soil and Bedrock class. This result is problematic as the Phoenix peri-urban area typically has cooler nighttime temperatures than the urban core and we would expect to see a negative correlation with these landscape metrics (Brazel et al. 2000; Hawkins et al. 2004). This result could be an artifact of the location of the nonurban grid cells in that they may include extensive regions of dark bedrock and soil. Further analysis of the entire Phoenix metropolitan region and comparison with other urban areas is needed to investigate this further. Negative correlation of night surface temperature for the Agriculture class in both urban and nonurban regions is obtained for the CA, ED, and IJI metrics. The distribution and configuration of agricultural lands in urban and peri-urban regions influences urban heat island and oases effects as discussed by Voogt and Oke (2003). The combined use of ASTER-derived landscape metrics and MODIS temperature data may facilitate further investigation of urban heat island/oasis effects and regional climatology. Initial modifications of land cover input to a regional climate model for Phoenix have already been accomplished using Landsat TM data (Grossman-Clarke et al. 2005; Zehnder 2002), and further improvements are possible using ASTER and MODIS data.

The strongest positive correlations are noted with NDVI and the nonurban Agriculture class for the CA, ED, and IJI metrics. Strong correlations are also observed with the urban and nonurban Built class for CA and IJI, and urban ED for the Undifferentiated Vegetation class. All of these aggregate land cover types include significant vegetation, and the correlations with the metrics reflect their complex distribution across the Phoenix urban and peri-urban landscape (particularly with regard to the Built and Undifferentiated Vegetation classes (Figs. 12.2–12.5). Relatively strong negative correlations of NDVI with the urban and nonurban Soil and Bedrock class in CA and MPS are the result of the generally low vegetation cover in both Phoenix urban mountain parks and the surrounding Sonoran desert (Whitford 2002).

Greenhill et al. (2003) calculate weighted mean patch size and lacunarity (a measure of the distribution of patches of pixels in a scene) for NDVI values obtained from the IKONOS sensor for a suburban area of southwest London. Their work demonstrates the usefulness of the landscape metric approach for urban planning applications and urban ecological research using high spatial resolution data. The availability of high spatial resolution data from sensors such as IKONOS and Quickbird is limited however both temporally and spatially compared to ASTER and MODIS (and similar) data. This is especially important with regard to comparative studies that include numerous urban centers and their surrounding regions.

The results presented here are of course scale-dependant (refer to Chapters 4 and 5), and could be expected to vary if different spatial scales or land cover classification schemes were used (Woodcock and Strahler 1987; Wu et al. 2000; Small 2003). The variance due to spatial scale was investigated in an allied study by Stefanov and Netzband (2005) using the same classified ASTER and MODIS data described here. Their study focused on gridded landscape metric analysis using 250 m/pixel,

500 m/pixel, and 1 km/pixel MODIS NDVI datasets, and obtained similar weak correlations with land cover type even at smallest spatial scale of analysis. In choosing to focus on relatively high spatial resolution ASTER data for landscape metrics, and high temporal resolution MODIS data for biophysical parameters, we highlight the particular strengths of each dataset for urban/peri-urban analysis and monitoring. We conclude however that standardized, remotely sensed datasets with high spatial, spectral, and temporal resolution will be required to meet the challenge of understanding, and perhaps more importantly predicting, local to global ecosystem change due to urban expansion and development.

Chapter Summary

New high spatial, temporal, and spectral resolution remotely sensed data have sparked a renewed interest in the investigation of physical, climatic, and social processes associated with human-dominated systems., i.e. cities. The interdisciplinary nature of such research likewise encourages the creative use of tools and data from different disciplines and sources. Integration of remotely sensed and ancillary geospatial data for highly-accurate land cover classification can be easily performed using an expert systems approach. An expert land cover classification system was built using ASTER data and land use information for the Phoenix, Arizona, USA metropolitan area. Landscape spatial structure for the Phoenix area was obtained using several landscape metric algorithms. Spatial metrics used include Class Area, Mean Patch Size, Edge Density, and the Interspersion and Juxtaposition Index. Linkages between urban spatial structure and biophysical parameters (albedo, fraction of photosynthetically active radiation, leaf area index, day/night surface temperature, and the normalized difference vegetation index) obtained from MODIS were investigated using linear regression of gridded landscape metric data. Our results indicate some control of these biophysical parameters by urban/peri-urban landscape structure. The correlations are not strong however, and may reflect both the spatial heterogeneity of the Phoenix metropolitan region and the relatively low variance of the MODIS data over the urban/peri-urban region at the 1 km/pixel scale.

The approach and results we present are of use to urban ecologists and land planners, as landscape structural analysis and measures of ecosystem function provide useful monitoring tools for regional habitat and climatic alteration associated with urbanization. Use of the uniform spatial reference systems provided by remotely sensed data in comparative studies permits quantitative evaluation of the configuration of existing developed and open space. This could improve the scope of usually small-scale and project-related analyses of local environmental change, and thus represents an important tracking system for regional planning and investigation of the ecological effects of increasing global urbanization.

LEARNING ACTIVITIES

Internet Resources

- MODIS home page (http://modis.gsfc.nasa.gov/).
- ASTER home page (http://asterweb.jpl.nasa.gov/).
- MOLAND Project (http://moland.jrc.it/).
- 100 Cities Project at Arizona State University (http://100cities.asu.edu/index.html).
- USGS Global Visualization Viewer (http://glovis.usgs.gov) – a useful site for browsing and ordering ASTER, Landsat, and MODIS data. For ASTER data, visible to near infrared and thermal infrared browse images can be examined at a minimum resolution of 155 m/pixel. Note that while Landsat and MODIS data can downloaded free of charge, there is a fee for ASTER data.

Study Questions

- Go to the GLOVIS website indicated above and compare 155 m/pixel VNIR browse images for Phoenix, Arizona, USA with Istanbul, Turkey; Baltimore, Maryland, USA; Madrid, Spain; Riyadh, Saudi Arabia; and Chongqing, China. Based on the observed physical structure, and regional physiography of each city, which would you expect to be similar to Phoenix in terms of ecological function and climate and why?
- Why is the spatial resolution of remotely sensed data of critical importance to urban studies?
- Is there an optimum scale of landscape metric analysis in urban systems? Why or why not?
- What are some of the difficulties inherent in using so-called "hard classifiers" in urban regions?
- This chapter presents research focused on questions of urban ecology; what other problems facing urban systems can be addressed using the datasets and approaches discussed here?
- The hypothesis that urban ecological function is directly related to urban structure is not strongly supported by the results presented in this chapter. Consider why this might be from the standpoint of:

  - The remotely sensed datasets used.
  - The land cover classification approach.
  - The landscape metrics used.
  - The spatial and temporal scales of the analysis.
  - How might further research explore some of these potential sources of uncertainty?

References

Abrams M (2000) The Advanced Spaceborne Thermal Emission and Reflection Radiometer (ASTER): data products for the high spatial resolution imager on NASA's Terra platform. Int J Remote Sens 21(5):847–859

Alberti M, Waddell P (2000) An integrated urban development and ecological simulation model. Integr Assess 1:215–227

Barnsley MJ, Barr SL (2000) Monitoring urban land use by Earth observation. Surv Geophys 21:269–289

Botkin DB, Estes JE, MacDonald RB (1984) Studying the Earth's vegetation from space. BioScience 34:508–514

Brazel AJ, Selover N, Vose R, Heisler G (2000) The tale of two climates: Baltimore and Phoenix LTER sites. Climate Res 15:123–135

Buyantuyev A, Wu J, Gries C (2007) Estimating vegetation cover in an urban environment based on Landsat ETM+ imagery: a case study in Phoenix, USA. Int J Remote Sens 28:269–291

Churchill P, Hubbard N (1994) Centre for Earth Observations (CEO). EARSeL Newsl 20:18–21

Dell'Acqua F, Gamba P (2001) Detection of urban structures in SAR images by robust fuzzy clustering algorithms: the example of street tracking. IEEE Trans Geosci Remote Sens 39(10):2287–2297

Dial G, Bowen H, Gerlach F, Grodecki J, Oleszczuk R (2003) IKONOS satellite, imagery, and products. Remote Sens Environ 88:23–36

Donnay J-P, Barnsley MJ, Longley PA (2001) Remote sensing and urban analysis. In: Donnay J-P, Barnsley MJ, Longley PA (eds) Remote sensing and urban analysis. Taylor & Francis, New York, pp 245–258

Elvidge CD, Hobson VR, Nelson IL, Safran JM, Tuttle BT, Dietz JB, Baugh KE (2001) Overview of DMSP OLS and scope of applications. In: Mesev V (ed) Remotely sensed cities. Taylor & Francis, New York, pp 245–258

Eurostat (1995) Pilot project delimitation of urban agglomerations by remote sensing: results and conclusions. Office for Official Publications of the European Communities, Luxembourg

Forster BC (1980) Urban residential ground cover using Landsat digital data. Photogramm Eng Remote Sens 46:547–558

Gamba P, Dell'Acqua F, Lisini G, Cisotta F (2006) Improving building footprints in InSAR data by comparison with a lidar DSM. Photogramm Eng Remote Sens 72(1):63–70

Gammage G Jr (1999) Phoenix in perspective: reflection on developing the desert. Arizona State University, Tempe, AZ

Geosystems L (2003) ERDAS field guide, 7th edn. Leica geosystems GIS & mapping, Atlanta, GA

Gong P, Howarth PJ (1990) The use of structural information for improving land – cover classification accuracies at the rural – urban fringe. Photogramm Eng Remote Sens 56:67–73

GP2100 (2003) Greater Phoenix regional atlas: a preview of the region's 50-year future. Arizona State University, Tempe, AZ

Greenhill DR, Ripke LT, Hitchman AP, Jones GA, Wilkinson GG (2003) Characterization of suburban areas for land use planning using landscape ecological indicators derived from Ikonos-2 multispectral imagery. IEEE Trans Geosci Remote Sens 41:2015–2021

Grimm NB, Grove JM, Redman CL, Pickett STA (2000) Integrated approaches to long-term studies of urban ecological systems. BioScience 70:571–584

Grossman-Clarke S, Zehnder JA, Stefanov WL, Yubao L, Zoldak MA (2005) Urban modifications in a mesoscale meteorological model and the effects on near-surface variables in an arid metropolitan region. J Appl Meteorol 44:1281–1297

Haack B (1983) An analysis of Thematic Mapper Simulator data for urban environments. Remote Sens Environ 13:265–275

Haack B, Bryant N, Adams S (1987) An assessment of Landsat MSS and TM data for urban and near-urban land-cover digital classification. Remote Sens Environ 21:201–212

Haff PK (2002) Neogeomorphology. Am Geophys Union EOS Trans 83(29):310–317

Hawkins TW, Brazel A, Stefanov WL, Bigler W, Safell EM (2004) The role of rural variability in urban heat island and oasis determination for Phoenix, Arizona. J Appl Meteorol 43:476–486

Herold M, Scepan J, Clarke KC (2002) The use of remote sensing and landscape metrics to describe structures and changes in urban land uses. Environ Plann A 34(8):1443–1458

Herold M, Couclelis H, Clarke KC (2005) The role of spatial metrics in the analysis and modeling of urban land use change. Comput Environ Urban Syst 29:369–399

Hope D, Gries C, Zhu W, Fagan WF, Redman CL, Grimm NB, Nelson AL, Martin C, Kinzig A (2003) Socioeconomics drive urban plant diversity. Proc Nat Acad Sci 1000(15):8788–8792

Huang J, Lu XX, Sellers JM (2007) A global comparative analysis of urban form: applying spatial metrics and remote sensing. Landscape Urban Plann 82:184–197

Imhoff ML, Lawrence WT, Stutzer DC, Elvidge CD (1997) A technique for using composite DMSP/OLS "City Lights" satellite data to map urban area. Remote Sens Environ 61(3):361–370

Irons JR, Petersen GW (1981) Texture transforms of remote sensing data. Remote Sens Environ 11:359–370

Jackson MJ, Carter P, Smith TF, Gardner W (1980) Urban land mapping from remotely-sensed data. Photogramm Eng Remote Sens 46:1041–1050

Jenerette GD, Harlan SL, Brazel A, Jones N, Larsen L, Stefanov WL (2007) Regional relationships between surface temperature, vegetation, and human settlement in a rapidly urbanizing ecosystem. Landscape Ecol 22:353–365

Jensen JR (1981) Urban change detection mapping using Landsat data. Am Cartographer 8:1237–1247

Jensen JR (1996) Introductory image processing: a remote sensing perspective, 2nd edn. Prentice-Hall, Upper Saddle River, NJ

Jensen JR (2000) Remote sensing of the environment: an earth resource perspective. Prentice-Hall, Upper Saddle River, NJ

Knyazikhin Y, Glassy J, Privette JL, Tian Y, Lotsch A, Zhang Y, Wang Y, Morisette JT, Votava P, Myneni RB, Nemani RR, Running SW (1999) MODIS Leaf Area Index (LAI) and fraction of Photosynthetically Active Radiation absorbed by vegetation (FPAR) product (MOD15) algorithm theoretical basis document. http://modis-land.gsfc.nasa.gov/pdfs/atbd_mod15.pdf. Accessed 8 Feb 2004

Kupel DE (2003) Fuel for growth: water and Arizona's urban environment. University of Arizona Press, Tucson, AZ

Lavalle C, Demicheli L, Turchini M, Casals CP, Niederhuber M (2001) Monitoring mega-cities: the MURBANDY/MOLAND approach. Dev Pract 11(2–3):350–357

Longley PA (2002) Geographic information systems: will developments in urban remote sensing and GIS lead to 'better' urban geography? Progr Hum Geogr 26(2):213–239

Maricopa Association of Governments (2000) Existing (year 2000) land use. Maricopa Association of Governments, Phoenix, AZ

Martin LRG, Howarth PJ, Holder G (1988) Multispectral classification of land use at the rural–urban fringe using SPOT data. Can J Remote Sens 14(2):72–79

McGarigal K, Marks BJ (1995) FRAGSTATS: spatial pattern analysis program for quantifying landscape structure. USDA For. Serv. Gen. Tech. Rep. PNW-351

Mesev V (2003) Remotely sensed cities: an introduction. In: Mesev V (ed) Remotely sensed cities. Taylor & Francis, London, pp 1–19

Musacchio L (2003) Landscape ecological classification and analysis of a 100-year floodplain corridor in the Phoenix metropolitan region. Central Arizona-Phoenix Long-Term Ecological Research (CAP LTER) Fifth Annual Poster Symposium, Tempe, AZ, 19 Feb

Narumalani S, Mishra DR, Rothwell RG (2004) Change detection and landscape metrics for inferring anthropogenic processes in the greater EFMO area. Remote Sens Environ 91:478–489

Nations D, Stump E (1996) Geology of Arizona, 2nd edn. Kendall/Hunt Publishing Company, Dubuque, IA

Netzband M, Kirstein W (2001) Landscape metrics as a tool for the comparison of different urban areas. Regensburger Geographische Schriften 35:222–231

Netzband M, Stefanov WL (2003) Assessment of urban spatial variation using ASTER data. The international archives of the photogrammetry, remote sensing, and spatial information sciences, vol 34, part 7/W9. Regensburg, Germany, pp 138–143

Netzband M, Stefanov WL, Redman CL (2007) Remote sensing as a tool for urban planning and sustainability. In: Netzband M, Stefanov WL, Redman C (eds) Applied remote sensing for urban planning, governance, and sustainability. Springer, Berlin, pp 1–23

Parkinson CL, Greenstone R (2000) EOS data products handbook: volume 2. NASA Goddard Space Flight Center, Greenbelt

Rainis R (2003) Application of GIS and landscape metrics in monitoring urban land use change. In: Hashim NM, Rainis R (eds) Urban ecosystem studies in Malaysia – a study of change. Universal Publishers, Parkland, pp 267–278

Ramsey MS (2003) Mapping the city landscape from space: the Advanced Spaceborne Thermal Emission and Reflectance Radiometer (ASTER) Urban Environmental Monitoring Program. In: Heiken G, Fakundiny R, Sutter J (eds) Earth science in the city: a reader. American Geophysical Union, Washington, DC, pp 337–361

Ramsey MS, Stefanov WL, Christensen PR (1999) Monitoring world-wide urban land cover changes using ASTER: preliminary results from the Phoenix, AZ LTER site. In: Proceedings of the 13th international conference, applied geological remote sensing, vol 2, Vancouver, BC, Canada, 1–3 Mar

Ridd M (1995) Exploring a V-I-S (Vegetation-Impervious Surface-soil) model for urban ecosystem analysis through remote sensing: comparative anatomy of cities. Int J Remote Sens 16:2165–2185

Ridd MK, Liu J (1998) A comparison of four algorithms for change detection in an urban environment. Remote Sens Environ 63:95–100

Robinson JA, McRay B, Lulla KP (2000) Twenty-eight years of urban growth in North America quantified by analysis of photographs from Apollo, Skylab, and Shuttle-Mir. In: Lulla KP, Dessinov LV (eds) Dynamic earth environments: remote sensing observations from Shuttle-Mir missions. Wiley, New York, pp 25–41

Running SW, Thornton PE, Nemani RR, Glassy JM (2000) Global terrestrial gross and net primary productivity from the Earth Observing System. In: Sala O, Jackson R, Mooney H (eds) Methods in ecosystem science. Springer, New York, pp 44–57

Sabins FF (1997) Remote sensing: principles and interpretation, 3rd edn. W.H. Freeman, New York

Sawaya KE, Olmanson LG, Heinert NJ, Brezonik PL, Bauer ME (2003) Extending satellite remote sensing to local scales: land and water resource monitoring using high-resolution imagery. Remote Sens Environ 88:144–156

Schaaf CB, Gao F, Strahler AH, Lucht W, Li XW, Tsang T, Strugnell NC, Zhang XY, Jin YF, Muller JP, Lewis P, Barnsley M, Hobson P, Disney M, Roberts G, Dunderdale M, Doll C, d'Entremont RP, Hu BX, Liang SL, Privette JL, Roy D (2002) First operational BRDF, albedo and nadir reflectance products from MODIS. Remote Sens Environ 83(1–2):135–148

Schneider A, McIver DK, Friedl MA, Woodcock CE (2003) Mapping urban areas by fusing coarse resolution remotely sensed data. Photogramm Eng Remote Sens 69:1377–1386

Schöpfer E, Moeller MS (2006) Comparing metropolitan areas – a transferable object-based image analysis approach. Photogrammetrie, Fernerkundung, Geoinformation 4:277–286

Seto KC, Fragkias M, Schneider A (2007) 20 years after reforms: challenges to planning and development in China's city-regions and opportunities for remote sensing. In: Netzband M, Stefanov WL, Redman C (eds) Applied remote sensing for urban planning, governance, and sustainability. Springer, Berlin, pp 249–269

Small C (2003) High spatial resolution spectral mixture analysis of urban reflectance. Remote Sens Environ 88:170–186

Small C (2007) Spatial analysis of urban vegetation scale and abundance. In: Netzband M, Stefanov WL, Redman C (eds) Applied remote sensing for urban planning, governance, and sustainability. Springer, Berlin, pp 53–76

Smith RL (1980) Ecology and field biology, 3rd edn. Harper & Row, New York

Stefanov WL (2002) Remote sensing of urban ecology at the Central Arizona-Phoenix Long Term Ecological Research site. Arid Lands Newsletter, 51. http://ag.arizona.edu/OALS/ALN/aln51/stefanov.html. Accessed 20 Feb 2009

Stefanov WL, Brazel AJ (2007) Challenges in characterizing and mitigating urban heat islands – a role for integrated approaches including remote sensing. In: Netzband M, Stefanov WL, Redman C (eds) Applied remote sensing for urban planning, governance, and sustainability. Springer, Berlin, pp 117–135

Stefanov WL, Netzband M (2005) Assessment of ASTER land cover and MODIS NDVI data at multiple scales for ecological characterization of an arid urban center. Remote Sens Environ 99:31–43

Stefanov WL, Christensen PR, Ramsey MS (2001a) Remote sensing of urban ecology at regional and global scales: results from the Central Arizona-Phoenix LTER site and ASTER Urban Environmental Monitoring Program. Regensburger Geographische Schriften 35:313–321

Stefanov WL, Ramsey MS, Christensen PR (2001b) Monitoring urban land cover change: an expert system approach to land cover classification of semiarid to arid urban centers. Remote Sens Environ 77:173–185

Stefanov WL, Ramsey MS, Christensen PR (2003) Identification of fugitive dust generation, transport, and deposition areas using remote sensing. Environ Eng Geosci 9:151–165

Stefanov WL, Netzband M, Möller MS, Redman CL, Mack C (2007) Phoenix, Arizona, USA: applications of remote sensing in a rapidly urbanizing desert region. In: Netzband M, Stefanov WL, Redman C (eds) Applied remote sensing for urban planning, governance, and sustainability. Springer, Berlin, pp 137–164

Stuckens J, Coppin PR, Bauer ME (2000) Integrating contextual information with per-pixel classification for improved land cover classification. Remote Sens Environ 71:282–296

Sutton PC (2003) Estimation of human population parameters using night-time satellite imagery. In: Mesev V (ed) Remotely sensed cities. Taylor & Francis, London, pp 301–333

Tucker CJ (1979) Red and photographic infrared linear combinations for monitoring vegetation. Remote Sens Environ 8:127–150

Vogelmann JE, Sohl T, Howard SM (1998) Regional characterization of land cover using multiple sources of data. Photogramm Eng Remote Sens 64:45–57

Voogt JA, Oke TR (2003) Thermal remote sensing of urban climates. Remote Sens Environ 86:370–384

Wan Z, Li Z-L (1997) A physics-based algorithm for retrieving land-surface emissivity and temperature from EOS/MODIS data. IEEE Trans Geosci Remote Sens 35:980–996

Weber C (1994) Per-zone classification of urban land cover for urban population estimation. In: Foody GM, Curran PJ (eds) Environmental remote sensing from regional to global scales. Wiley, Chichester, pp 142–148

Weber C (2001) Urban agglomeration delimitation using remote sensing data. In: Donnay J-P, Barnsley MJ, Longley PA (eds) Remote sensing and urban analysis. Taylor & Francis, New York, pp 245–258

Weber C, Puissant A (2003) Urbanization pressure and modeling of urban growth: example of the Tunis metropolitan area. Remote Sens Environ 86:341–352

Wentz EA, Stefanov WL, Gries C, Hope D (2006) Land use and land cover mapping from diverse sources for an arid urban environments. Comput Environ Urban Syst 30:320–346

Wentz EA, Nelson D, Rahman A, Stefanov WL, Roy SS (2008) Expert system classification of urban land use/cover for Delhi, India. Int J Remote Sens 29(15):4405–4427

Wentz EA, Stefanov WL, Netzband M, Möller M, Brazel A (2009) The urban environmental monitoring/100 cities project: legacy of the first phase and next steps. In: Gamba P, Martin H (eds) Global mapping of human settlement: experiences, data sets, and prospects. Taylor & Francis, New York

Whitford WG (2002) Ecology of desert systems. Academic, New York

Whitford V, Ennos AR, Handley JF (2001) City form and natural process – indicators for the ecological performance of urban areas and their application to Merseyside, UK. Landscape Urban Plann 57:91–103

Woodcock CE, Strahler AH (1987) The factor of scale in remote sensing. Remote Sens Environ 21:311–332

Wu J, Jelinski DE, Luck M, Tueller PT (2000) Multiscale analysis of landscape heterogeneity: scale variance and pattern metrics. Geogr Inf Sci 6:6–19

Yu XJ, Ng CH (2007) Spatial and temporal dynamics of urban dynamics of urban sprawl along two urban–rural transects: a case study of Guangzhou, China. Landscape Urban Plann 79:96–109

Zehnder JA (2002) Simple modifications to improve fifth-generation Pennsylvania State University-National Center for Atmospheric Research mesoscale model performance for the Phoenix, Arizona metropolitan area. J Appl Meteorol 41:971–979

Zhu G, Blumberg DG (2002) Classification using ASTER data and SVM algorithms: the case study of Beer Sheva, Israel. Remote Sens Environ 80:233–240

Zipperer WC, Wu J, Pouyat RV, Pickett STA (2000) The application of ecological principles to urban and urbanizing landscapes. Ecol Appl 10(3):685–688

Chapter 13
Remote Sensing of Desert Cities in Developing Countries

Mohamed Ait Belaid

This chapter focuses on remote sensing (RS) of desert cities, with a special focus on the developing countries context. It introduces the reader to the characteristics of urban areas in the desert environment, and discusses the potential of satellite imagery and how they are used to map and monitor changes in these areas over space and time.

Learning Objectives

Upon completion of this chapter, you should be able to:

❶ Distinguish the unique character of urban areas in desert environments
❷ Articulate thorough examples of how urban areas can be mapped using remotely sensed data
❸ Discuss a range of applied change detection and modeling techniques used in urban areas monitoring

---

13.1 Urban Areas in Desert Environment

Urban areas in desert cities are characterized by many constraints and problems, primarily desertification, climate change, drought, urbanization, sand dune formation, water scarcity, pollution, vegetation degradation, land management, waste management, wind and water erosion, demographic explosion and immigration. This section discusses the characteristics of urban areas particularly the desertification phenomenon and its impact on desert cities in Africa and West Asia. We will also review how remote sensing is used to tackle desertification related issues in these cities.

M. Ait Belaid (✉)
College of Graduate Studies, Arabian Gulf University, P.O. Box 26671,
Manama, Kingdom of Bahrain
e-mail: belaid@agu.edu.bh

T. Rashed and C. Jürgens (eds.), *Remote Sensing of Urban and Suburban Areas*,   245
Remote Sensing and Digital Image Processing 10,
DOI 10.1007/978-1-4020-4385-7_13, © Springer Science+Business Media B.V. 2010

13.1.1   Status and Extent of Desertification

Desertification is defined by the United Nations Convention to Combat Desertification (UNCCD) as a decline in land productivity in arid, semi-arid

> **desertification is a world-wide phenomenon defined as a decline in land productivity in arid, semi-arid and dry sub-humid regions due to such factors as human activities, population growth, climate changes, agricultural mismanagement, and fuel wood consumption**

and dry sub-humid regions, resulting from many factors, such as human activities and climate changes (UNCCD 1994). It is a world-wide phenomenon that has a profound effect, especially in Africa and West-Asia. According to a global assessment undertaken by the United Nations Environment Program (UNEP) in 1991, around 70% of the arid and semi-arid areas (excluding hyper-arid deserts) are affected by desertification. The economic losses of such phenomenon are estimated at US$42.3 billion per year. These consequences have affected more than 100 countries, 80 of which are developing countries that account for approximately one sixth of the world population. The main causes beyond increased desertification are: deforestation, overgrazing, fuel wood consumption, agricultural misman-agement, wind and water erosion, urbanization, and industry. The severe climatic conditions and the demographic explosion have pushed the rural population in poverty, and forced them to immigrate to the near cities and live in diffi-cult socioeconomic conditions. These immigrants put a huge strain on services (e.g., housing, water, waste, health) and infrastructure (e.g., transport, education) resources that are already overstretched. As the demand of housing increases, cities begin to expand into new areas (Chapter 2). This in turn causes urban development to expand into natural areas such as deserts, rangelands, and agricultural lands. Figure 13.1 shows a flowchart designed for desertification monitoring and control using both conventional and remotely sensed data (Ait Belaid 1999).

13.1.2   Urban Areas in Africa

> **Africa's high urban growth rate is a result of rural–urban immigration, population growth, and in some areas, conflict and environmental disasters**

The majority (62.1%) of African population is still rural, but the urban population is expected to increase from 10% to 17% between 2000 and 2013. North Africa is the most urbanized sub-region, with an average urban popu-lation of about 54% of the total populations in countries located in this region. The least urbanized African sub-region is Eastern Africa with only 23% of the popula-tion living in urban areas (UNPD 2001).

   The number of African cities is increasing and exist-ing cities are expanding in coverage. About 43 African cities have population greater than one million inhabitants

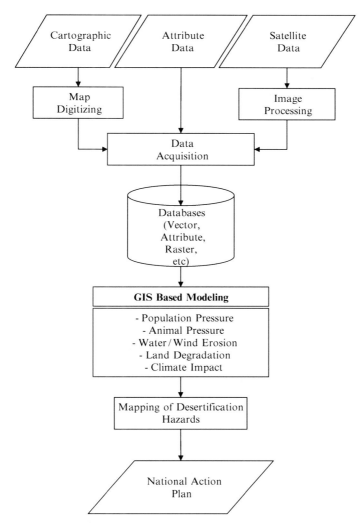

Fig. 13.1 Flowchart for desertification monitoring and control (Adapted from Ait Belaid 1999)

(UNPD 2001). Africa's high urban growth rate is a result of rural–urban immigration, population growth, and in some areas, conflict. People migrate from
rural areas because of declining agricultural productivity, lack of employment
opportunities, and lack of access to basic physical and social infrastructures. In
Southern Africa, up to 45% of urban households grow crops or raise livestock in
urban environments in order to support their livelihoods (UNPD 2001).
Environmental disasters and conflicts have also caused many people to flee rural
areas and seek refugee in urban centers. For example, about 4.5 million rural

people in Mozambique were displaced to urban areas in the 1980s (Chenje 2000) and the world's third largest refugee camp exists in urban Sierra Leone (UNCHS 2001).

About 85% of African urban population has access to water sources and sanitations (WHO and UNICEF 2000). In large cities, the air pollution is a growing concern, particularly in Cairo, Egypt, where pollutants create a permanent haze over the city, in addition of a range of high risk of respiratory disorders to its 12+ million inhabitants (UNCHS 1996).

13.1.3   Urban Areas in West Asia

oil boom, wars and globalization are the main factors fueling urban growth of cities in West Asia

The majority of Western Asian population lives in urban areas with the notable exception of Yemen. The past 30 years brought about significant economic, political and technological changes, which have influenced the structure and the functions of urban areas. Three main factors have shaped the urban landscapes of west Asia (UNESCWA 1999): the 1970 oil boom, the large-scale movement resulted from the Gulf Wars, and globalization.

Rapid economic growth, which occurred in most countries in the West Asia region over the past 3 decades, was accompanied by population growth and increased urbanization. There has been a massive migration of the population from rural to urban areas, as well as migration of foreign labor into urban areas, especially in the Gulf Cooperation Council (GCC) countries. Urbanization has continued to increase at a faster rate than the total population (UNPD 2001). All the countries of the Arabian Peninsula currently have an 84% average level of urbanization. With the exception Yemen which is only 25% urbanized, almost the entire population of Bahrain (92.2%), Kuwait (97.6%) and Qatar (92.5%) are living in urban areas (UNDP 2001). The number of large cities with more than one million residents is estimated at 12 cities in this part of the world.

urbanization forms in West Asia cities involve transformation of prime agricultural land, coastal habitats and forests into land for housing, roads, and industry

Economic development has brought dramatic improvement in the well-being of the western Asian population (UNDP 2001). Despite these positive impacts, several cities are negatively influenced by many problems such as lacking adequate services for the urban poor, placing human health and well-being at risk (UNDP, UNEP, WB, WRI 1998). Most of the large cities in the West Asia have higher levels of air pollution. As urban areas expand, prime agricultural land, coastal habitats and forests are transformed into land for housing, roads and industry. Land conversion activities range from draining and filling of marshes and wetlands to large-scale reclamation projects that extend shorelines into the sea. In Lebanon and most of the GCC countries, these activities have been carried out for decades. Dubai city, for example, increased in size by 92 km^2 in 15 years (Doxiadis Associates 1985). Likewise, the area of Bahrain increased by 47.3 km^2 in 23 years, by continuing land reclamation along the coastal zones at the Persian Gulf for urban development (CSO 1999). It is estimated that municipal waste generation in the region increased to 25 t/year in 1995 (Kanbour 1997), which is double of the annual waste generation in Mashriq countries of Iraq and Syria. In the GCC countries, waste collection and disposal systems are fairly efficient compared to those in the Mashriq. Plants for composting municipal solid wastes and sewage sludge have been established in several countries (Kanbour 1997).

Remote Sensing Infrastructure in Africa and West Asia

Remote sensing techniques, geographic information systems (GIS), Global Positioning System (GPS), and telecommunications are valuable tools that provide information and data analyses capabilities for natural resources management, environmental monitoring and human activities assessment. In developing countries, especially in Africa and West Asia, many research institutions, centers for studies, and universities have been introduced to these technologies and tools to support decision making processes.

Remotely sensed data are gathered throughout national agencies or regional providers of imagery. The United Nations Regional Center for Space Sciences and Technology Education in Asia has been in operation since November 1995. This was followed by the establishment of two similar centers; one in Morocco (for French speaking countries), and one in Nigeria (for English speaking countries). There is also another international initiative for establishing a new regional center in Western Asia, which currently involves Jordan, Saudi Arabia and Syria. These centers are expected to enhance the academic and professional capabilities and technical infrastructure in each region (UNISPACE-3 1999). In addition, there are several remote sensing centers in the GCC countries. The first center of remote sensing was established by the Institute of Space Research (ISR) at King Abdulaziz City for Science and Technology (KACST). It has provided remote sensing imagery and data since 1987 for the public and private sectors (KACST 2002). A second center was set-up in 1997 by United Arab Emirates and is operated by Dubai Space Imaging (DSI). DSI sells information products and services, as well as imagery and will provide training in imagery analysis and GIS tools and applications (Baker et al. 2001). A number of GIS and RS forums, are being held in the region, at bi-annual basis; for example: the MARISY Symposium in Morocco, AFRICAGIS in Africa, and a GIS Conference at Bahrain.

13.2   GIS and RS in Desert Cities

13.2.1   A General Framework for Mapping Desertification Hazards

Figure 13.1 below depicts a framework for the mapping and inventory processes that are typically used in the context of desert cities. It shows also the contribution of remote sensing and GIS techniques to these processes, which are illustrated with a number of applications ranging from the delineation of location and extent of urban areas, to surface temperature estimation and green belt implementation.

The framework shown in Fig. 13.1 shows how Remote sensing (RS) along with GIS tools, which are used to gather, display, store, analyze and output data related to the urban and sub-urban environment, can provide planners with necessary information that is suited for the management of urban and sub-urban areas, including (Donnay et al. 2001; Bahr 2001):

- Analyzing location and extent of urban areas
- Delineating the spatial distribution of different land use categories
- Managing primary transportation network and related infrastructure
- Collecting and deriving various census-related statistics and socio-economical indicators
- Depicting the 3D structure of urban area for telecommunications
- Conducting Environmental Impact Assessment (EIA) studies
- Monitoring changes in urban features over time

Some of the applications stated above can be tackled through land use/cover mapping and land use change detection, using the appropriate techniques of image classification, and change detection and analysis. Other application areas (e.g., transportation analysis, 3D modeling, EIA studies) still require further development to be fully operational in desert cities of developing countries.

The increasing demands of urban planning and management necessitate the applications of remote sensing and GIS for the sustainable development of urban areas. The implementation of urban development plans must incorporate an integrated approach for spatial modeling using RS data, GIS databases and GPS solutions. All these datasets and information can be integrated into a form of Urban Information System (UIS). An UIS can help develop efficient and economical models for the development and location of industries, education, housing, water supplies, service facilities, and disposal systems at both regional and national levels. The following is a number of application examples of how RS has been contributing to mapping desertification hazards and the sustainable development and management of desert cities.

13.2.2   Fusion of Optical and Radar Imagery

The city of Algiers, Algeria has developed new techniques of fusion of radar and optical imagery, particularly ERS/SAR and Spot-HRV imagery. For this purpose, two methods were developed and tested. They were mainly based on the fusion before and after operating fuzzy classification on satellite images. The results are comparable to those obtained by principal component analysis and ground truthing (Smara et al. 2003a).

13.2.3 Surface Temperature Estimation

Asmat et al. (2003) attempted to estimate the surface temperature above ground, based on the classification of Landsat-ETM+ imagery acquired in May 2001 for CyberJava (Malaysia). The land surface temperature and the classified land cover categories were generated and exported into a GIS environment for urban heat mapping analysis. The correlation between land use categories and surface temperature showed that urban areas exhibited higher surface temperature than forests, agriculture areas and water. A polygon-based classification process utilizing fuzzy logic has been adopted to integrate a broad spectrum of different object features, such as spectral values, shape, and texture.

13.2.4 Green Belt Implementation

Sand drift and movement in the Errachidia Oasis in the Moroccan desert have been a source of major environmental and socioeconomic changes. A huge quantity of sands is accumulated near urban areas, across roads, and inside palm trees, resulting in an out-migration of rural people. For this purpose, Lahraoui (2003) investigated a new approach to determine the shape and the location of a "green belt" around the oasis. This belt has been designed, in order to combat against desertification, especially sand movement, dune formation, and to protect the main infrastructures (e.g., airport, roads, irrigation channels, crops, and biodiversity). One single date Spot-HRV imagery of July 2002 was processed to delineate the potential path of the green belt, taking into account many physical, biological and social parameters including: soil, climate, surface stage, topography and socio-economy.

13.2.5 GIS Implementation at Municipalities

The GCC countries started the development of GIS applications since early 1990s. Since that time many GIS have been implemented in different areas related to base maps, municipality, utilities, business, and land registration. Below are some examples of currently operational GIS in GCC countries:

- United Arab Emirates has developed four GIS applications within various municipalities, including Dubai, Abu Dhabi, Sharjah, and Fujairah.
- The Kingdom of Bahrain has developed several applications related to geocoding and address locations, and crime analysis in all municipalities of Bahrain. Durrat Al-Bahrain is an enterprise real estate project based on Internet -based GIS.

- The Kingdom of Saudi Arabia has developed databases for the cities of Madina, Macca and Riyadh.
- The State of Qatar has a unique country-wide GIS that has been in operation for more than 10 years. This large GIS was designed efficiently to integrate data from 16 different government agencies.

13.2.6 Other Urban Application Examples

There are other applications of remote sensing related to desert cities such as:

- Extraction of primary transportation network
- Census related statistics (population estimation)
- Natural hazards and crisis management, flood, desertification, war, earthquake
- Site selection of landfills for industrial and domestic waste in urban areas
- Suitability analysis of tourist development sites in coastal environments

Currently, however, we do not know of any operational applications of these issues in developing countries. However, there are examples of other urban remote sensing applications in developing countries, such as the use of RS in the monitoring of floods in the Syria during the recent years, two wars in Iraq, and earthquakes in the cities of Algiers (Algeria), Pam (Iran) and Taza (Morocco)

13.3 Urban Areas Monitoring and Modeling

This section presents a set of techniques used to detect changes over time in urban areas, and drivers of such changes. The techniques are illustrated in brief through case study examples from the literature.

13.3.1 Land Use Change Detection

There is a wide range of techniques used for land use change detection in urban areas. These techniques can be subdivided into the following categories:

Multi-Date Composite Image Method: Images of different periods can be combined to form a new composite single image, in which changes may be visually inferred from variations in gray tone or color hue. Enhancement procedure most frequently employed for change detection includes image overlay, image differencing, image ratioing, and vegetation indices. The composite image is then classified (Fung and Zhang 1989).

Image Comparison Method: A comparison can be made between multi-date satellite images. These images may be single band gray tone or multiple band color composite images. Image variation between dates signals the occurrence of land cover change (Martin 1989).

Comparison of Classified Imagery: A study was conducted in North Eastern Cairo, in Egypt, using multi-source data, Landsat-TM, Spot-HRV and KVR-1000, with spatial resolution of 30, 20 and 5 m, respectively. Overall, the merged images TM-KVR and HRV-KRV provided higher information content. It was possible to map out land use changes over 48 years (1945, 1986, 1991, and 1993), which offers a valuable indicator for urban growth (CEDARE 1998). Two similar studies were conducted respectively in the cities of Casablanca (Benchekroun 1993) and in Rabat-Salé (Benchekroun and Layachi 1993).

Combination of Classified Imagery: A methodology was developed by Mongkolsawat and Thirangoon (1990) to develop land cover change detection methodology for the Yasothon province of Thailand. Two Landsat-TM imagery acquired during the wet and dry seasons (September and April respectively) were co-registered and classified based on land cover and terrain types. The September image was classified into 11 classes, while the April image was classified into 14 classes. The two classified images were then mathematically combined resulting in a unique linear combination of 154 possible gray levels image. Through a process of regrouping these gray levels into classes, the resultant image provided a meaningful land cover dynamics with respect to the terrain types.

Comparison of the Classified Radar Imagery: Smara et al (2003b) describes a radar based change analysis methodology in which two Radar SAR images of ERS satellite were acquired for the periods of 1992 and 1996. The techniques proposed were applied and the results were compared against the principal component analysis method. The quantitative study revealed some changes in buildings, vegetation and other related seasonal variations (Smara et al. 2003b).

Comparison of Conventional Data: Johnson (2003) investigated the impact of land degradation along a coastal strip in Togo, both under the urbanization encroachment and water erosion. The study produced two land use maps, based on 23 aerial photographs at a scale of 1:10,000 taken in 1985 and the topographic map of 1960 at a scale of 1:50,000. These two maps were analyzed by GIS. Changes of a period of 25 years were found to be related mainly to the retreat of the shoreline (amplitude ranging from 45 to 230 m depending on the locations) and the great entropic pressure on the other land use categories, especially the coconut palm.

13.3.2   *Land Use Change Analysis and Modeling*

Land Use Analysis and Modeling: The integration of time series econometrics and artificial neural networks improves the monitoring of change using a series of images. This kind of time series analysis typically requires an understanding of the drivers of change. This understanding could be built using a multivariate regression model, which links the changes in dependant variables (i.e., land use changes) to the changes in independent variables (i.e., human activities). This postulates a linear relationship between the dependent and independent variables, which can be represented and processed mathematically (Weicheng et al. 2002). Nagai et al. (2002) developed a new methodology to reconstruct the long-term land cover changes from fragmentary observational data and knowledge of the changes. Genetic algorithm has also been used as an interpolation method to create most probable spatial-temporal distribution of land cover categories. A convenient classifier such as maximum likelihood could then be applied to the output of the genetic interpolation algorithm to construct a long-term land cover changes during.

Urban Information Systems (UIS): Cities typically manage considerable collection of land-related information. However, the traditional separation of this information into different components themes, combined with disjointed information related to management regimes, leads to a considerable loss in the value of the information as a resource. City-wide Land Information Management (LIM) provides the means to integrate these components of land information into truly corporate information resource. Figure 13.2 below illustrates how city-wide LIM can add value by combining information concerning use, condition, value and tenure of land and disseminating this to the decision makers (UN-Habitat and FIG 2002).

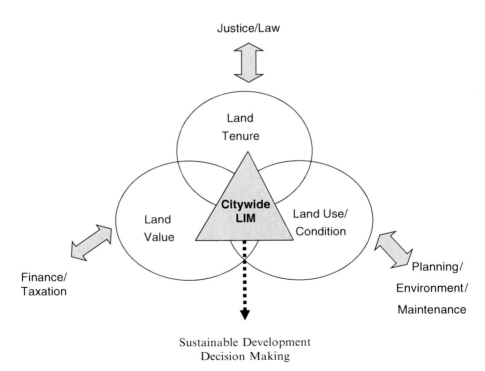

Fig. 13.2 Citywide LIM supporting sustainable development decision making (Adapted from UN-Habitat and FIG 2002)

13.4   Case Studies

This section briefly presents an application example of how remote sensing has been used to investigate urbanization processes in desert cities. The discussed example includes two multi-temporal studies conducted in two different sites in Morocco and Saudi Arabia respectively. The methodologies adopted in the two studies are slightly different, but the main steps are similar and are summarized below (Fig. 13.3).

13.4.1   Adopted Methodology

Geographical Study: An intensive geographical study was conducted through analysis of the existing and available literature. It included a description of physical conditions, i.e., climate, erosion, land, water, and vegetation; but also the character-ization of the impact of human and animal pressures and severe climate conditions.

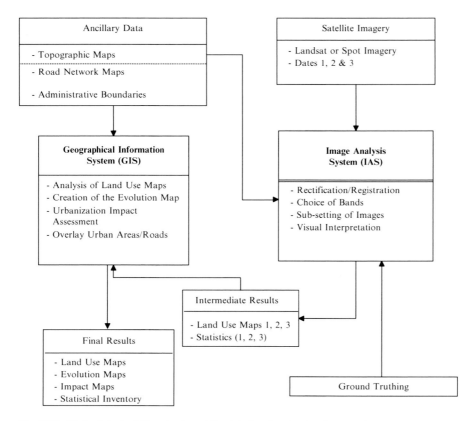

Fig. 13.3 Methodological flowchart used for data input, output and processing

In addition, the study focused on urbanization encroachment, population/demography, and agricultural practices on the urban–rural fringe.

Satellite and Ancillary Data Acquisition: In order to perform the study, satellite imagery was required at the appropriate period of time (year, season). Existing cartographic data (topographic maps, road network maps, boundaries) are also necessary to be used in the process of registration and integration with satellite imagery and/or to supplement the analysis with the necessary input in dates for which satellite imagery could not be acquired. Ground observations and existing statistics could also be used in visual interpretation and the validation of remote sensing results.

Land Use Mapping Using Satellite Imagery: Two multi-date satellite images were used for each city. The adopted interpretation technique was based on visual interpretation assisted by computer. The process consisted of displaying of the rectified imagery on the screen and the digitizing of polygons, which represent different land use categories, based on image characteristics such as tone, texture, color and pattern that can be translated into land use attributes. The whole process was guided by ground observations and local knowledge. These polygons were identified by independent labels attached to the polygon centeroids. The ground truthing aimed at localizing and characterizing land use categories using GPS instruments. The result of the visual interpretation technique was a digital coverage layer, in which polygons represented different land use categories. A full database is attached to this coverage, in which every polygon is characterized by many attributes like, the category number, area and perimeter of polygons.

Data Analysis and Interpretation Using GIS: We utilized the pair-wise comparison of land use maps, as well as their statistical inventories, in order to detect changes both in terms of area and geographic location of change. Urban and sub-urban areas were identified and assessed, in order to produce a map between two periods depicting changes in only urban and sub-urban areas and to investigate the impact of urbanization on other land uses and vise-versa. An "evolution matrix" was established for the two periods to provide all possible paths of change.

13.4.2   The Case Study of Morocco

Context: The main objective of this study was the assessment of urbanization impact on agricultural lands in the three cities of Ksar El Kébir, Béni Mellal, and Khémiset, during the last 3 decades (1970s, 1980s and 1990s). This multi-temporal study has been conducted in order to inform the urban planning scheme (Shéma Directeur d'Aménagement Urbain). Each study site covered a square of 30 km by 30 km centered on each of the cities. Two panchromatic Spot-HRV images with 10 m spatial resolution were used to cover the last 2 decades. In addition, topographic city maps were used to cover the first decade (1970s). Ground observations were used to complete the process of validation and control of land use mapping.

Data Analysis: For each city, we performed the following operations: (1) land use mapping using ancillary data and GIS for digitizing of city maps at a scale of 1:10,000 from the 1970s; (2) pre-processing of two panchromatic Spot-HRV imagery, which consisted of image subsetting, geometric correction according to the parameters of the national reference system (Clarck's 1880 projection and Merchich Datum); (3) land use mapping based on visual interpretation, using two panchromatic Spot-HRV imagery acquired during the last 2 decades; and (4) analysis and interpretation of the three land use maps, using GIS for analyzing the changes of land uses and modeling their mutual impacts.

Results and Discussion: The results and findings of the case study in Morocco included detailed land use maps, at a scale 1:50,000. These maps offered an extensive diagnosis of land use categories in terms of areas that were developed during each decade (1970s, 1980s and 1990s) (Anys et al. 1999a, b; Ait Belaid 2003). The adopted typology included eight thematic classes, namely: urban areas, rural settlements, fruit trees, rain-fed crops, irrigated crops, rangelands, forests and water bodies. The total area of each site is 90,000 ha.

Evolution maps for three intervals of time (1970–1980, 1980–1990 and 1970–1990) revealed the temporal evolution and trends of land use categories in the study sites. In 20 years, the area of urban areas increased by 598 ha (344%) in the city of Ksar El Kébir (Fig. 13.4), and by 981 ha (180%) in Béni Mellal, while the area of rural settlements increased by 1,945 ha (160%) in Ksar El Kébir and 756 ha (200%) in Béni Mellal. For the city of Khémisset (for which there was no city map available for the 1970s), the 10 years extension was 648 ha (70%) for urban areas and 363 ha (60%) for rural settlements. It is noticed that most urban changes occurred during the 1980s, compared to the 1970s.

The evolution matrix generated for the city of Ksar El Kébir, for the whole interval 1975–1996 signals the major land use changes that occurred (Table 13.1). Furthermore, it quantitatively illustrates the land use categories, which are impacted by any change and vise-versa. The zero value in the matrix implies no land use change in corresponding classes. During the period 1975–1996, urban areas expanded 598 ha, of which 501 ha were converted from rain-fed crops, 62 ha from rural settlements and 19 ha from fruit trees. The area of rural settlements increased tremendously by 1,945 ha, of which 1,372 ha were taken from rain-fed crops, 520 ha from rangelands and 61 ha from fruit trees. In sum, urban growth and rural settlements were severely impacting agricultural lands (rain-fed crops, fruit trees, rangelands). We can notice as well the increase of the area of water bodies by 1,457 ha, and the area of irrigated crops by 12,917 ha, which correspond respectively to the construction of a new dam and the development of a new irrigation plan in the region between 1970 and 1980.

The cartographical and statistical results are powerful tools for depicting the areas of agricultural lands that need to be protected. Furthermore, these results generated from this analysis were used to inform the master plan of urban planning (Urban Planning Director Scheme). The cost of the study estimated at US$18/km^2. The duration of this analysis was 8 months. Given the scale of the analysis conducted,

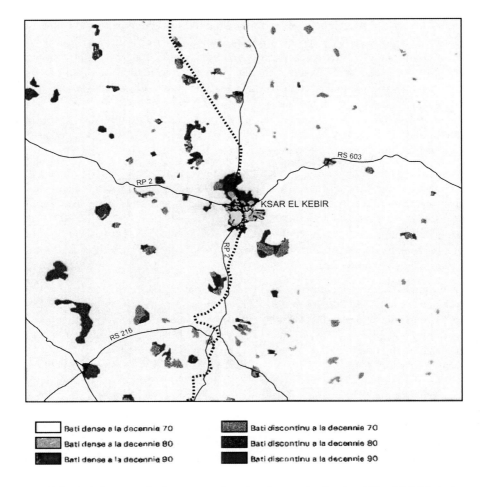

Fig. 13.4 Evolution map of urban areas and rural settlements for the city of Ksar El Kébir during 1975, 1987 and 1996 © CNES/Spot Image (1987–1996)/processed at CRTS

this relatively short time should encourage and facilitate the application of the methodology and extend studies to other cities in Morocco.

13.4.3   The Case Study of Saudi Arabia

Context: Al-Ahsa is one of the biggest oases in Saudi Arabia. It includes three cities, namely, Hufuf, Mubarraz and Al-Youn. The design of the three cities has changed consistently during the last 30 years, especially with the development of oil industry during the 1970s (Mufareh 2002; UNEP 2002). The population of Al-Ahsa

Table 13.1 Land use changes (evolution matrix) for Ksar El Kébir between 1975 and 1996

1996	1975									
	A	B	C	D	E	F	G	H	Ext.	Evol.
A	–	61.7	0	18.8	0	0	501.4	16.2	598.1	+598.1
B	0	–	5.1	61.7	52.4	0	1,371.6	48	2,006.8	+1,945.2
C	0	0	–	9.2	176.5	0	1,051.8	0	1,237.5	+763.1
D	0	0	469.3	–	218.9	0	510.4	0	729.3	–267.1
E	0	0	0	313.2	–	0	650.9	0	1,120.2	–82.9
F	0	0	0	252.5	0	–	1,141.4	0	1,456.6	+1,456.6
G	0	0	0	339	287.3	0	–	0	252.5	–17,329.6
H	0	0	0	0	0	0	12,354.6	–	12,980.9	+12,916.6
Reg.	0	61.7	474.4	996.4	1,203.1	0	17,582.1	64.2		

A: urban areas; B: rural settlements; C: forest areas; D: fruit trees; E: rangelands; F: water bodies; G: rain-fed crops; H: irrigated crops;
Ext. (extension): increase of the area; Reg. (regression): decrease of the area; Evol. (evolution): extension–regression; unit: hectares

oasis is estimated at 800,000 inhabitants, of which 78% live in urban areas and 22% in rural areas and villages. The main activity is concentrated on the agricultural sector (palm trees), but the overall agricultural area is declining due to climatic and anthropogenic factors such as sand movement, urban encroachment, decrease of the level of ground water tables, lack of drainage (sabkha), the small size of agricultural parcels which are not exploited cost-effectively. The climate is arid and the precipitation ranges from 75 to 110 mm.

The site covered in this study was a 40 × 55 km rectangle centered on Al-Ahsa oasis. Two multi-temporal multispectral Landsat-TM imagery with 30 m spatial resolution, were used to analyze two points of time: October 1987 and October 2001. Topographic maps at a scale of 1/25,000 and road network map were also used, along with ground observations.

Data Analysis: We performed the following operations: (1) pre-processing of the two Landsat-TM images, which consisted of image subsetting, geometric correction according to the parameters of the national grid reference (UTM projection, WGS'84 Datum and Zone 39); (2) land use mapping based on visual interpretation, using ground observations and the two Landsat images; and (3) analysis of the two land use maps using GIS in order to investigate the impact of urbanization on other land uses.

Results and Discussion: The results of this case study included land use maps at a scale of 1:100,000, which offered a diagnosis of urban and sub-urban areas at each period (1987 and 2001). The land use categories include seven thematic classes, namely: urban areas, urban planning, agricultural lands, sabkha, sand dunes, hills and water bodies (Mufareh 2002; Ait Belaid and Mufareh 2002; Ait Belaid 2003). The total area of the study site is approximately 222,000 ha.

One evolution map for the period (1987–2001) was created to illustrate the temporal evolution of land use categories. During 14 years, urban areas (Fig. 13.5) have expanded by 5,326 ha (75%) and agricultural areas by 4,430 ha (22%). During the same period of time, there was a decrease in the areas of sabkha by 8,063 ha (12%), urban planning zones by 1,341 ha (27%), sand dunes by 206 ha and hills by 146 ha.

The evolution matrix generated for Al-Ahsa oasis is shown in Table 13.2. It illustrates the major land use changes that occurred between 1987 and 2001. It signals also the land use categories, which are not impacted by any change and vise-versa. The zero value in the matrix implies no land use change in corresponding classes.

During 14 years, urban areas increased by 5,326 ha, of which 2,746 ha were converted from sabkha, 1,500 ha from urban planning zones, 972 ha from sand dunes, and 513 ha from agricultural lands. In contrast, the area of agricultural lands increased by 4,430 ha, of which 4,480 ha converted from sabkha, 1,654 ha from sand dunes, 184 ha from urban planning zones, and 119 ha from urban areas. The agricultural lands experienced many transformations. Overall, urban areas and agricultural lands show an explicit evolution of change during the study period. This expansion was compensated by a regression of other land use categories, implying that new sand dunes, sabkha, urban planning uses were developed between 1987 and 2001 but consumed by urban areas and agricultural lands. In conclusion, Al-Ahsa oasis is

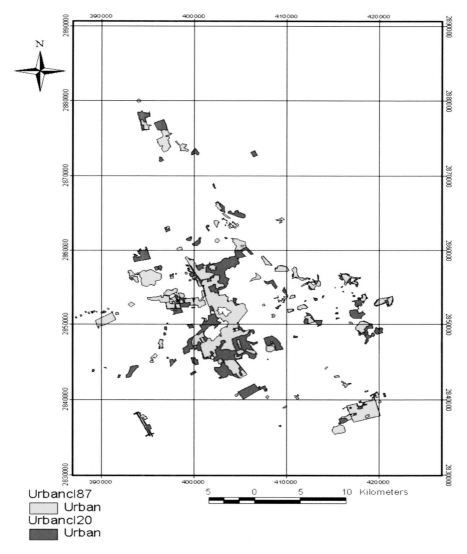

Fig. 13.5 Evolution map of urban areas in Al-Ahsa oasis during 1987 and 2001. © EOSAT (1987–2001) – KACST/processed at AGU

facing many problems related to urbanization extension, deterioration of agricultural lands, formation of new sabkha and new sand dunes. These findings confirm the original hypothesis of the study.

Overlay of the evolution map and the road network map to localize the urban changes and directions of such change was also performed. The cost of the study was estimated at about US$9/km² and the delivery time of the analysis took about 4 months.

Table 13.2 Land use changes (evolution matrix) for Al-Ahsa oasis between 1987 and 2001

| 2001 | 1987 | | | | | | | Ext. | Evol. |
	A	B	C	D	E	F	G		
A	–	513.1	1.5	21.5	2,745.9	1,499.8	972.3	5,754.1	+5,325.8
B	1,193	–	0	82.4	4,480.4	184.3	1,654.1	6,520.7	+4,430.3
C	6.4	0	–	0	0	0	1.9	8.3	+2.0
D	13.7	1.4	0	–	197.9	0	0	215	–146.4
E	221.2	1,396	0	257.5	–	533.3	4,368.5	6,776.5	–8,063.4
F	14.1	22.1	0	0	193.1	–	827	1,056.3	–1,341.7
G	51.6	157.8	4.8	0	7,222.6	180.6	–	7,617.4	–206.4
Reg.	428.3	2,090.4	6.3	361.4	14,839.9	2,398	7,823.8		

A: urban areas; B: agricultural lands; C: water bodies; D: hills; E: sabkha; F: urban planning; G: sand dunes
Ext. (extension): increase of the area; Reg. (Regression): decrease of the area; Evol. (evolution): extension–regression; unit : hectares

Chapter Summary

Urban and sub-urban landscapes of desert cities are shaped by various factors such as desertification, economic development, wars and conflicts. Furthermore, there are remarkable differences in the level and pace of urbanization between the sub-regions and among developing countries.

Urban areas of desert cities can be mapped out and monitored using remote sensing data and appropriate change detection and analysis techniques. There are several techniques available, of which we concentrated on those more suitable to the technical context and infrastructure of developing countries. These include photo-interpretation techniques assisted by computer to produce the classified imagery maps of land use categories and the comparison of the classified imagery land use changes in urban areas. GIS tools were used to produce the evolution matrix, which presents the whole changes occurred over time and allows for the analysis of these land use changes.

LEARNING ACTIVITIES

Internet Resources

- 2nd FIG Regional Conference 2003, Marrakech, Morocco http://fig.net/conference
- 6th Annual ESRI Middle-East and North Africa User Conference, 2002, Dubai. http://www.gistec.com/mea2002
- Asian Conference for Remote Sensing. http://www.gisdevelopment.net/aars/acrs/2002/luc/luc015shtml
- Third United Nations Conference on the Exploration and Peaceful Uses of Outer Space (UNISPACE-3) 1999, Vienna, Austria. http://www.un.org/events/unispace3/bginfo/activities.htm
- Durrat Al Bahrain Real Estate Project. http://217.17.227.19/sales/

Study Questions

- Discuss with examples some significant environmental and socio-economical problems related to desert cities.
- Discuss the range of potential applications of remote sensing in desert cities.
- There are several land use change detection techniques available. Which one was discussed in this chapter? Which ones may have also proven to be suitable for the case studies presented in this chapter? Why?

- How can you utilize the confusion matrix to understand urban extension, regression and neat evolution?
- What is the difference between the two Figs. 13.4 and 13.5 in terms of the nature of classes mapped?

References

Ait Belaid M (1999) Remote sensing systems for drought and desertification monitoring, the case of Morocco. International archive of photogrammetry and remote sensing, vol XXXII(7C2). Vienna, Austria, pp 104–109

Ait Belaid M (2003) Urabn–rural land use change detection and analysis using GIS and RS technologies. 2nd FIG Regional Conference, Marrakech, Morocco. http://fig.net/conference. Accessed Jan 2004

Ait Belaid M, Al_Rowali M (2002) Investigation of the impact of urbanization on agricultural lands in AL-Ahsa Oasis using geographic information systems and remote sensing technologies. 6th Annual ESRI Middle East & Africa User Conference, Dubai. http://www.gistec.com/mea2002. Accessed Dec 2003

Al Shaikh MN (2002) Application for managing addresses spatially in Kingdom of Bahrain. 6th Annual ESRI Middle East & Africa User Conference, Dubai. http://www.gistec.com/mea2002. Accessed Dec 2003

Anys H, Ait Belaid M, Bijaber N (1999a) Étude de l'Environnement Urbain et Évaluation de l'Impact de l'Urbanisation sur les terres Agricoles. 67ème Congrès de l'Association Canadienne Francophone pour l'Avancement des Sciences (ACFAS), Canada

Anys H, Ait Belaid M, Bijaber N (1999b) Cartographie de l'Évolution du Tissu Urbain et Évaluation de l'Impact de l'Urbanisation sur les terres Agricoles. GéoObservateur, Rabat, Morocco 10:3–11

Asmat A, Mansour S, Hong W (2003) Rule based classification for urban heat island mapping. 2nd FIG Regional Conference, Marrakech, Morocco. http://fig.net/conference. Accessed Jan 2004

Bahr H-P (2001) Remote sensing and urban analysis. In: Donnay J-P, Barnsley MJ, Longley PA (eds) Remote sensing and urban analysis. Taylor & Francis, New York, pp 96–113

Baker JC, O'Connell KM, Williamson RA (2001) Commercial observation satellites at the leading edge of global transparency. A joint Publication of RAND and the American Society for Photogrammetry and Remote Sensing (ASPRS), USA

Benchekroun H (1993) Utilisation de la Télédétection pour le suivi et l'Aménagement de la Ville de Casablanca. GéoObservateur, Rabat 3:61–69

Benchekroun H, Layachi A (1993) Utilisation de la Télédétection et des Systèmes d'Information Géographique dans le Domaine Urbain: Cas de la Wilaya de Rabat-Salé. GéoObservateur, Rabat 3:71–81

CEDARE (1998) Integrating multi-source data to determine urban pattern and changes in north-eastern Cairo. Report CTM-RAC/ERS, MAP/UNEP, Palermo, pp 11–16

Chenje M (ed) (2000) State of environment Zambezi Baasin 2000. SADC, IUCN, ZRA and SARDC, Maseru, Lusaka and Harare, Global Environmental Outlook (GEO-3)

CSO (1999) Statistical abstract 99. Directorate of Statistics-Central Statistics Organization, Bahrain, Global Environmental Outlook (GEO-3)

Donnay J-P, Barnsley MJ, Longley PA (2001) Remote sensing and urban analysis. In: Donnay J-P, Barnsley MJ, Longley PA (eds) Remote sensing and urban analysis. Taylor & Francis, New York, pp 245–258

Doxiadis Associates (1985) Comprehensive development plan for Dubai Emirate, vol 2. Doxiadis Associates, Athens, Global Environmental Outlook (GEO-3)

Fung T, Zhang Q (1989) Remote sensing and methodologies of land use change analysis, vol 6. Occasional Paper, University of Waterloo, pp 135–153

Johnson DB (2003) Un Exemple d'Approche Multisource de l'Etude de l'Occupation du Sol pour l'Analyse de la Dynamique Spatiale sur la Bande littorale du Togo. 2nd FIG Regional Conference, Marrakech, Morocco. http://fig.net/conference. Accessed Jan 2004

KACST (2002) King Abdulaziz City of science and technology. Third Saudi Satellite Launched, KACST news. http://www.kacst.edu.sa/en/script/displaynewsE.asp?news_id=115. Accessed Dec 2003

Kanbour F (1997) General status of urban waste management in West Asia. UNEP Regional Workshop on Urban Waste Management in West Asia, Manama, Bahrain

Lahraoui L (2003) Trace d'une Ceinture Verte pour Lutter Contre l'Ensablement à partir de l'Analyse d'une Image Spot-5. 2nd FIG Regional Conference, Marrakech, Morocco. http://fig.net/conference. Accessed Jan 2004

Martin LRG (1989) Remote sensing and methodologies of land use change analysis, vol 6. Occasional Paper, University of Waterloo, pp 101–113

Mongkolsawat C, Thirangoon P (1990) Land cover change detection using digital analysis of remotely sensed satellite data: a methodological study, ACRS. http://www.gisdevelopment.com/aars/acrs/1990/G/lclu003.shtml. Accessed Dec 2003

Mufareh AR (2002) Investigation of the impact of urbanization on land use in AL-Ahsa oasis using remote sensing and geographic information systems technologies. Master Thesis, Arabian Gulf University, Manama

Nagai M, Shibasaki R, Shaobo H (2002) Reconstruction of long term land cover changes by a maximum likelihood interpolation method using genetic algorithm (ACRS). http://www.gisdevelopment.net/aars/acrs/2002/luc/luc013.shtml Accessed Dec 2003

Smara Y, Ouarab N, Laama S, Cherifi D (2003a) Techniques de Fusion et de Classification Floue d'Images Satellitaires Multisources pour la Caractérisation et le Suivi de l'Extension du Tissu Urbain de la Région d'Alger (Algérie). 2nd International Federation of Surveyors (FIG) Regional Conference, Marrakech, Morocco. http://fig.net/conference. Accessed Jan 2004

Smara Y, Ouarab N, Laama S, Cherifi D (2003b) Detection de la Zone Urbaine et Peri-urbaine de la Région d'Alger (Algérie) a l'Aide de l'Imagerie Radar SAR. 2nd FIG Regional Conference, Marrakech, Morocco. http://fig.net/conference. Accessed Jan 2004

UNCCD (1994) United Nations Convention to Combat Desertification, Paris, France

UNCHS (1996) An urbanizing world: global report on human settlements 1996. Oxford University Press, New York/Oxford, Global Environmental Outlook (GEO-3)

UNCHS (2001) State of the world cities 2001. United Nations Centre for Human Settlements (Habitat), Nairobi, Global Environmental Outlook (GEO-3)

UNDP, UNEP, WB, WRI (1998) World resources 1998–1999. Oxford University Press, New York, Global Environmental Outlook (GEO-3)

UNEP (1991) Status of desertification and implementation of the United Nations of action plan to combat desertification. Report of the Executive Director, Nairobi

UNEP (2002) Global Environmental Outlook (GEO 3), past, present and future perspectives. Earthscan, London

UNESCWA (1999) Survey of economic and social developments in the ESCWA region. United Nations Economic and Social Commission for Western Asia, New York, Global Environmental Outlook (GEO-3)

UN-Habitat and FIG (2002) Land information management for sustainable development of cities: best practices guidelines. In the City-Wide Land Information Management. FIG and UN-Habitat, Publication, N. 31

UNISPACE-3 (1999) Report of the Third United Nations Conference on the Exploration and Peaceful Uses of Outer Space (UNISPACE III). Vienna, Austria, 19–30 July. A/CONF.184/6

UNPD (2001) World urbanization perspects 1950–2050 (the 2000 revision). United Nations Population Division, New York, Global Environmental Outlook (GEO-3)

Weicheng W, Lambin EF, Courel M-F (2002) Land use and cover change detection and modeling for North Ningxia, China, Map Asia Conference. www.gisdevelopment.net/updates/biweekly/bwu081002.htm. Accessed Dec 2003

WHO, UNICEF (2000) Global war supply and sanitation assessment 2000 report. World Health Organization and United Nations Children's Fund, New York, Global Environmental Outlook (GEO-3)

Chapter 14
Remote Sensing of Urban Environmental Conditions

Andy Kwarteng and Christopher Small

Surface temperature and vegetation abundance are two environmental conditions that can be accurately measured from satellites. This chapter gives an overview of the following: (1) urbanization and the urban environment; (2) urban vegetation, surface temperature and public health issues; (3) techniques for urban vegetation mapping; (4) urban thermal mapping; and (5) comparison of urban vegetation and surface temperature and their impact on environmental conditions in New York City and Kuwait City.

Learning Objectives

Upon completion of this chapter, you should be able to:

❶ Speculate on the environmental problems associated with urbanization, including those related to urban heat island
❷ Explain the kind of environmental conditions that can be measured accurately from satellites
❸ Describe and compare some remote sensing techniques for urban vegetation and surface temperature mapping

---

A. Kwarteng (✉)
Remote Sensing and GIS Center, Sultan Qaboos University, P.O. Box 33, Al-Khod, PC 123, Muscat, Oman
e-mail: kwarteng@squ.edu.om

C. Small
Lamont-Doherty Earth Observatory, Columbia University, 108 Oceanography, 61 Route 9W, Palisades, NY 10964-8000, USA
e-mail: small@ldeo.columbia.edu

T. Rashed and C. Jürgens (eds.), *Remote Sensing of Urban and Suburban Areas*,   267
Remote Sensing and Digital Image Processing 10,
DOI 10.1007/978-1-4020-4385-7_14, © Springer Science+Business Media B.V. 2010

14.1 Introduction

Nearly half of the world's human population is now believed to live in urban areas (United Nations 2001). The rate of global population growth is currently decreasing but demographic momentum implies that population growth will continue until at least the year 2100 (O'Neill et al. 2001). Almost all of this population growth is expected to occur in urban areas (United Nations 2001). Widespread urbanization also tends to concentrate this growing population into dense settlements at rates often exceeding 3% per year. Contrary to popular belief, short-term (<50 years) population growth is expected to occur not in megacities (>10 million inhabitants) but rather in moderate sized urban areas of developing countries maintaining high birth rates (United Nations 2001). Furthermore, most of this growth is expected to occur in developing countries where economic conditions may not provide the resources necessary to maintain the integrity of the physical environment.

Urban sprawl and associated large-scale alteration of the natural landscape (Chapter 2) will continue to escalate and have a profound effect on environmental conditions and processes. In addition to challenges presented in the area of land use planning, housing, pollution and development, urbanization has received much of attention worldwide due to implications for changes in microclimate, regional scale climates, and impact of potential sea level rises. There is much ongoing and future research of this phenomenon as a major component of anthropogenic climatic changes. The understanding of relations among urban systems, microclimate, and global scale climates has lead to the creation of several interest groups and much information continues to appear in the literature and on the internet; however coherent syntheses are relatively rare. A major objective of city planners is to ensure a healthy and pleasant environment for inhabitants and avoid any harmful repercussion from any large-scale changes. Any effective mitigation techniques should be based on an effective long-term environmental monitoring because of the constant change of urban morphology and environmental conditions.

the environmental implications of urbanization include effects on microclimate, regional scale climates, and impact of potential sea level rises

Remote sensing provides an invaluable tool for the long-term monitoring of urban growth and surface conditions. The synoptic view of urban land cover provided by satellite-based sensors is an important complement to in situ measurements of physical and environmental conditions in urban settings. Other advantages of remote sensing applications include the combination of cost-effectiveness, non-destructiveness and relative high spatial resolution, repetitive data at the same scale for a particular satellite, digital format, and acquisition of imagery from inaccessible areas without the hindrance of political or security restrictions. Forster (1983) provides a thorough summary of the early evolution of urban remote sensing and introduces a methodology with which some socioeconomic parameters may be predicted using reflectance based estimates of land cover classes. Compared to other land use categories however, application of remotely sensed observations to studies of the urban environment has been very limited. In part, this is because accurate identification of most built components of the urban environment requires finer spatial resolution (see Chapters 5 and 6) than has traditionally been available from operational satellites such as the Landsat or Satellite Pour d'Observation de la Terre (SPOT) sensors. The 30 m spatial resolution of the Landsat Thematic Mapper (TM) sensor is comparable to the characteristic scale of urban land cover (Welch 1982), but is generally too coarse for identification of individual structures. While this resolution has limited Landsat and similar satellites' use for studies of the built urban environment, it is sufficient to detect significant spatial and temporal variations in urban land cover and surface conditions – specifically vegetation, albedo, and surface temperature.

the synoptic view of urban land cover provided by satellite-based sensors complements in situ measurements of physical and environmental conditions in urban settings

The most successful applications of remote sensing to the urban environment generally involve measurement of physical quantities related to environmental conditions such as vegetation abundance and surface temperature. In spite of the great value that remotely sensed measurements can provide to the study of urban environments, it is important to acknowledge the limitations of these tools. While remotely sensed imagery provides an invaluable reconnaissance and monitoring tool, there are often other sources of environmental information that are more accurate and informative (Miller and Small 2003). This chapter explains the rationale and techniques for urban vegetation and surface temperature mapping from satellite imagery, and illustrates the differences in two cities, New York and Kuwait City, by comparing their surface properties and their influence on energy flux using Landsat 7 imagery.

14.2   Urban Physical Environment

The physical environment of a large city is profoundly different from the physical environment of small settlements in rural settings. The spatial agglomeration of built surface in most cities has different physical characteristics from most naturally occurring environments. These physical characteristics influence environmental conditions by changing the flux of mass and energy through the environment. Most of the inhabited areas on Earth are characterized by soil and vegetation but built environments are dominated by more impervious surface and have relatively small amounts of vegetation. Soil retains moisture and allows for continuous evaporation to the atmosphere while impervious surface increase runoff and significantly reduce evaporative moisture flux in the urban environment.

built environments are dominated by more impervious surface and have relatively small amounts of vegetation

14.2.1   Importance of Urban Vegetation

The abundance and spatial distribution of vegetation has a strong influence on the urban and suburban environmental conditions. Vegetation influences energy fluxes by selective reflection and absorption of solar radiation. **urban trees can** Vegetation has a significant cooling effect because it absorbs **be thought of as** much of the incoming solar radiation and dissipates the **in situ filters as** energy by transpiring water rather than converting the energy **they lower** to heat and reradiating it as built surfaces do (Goward et al. **temperature,** 1985; Gallo et al. 1993; Price 1990; Carlson et al. 1994; **improve the air** Gillies et al. 1997; Owen et al. 1998). This results in differ- **quality, and** ent fluxes of moisture and solar radiation, influences com- **reduce the** fort levels, and ultimately result in energy savings from **amount of** cooling. In addition to providing shade and wind shelterbelts, **urban smog** urban trees can lower the ambient temperature around a building or in a park, improve the air quality as well as reduce the formation of urban smog (Akbari 2002). In cold climates, trees shield buildings from cold winter wind and thereby save energy on heating energy. In warm weather, well-planned landscaping of trees and shrubs can significantly reduce the daily air-conditioning electricity consumption by as much as 50% (Parker 1981). Well-maintained urban vegetation in the form of trees, grasses, and flowers is visually pleasing, particularly in arid environments where native vegetation is limited by harsh environmental conditions.

14.2.2   Urban Microclimates

Although urban areas currently occupy only ~2% of the Earth's habitable land area (Small et al. 2005), their energy fluxes have a considerable impact on microclimate and ecosystem function at local and regional scales (Berry 1990; Landsberg 1981). Urban nitrogen emissions (NO_x) have been shown to have an adverse impact on regional agricultural productivity (Chameides et al. 1994). There is some evidence to suggest that urban heat island effects are responsible for inducing atmospheric convergence sufficient to influence thunderstorm formation and movements observed near large urban areas (Bornstein and LeRoy 1990; Bornstein and Lin 2000). Regional climate models also indicate a strong sensitivity to land cover variations at scales of kilometers (Pielke et al. 2002; Roy and Avissar 2000; Li and Avissar 1994) suggesting that urban land cover conversion may significantly modify the physical environment beyond the city itself.

14.2.3   Public Health Issues

The land surface modifications associated with urban growth impact a variety of public health issues. We focus here on those related to the thermal environment and air quality.

Reduction of vegetation cover impacts both of these environmental conditions. Removal of vegetation perturbs both mass and energy fluxes through the urban environment. The net results are generally higher nighttime temperatures and higher ozone and particulate concentrations in the air. Higher nighttime temperatures have a well-documented effect on heat stress during heat events (Curriero et al.

higher nighttime temperatures may cause heat-induced illnesses and deaths in large cities

2002). Greater absorption of solar radiation by solid surfaces results in time delayed re-radiation of the energy as heat. This, in turn, results in increased nighttime temperatures. High nighttime temperatures are believed to contribute to the physical stress of high daytime temperatures by forcing the human body to devote energy to cooling for sustained periods of time. Heat-induced illnesses and deaths in large cities have been noted at least since the early part of this century (Gover 1938).

The presence and abundance of vegetation in urban areas may also influence air quality and human health (Wagrowski and Hites 1997). Air pollutants, such as ozone and fine (~2.5 µm) particulate matter ($PM_{2.5}$), have measurable adverse public health impacts (Kinney 1999). The primary sources of these pollutants are motor vehicles, power generation plants and heavy industry. Meteorology also influences the production of some pollutants, with higher temperatures promoting the atmospheric reactions which form ozone and secondary fine particles. Because vegetation provides abundant surface area in contact with the atmosphere, the leaves serve as sinks for these surface reactant pollutants. Through leaf stomata, trees can directly remove pollutant gases such as CO, NO_x, O_3 and SO_2 (Smith 1984; Fowler 1985; Nowak 1994; Taha et al. 1997). Urban trees play contributing roles in sequestering CO_2 and consequently reduce global warming (Akbari 2002). In this sense, urban vegetation serves as in situ air filters. On the other hand, urban vegetation experiences both short- and long-term phenological changes and may itself be sensitive to subtle changes in environmental conditions. While changes in the built component of the urban environment are often documented at some level of detail, phenological changes in urban vegetation are not under direct human control and are not generally monitored.

Remote Sensing and Urban Environmental Conditions

Information derived from airborne and/or satellite remote sensing, which is instantaneous and does not require time synchronization, could be valuable to city planners in establishing effective landscape policies, zoning, and greenification. Thermal infrared remote sensing offers an unparalleled technique to measure radiant temperature, which is essential for the understanding of all physical, biological, and chemical systems on the Earth, including urban climate (Norman et al. 1995). The use of thermal observations have been limited to some extent due to the difficulty in obtaining emissivity values for the various surface materials as well as the incorporation of other variables that affect thermal radiance and energy partitioning at the surface in thermal models. Nevertheless, several useful studies on temperature and urban areas have been conducted (Oke 1987; Voogt and Oke 2003). From such studies, urban heat island and any amelioration methods could be studied and monitored continuously and linked to land use patterns. For example, city planners would like to know the effect of temperature and air pollution from the layout of parks, water bodies, industrial and commercial areas. The technology of using airborne and spaceborne thermal infrared sensors to measure surface temperature of urban areas has been available since the 1970s. Such studies entail the conversion of thermal infrared data into surface temperature.

A major obstacle encountered with the use of satellite imagery in environmental studies has been the relatively coarse spatial resolution (see Chapter 5). However, as discussed in Chapter 7, the spatial resolution of satellite imagery has increasingly become finer over the last three decades. Starting from Landsat Multispectral Scanner (MSS) with a resolution of 79 m per pixel in the early 1970s, presently QuickBird satellite has a spatial resolution of 0.61 m. In parallel with the improved spatial resolution is the improved spectral resolution. Landsat MSS has five bands in the visible and near-infrared regions, but presently the Hyperion satellite has 220 bands between 0.4 and 2.5 μm. However, the improvement in the spatial resolution of thermal bands do not parallel the advances in the reflective bands and has been relatively lower than other regions – due to different technologies used for the visible, near-infrared, shortwave and thermal infrared regions. At the moment, the highest spatial resolution for satellite thermal infrared imagery is 60 m per pixel from Landsat Enhanced Thematic Mapper Plus (ETM+). The Advanced Very High Resolution Radiometer (AVHRR) has a spatial resolution of 1.1 km per pixel (at nadir) and has been utilized in several large-scale urban analyses (Roth et al. 1989; Owen et al. 1998).

14.3   Urban Vegetation Mapping

Although vegetation mapping is one of the primary uses of optical remote sensing data, relatively little work has focused on urban vegetation. In part, this is a result of the greater diversity of natural vegetation mapping applications but it is also a result of the characteristic scales of urban vegetation relative to the spatial resolution of operational sensors. Recent comparative analysis of IKONOS 1 m resolution imagery in several urban areas worldwide indicates that the characteristic scale of individual features in urban mosaics is between 10 and 20 m (Small 2004). Until relatively recently, SPOT and Landsat with 20 and 30 m multispectral spatial resolutions, respectively, served as the primary tools for fine scale vegetation mapping. As a result of the low resolutions, SPOT and Landsat image urban areas primarily as spectrally mixed pixels. Although vegetation has a strong, and measurable influence on the mixed spectra that these sensors image, it is not amenable to the thematic classification tools that are usually used for mapping more spectrally homogeneous land covers like forests and agriculture.

14.3.1   Normalized Difference Vegetation Index

The most commonly used technique for vegetation analysis is the normalized difference vegetation index (NDVI), which is computed as:

$$NDVI = \frac{\rho_{NIR} - \rho_{red}}{\rho_{NIR} + \rho_{red}} \qquad (14.1)$$

where ρ_{NIR} and ρ_{red} are reflectance values derived from the near-infrared and red channels, respectively. The ratio is a measure of the deviations between chlorophyll absorption minimum and the infrared plateau, and thus, an indirect proxy for the amount of photosynthetically active green biomass (Tucker and Seller 1986). NDVI values can be computed for Landsat TM and MSS, SPOT and AVHRR imagery. In spite of its popularity, the NDVI suffers from several shortcomings. The index generally saturates for areal vegetation fractions higher than 40–50% (Elmore et al. 2000; Small 2001) and is non-associative (Price 1990) – limiting its utility for characterizing scale dependent vegetation abundance. It is also sensitive to spectral band definitions – complicating comparisons of indices measured by different sensors (Price 1987). Sensors, such as Landsat TM and ETM+, Advanced Spaceborne Thermal Emission and Reflection Radiometer (ASTER) and Moderate Resolution Imaging Spectroradiometer (MODIS) provide sufficient spectral resolution to be used for spectral mixture analysis as described in the next section.

vegetation indices are not well suited to mapping vegetation within the urban mosaic because of the heterogeneity of urban vegetation distribution and mixed pixels of coarse multispectral images

NDVI and other vegetation indices usually map vegetation cover as a continuous quantity (as opposed to a thematic class) but these indices are not well suited to mapping vegetation within the urban mosaic because of the heterogeneity of urban vegetation distribution. In addition to the difficulty in establishing a quantitative calibration between areal vegetation abundance and NDVI at sub kilometer scales, it has been shown that NDVI increases nonlinearly with increasing vegetation fraction (Small 2001).

14.3.2   Spectral Mixture Analysis

Spectral mixture analysis (SMA) provides a physically based methodology to quantify spectrally heterogeneous urban reflectance. SMA is based on the observation that, in many situations, radiances from surfaces with different "endmember" reflectance mix linearly in proportion to area within the instantaneous field of view (IFOV) (Singer and McCord 1979; Singer 1981; Johnson et al. 1983). This observation has made possible the development of a systematic methodology for SMA (Adams et al. 1986, 1989; Smith et al. 1990; Gillespie et al. 1990) that has proven successful for a variety of quantitative applications with multispectral imagery (e.g. Adams et al. 1995; Pech et. al. 1986; Smith et al. 1990; Elmore et al. 2000; Roberts et al. 1998). The linear mixing model can be expressed as:

$$R_i = \sum_{k=1}^{n} f_k R_{ik} + \varepsilon_i \text{ and } 0 \le \sum_{k=1}^{n} f_k \le 1 \qquad (14.2)$$

where R_i is the spectral reflectance for band i for each pixel; n is the number of endmembers; f_k is the fraction of endmember k; R_{ik} is the reflectance of endmember k in band i; and ε_i the difference between measured and modeled digital number in band i. A unique solution for Eq. 14.2 is obtained by minimizing the residual error, ε_i, in a least-square solution, given the constraints on f_k.

Endmembers can be selected from either a spectral library derived from laboratory or field measurement, or from representative homogeneous pixels in a satellite image. If a limited number of distinct spectral endmembers is known *a priori* it is possible to define a "mixing space" within which mixed pixels can be described as linear mixtures of the endmembers. Given sufficient spectral resolution, a system of linear mixing equations can be defined and the best fitting combination of endmember fractions can be estimated for the observed reflectance spectra. The strength of the SMA approach lies in the fact that it explicitly takes into account the physical processes responsible for the observed radiances and therefore accommodates the existence of mixed pixels.

spectral mixture analysis provides a physically based methodology to quantify spectrally heterogeneous urban reflectance within mixed pixels

A comparative SMA of Landsat ETM+ imagery for a diverse collection of 28 cities highlights several consistencies in the spectral properties of these cities (Small 2004). The analysis demonstrates that all of these cities have similar triangular mixing spaces in which urban reflectance can be accurately represented as linear mixtures of high albedo substrate, vegetation, and dark surfaces. An analysis of high-resolution IKONOS imagery for 14 cities yields very similar results suggesting that a simple three-endmember spectral mixture model provides a multiscale physical representation of the reflectance of urban mosaics. Areal vegetation fraction estimates for these urban areas agree with high-resolution vegetation fraction measurements to within 10% in New York (Small 2001, 2004).

Several recent studies have used physical rather than statistical classifications of urban land cover in individual cities with some degree of success (e.g., Kressler and Steinnocher 2001; Rashed et al. 2002; Small 2001, 2004; Wu and Murray 2003; Lu and Weng 2004). SMA is quite useful in the analysis of Landsat TM and SPOT images of urban areas, however it is important to acknowledge that with the significant increase of high spatial resolution satellite imagery during the last couple of years, textural and image segmentation approaches will play important roles in urban land cover classification.

14.4   Urban Thermal Mapping

14.4.1   *Principles of Thermal Radiation*

A basic understanding of thermal radiation is necessary for the interpretation of thermal imagery acquired by scanners. The region beyond 3 μm of the electromagnetic spectrum is characterized by emitted electromagnetic energy from natural materials which can be related to temperature according to the Stefan–Boltzmann and Wien's laws. The total energy emitted by a blackbody is a function of the surface temperature as expressed by Stefan–Boltzmann law, which states that:

$$M = \sigma T^4$$

(14.3)

where M is total radiant emittance energy from the radiant surface of a material; watts (W) m^{-2}; σ is Stefan-Boltzmann constant, 5.6697×10^{-8} W m^{-2} K^{-4}; and T is the absolute temperature in K of the emitting material. Wien's displacement law establishes the relations between the wavelength at which a blackbody radiation reaches a maximum and its temperature in the following equation,

$$\lambda_m = \frac{2898}{T}$$

(14.4)

where λ_m is the dominant wavelength in micrometers, T is temperature in K. Both laws (Eqs. 14.3 and 14.4) are related to the ideal behavior of a blackbody; however, natural materials are not perfect and only approximate blackbodies.

Due to the Earth's absorption of certain wavelength by gases, or the so-called absorption window, remote sensing of the thermal infrared region is limited to 3–5 μm and/or 8–14 μm of the electromagnetic spectrum. The maximum Earth's energy at 300 K is emitted at 9.7 μm. Radiometers and scanners can be used to measure thermal energy from the Earth's surface at anytime of the day or night. All objects above 0 K, or −273°C, emit energy whose spectral composition and intensity are characteristic of the composition and kinetic temperature of the target. The digital numbers in a thermal image are measures of the radiant energy which is a function of the emissivity of the different targets. A blackbody has an emissivity of 1 whereas all natural materials have values between 0 and 1. Consequently, the temperature derived from the thermal infrared using the Wien's displacement law and Stefan–Boltzmann law is less than the true surface temperature. The ability to derive the 'true' temperature of an object depends on the applying the approximate emissivity values from the blackbody laws. Materials that have high emissivity absorb and emit relatively larger proportions of incident energy.

14.4.2 Retrieval of Thermal Data from Satellite Imagery

Brightness temperatures (also referred to as blackbody temperatures) can be derived from satellites' thermal infrared measurements through Planck's law (Flynn et al. 2001; Dash et al. 2002). The digital numbers of Landsat TM and ETM+ thermal infrared band 6 (10.4–12.5 μm) are converted into radiance using the equation:

$$L_\lambda = gain * DN + offset \tag{14.5}$$

where L_λ is at sensor radiance, DN is the digital number of a pixel, $gain$ is slope of the radiance/DN conversion function in Wm^{-2} sr^{-1} μm^{-1}, offset is the rescaled bias which is the intersection of the radiance/DN conversion function in Wm^{-2} sr^{-1} μm^{-1} (Landsat Project Science Office 2004). Each Landsat TM scene is accompanied by gain and offset values as part of the metadata. The TM band 6 spectral radiance values are subsequently transformed to surface temperature values using the relationship:

$$T_s = \frac{K2}{In\left(\dfrac{K1}{L_\lambda} + 1\right)} \tag{14.6}$$

where T_s the radiant surface temperature in $K1$ and $K2$ are thermal calibration constants in Wm^{-2} sr^{-1} μm^{-1} supplied by the Landsat Project Science Office (2004), and L_λ is spectral radiance of thermal band pixels in Wm^{-2} sr^{-1} μm^{-1}. For Landsat 7, K1 is 666.09 and K2 is 12822.71, and for Landsat 5 K1 is 607.76 and K2 is 1260.56.

The composite emissivity values for urban mosaics are rarely known, but usually assumed to be near 1. The typical emissivity values for man-made surfaces such as concrete and asphalt range from 0.95 to 0.97 (Buettner and Kern 1965). Here, it is important to acknowledge the ambiguity introduced in urban surface temperature distribution by incomplete knowledge of surface emissivity.

14.4.3 Urban Heat Island

The phenomenon commonly referred to as urban heat island, results from the inadvertent urban climate modification from anthropogenic activities (Oke 1987). Urban materials such as construction material, roofs, asphalt, concrete and roads absorb more heat from the sun. The subsequent release of the energy causes urban areas to be warmer compared to the surrounding non-urban areas giving rise to the urban heat effect (Oke 1987; Gallo et al. 1995; Lo et al. 1997; Owen et al. 1998; Nichol 2003; Voogt and Oke 2003). In addition, the non-porous human-made materials give rise to reduced evapotranspiration and more rapid runoff of rainwater (Kim 1992). Urban heat island has been observed and documented for more than one and half centuries (Howard 1833; Oke 1987). Depending on the location, the urban heat island effect can be exacerbated by the absence or removal of vegetation, which provides shade and evapotranspiration to cool the air on warm days.

urban materials such as construction material, roofs, asphalt absorb more heat from the sun and subsequently release this energy causing urban areas to be warmer compared to the surrounding non-urban areas giving rise to the urban heat effect

The dynamics of the heat island effect are a function of the time, meteorological conditions, local and urban characteristics, and consequently, could be unique for particular urban areas. In general, parks, lakes and open areas, appear relatively cooler compared to commercial, industrial or dense buildings (Roth et al. 1989). The urban heat intensity, $\Delta T(u - r)$, is the difference between the urban maximum temperature (u) and the non-urban low temperature (r), and is controlled by the unique characteristics of particular urban and suburban areas. For example, in cities with tall buildings, the three-dimensional structures alter the airflow that could reduce heat loss resulting in higher temperatures (Oke 1987; Nichol 1996, 1998).

Why should urban planners, environmentalists, and residents worry about urban heat island? There are positive as well as negative effects from urban heat island. Positive contributions include the potential for early budding and blooming of flowers and trees in urban compared to non-urban areas, and the reduction in winter heating. The adverse effects, which appear to outweigh any positive gains, include stress for humans as the summer discomfort levels rise, higher demand for air-conditioning in the summer, and the speeding up of the process of chemical weathering of building of materials, especially in the tropical and other low latitude areas. Air conditioners use energy in the form of fuel and electricity and result in increases in the emission of SO_2, CO, CO_2 and NO_2, which contribute to global warming and climatic change (Akbari 2002). In temperate climates, the use of air-conditioning could result in the formation of fog, which could raise the level of pollution in the air (Rosenfeld et al. 1995).

Traditionally, urban heat island effects are observed from the measurement of air temperatures from fixed weather observation stations and sensors mounted on moving vehicles. The urban heat island effect is best developed in the night under cloudless skies and calm winds. Any urban heat intensity gradients are obliterated by strong winds. In ideal conditions, a village or small town with a population of over 1,000 inhabitants exhibit a heat island effect (Oke 1987). Large cities may exhibit higher heat intensities than smaller ones, but in general the morphology of a city ultimately

the dynamics of the heat island effect are a function of the time, meteorological conditions, local and urban characteristics

determines the nature for an area. Urban heat island intensity, $\Delta T(u - r)$, of up to 12°C has been recorded under ideal conditions (Roth et al. 1989).

One of the earliest applications of spaceborne measurements was for surface temperature and its relationship to the urban heat island effect and urban climate. Rao (1972) is credited with the first study of urban heat islands from an environmental satellite. Since then, several other studies have utilized the thermal infrared data from the AVHRR, Heat Capacity Mapping Mission (HCMM), Landsat TM, TIROS Operational Vertical Sounder (TOVS). Carlson et al. (1977) used satellite-derived measurements of surface temperature to investigate the relationship between urban land use and heating patterns. Other studies have used satellite measurements of both reflected and emitted infrared radiation to quantify the relationship between urban land use and the urban heat island effect (e.g., Owen et al. 1998; Roth et al., 1989; Gallo et al. 1993; Gillies et al. 1997). At finer scales, Nichol (1994) used Landsat TM thermal imagery to quantify the effect of solar radiation on the microclimate in Singapore. Comprehensive summaries of remote sensing application to urban heat island are presented in Gallo et al. (1995), Roth et al. (1989), and Voogt and Oke (2003).

Within the urban environment, the abundance and distribution of vegetation plays an important role in controlling temperatures and air quality (Akbari et al. 1996, 2001; Nowak et al 2000). To ameliorate the effect from urban heat island, several urban areas have recommended the planting of trees and grasses and the use of reflective building and roofing material and pavement surfaces. Taha et al. (2000) observed that through the impact of vegetation and reflected roofs and pavement, the daytime ambient temperature could be reduced between 1 and 2 K.

14.4.4   Relations Between Satellite Thermal Measurements and In Situ Air Temperatures Observations

A major objective of researchers is to derive measurements that are meaningful or familiar to human interpreters or city planners, and thus, the tendency is to compare satellite-derived measurements to those acquired using *in situ* techniques. However, the fundamental differences between the two datasets cannot be over-looked and any attempt to establish direct correlations is not always straightforward (Price 1979; Vukovich 1983; Roth et al. 1989). Satellite measurements are unique and have some attractive features, which include dense grid synoptic view that eliminates synchronizing of data. The differences between traditional and satellite-derived thermal measurements of urban areas have been discussed extensively in the literature (Roth et al. 1989; Voogt and Oke 2003; Nichol 2003). While attempts have been made by some researchers to compare the two datasets, others have warned against using the two datasets as surrogates due the inherent differences and the measured surfaces

in situ and satellite-derived thermal measurements of urban areas inherent fundamental differences that must be acknowledged in the analysis

(Carlson and Boland 1978; Roth et al. 1989). Nichol (1996) recorded a strong correlation between air measurements and satellite measurements in low wind environments in Singapore.

The majority of satellite measurements are restricted to the late mornings to optimize conditions for the visible and near infrared regions. One exception is NOAA AVHRR which records temperature in the night, but has a coarse resolution of 1.1 km per pixel at nadir. During mornings, urban materials will slowly absorb heat. Urban heat islands are best observed in the night when maximum radiation from urban materials occurs. The timing for satellites is therefore not optimized for detecting urban heat islands, and perhaps more prone to the observation of heat sinks because of the heat absorption behavior of urban surfaces (Nichol 1996). The fact that satellite measurements are recorded only at particular times makes the data suitable for comparative studies over long periods. However, the behavior of urban heat or any environmental parameter during other times of the day, apart from mornings, cannot be obtained from satellite data. Studies have shown that the urban heat effect is dynamic over a 24-h period (Roth et al. 1989; Jauregui 1997).

14.5   Comparison of Environmental Conditions in New York City and Kuwait City

In this section, we present a case study to demonstrate environmental analysis of two cities using satellite imagery. We illustrate differences in surface properties and their influence on energy flux by comparing Landsat 7 imagery acquired for New York City and Kuwait City. Surface temperature distribution and vegetation fraction, or the areal proportion of vegetation within a pixel, distribution can be measured accurately and provide a synoptic view of spatial variations in environmental conditions related to albedo, emissivity and evapotranspiration. These spatial variations are a primary determinant of the environmental conditions that influence human comfort levels. Composite surface temperature measurements are influenced by the heterogeneity of materials within the sensor IFOV just as optical measurements are. While spectrally mixed pixels can be "unmixed" if spectral endmembers are known, the same procedure cannot generally be applied to thermal data as most satellite sensors collect only a single thermal band and there is no evidence that thermal spectra mix linearly. Consequently, most analyses of thermal data assume that the target within the IFOV is thermally homogeneous and has uniform emissivity, which is assumed to be near 1. However, when thermal imagery is used in conjunction with optical multispectral imagery it is possible to interpret the distribution of surface temperatures in the

spectral mixture analysis cannot generally be applied to thermal data as most satellite sensors collect only a single thermal band and there is no evidence that thermal spectra mix linearly

context of endmember abundances estimated by the spectral mixture analysis. This is the approach used here.

Kuwait City and most parts of the Arabian Shield are characterized by a desert environment with scanty rainfall, and a dry, hot climate. Spring (January–March) temperatures in Kuwait are generally low and quite pleasant compared to the summers, especially July with a mean temperature of 37.4°C and a maximum mean temperature of 45°C. The March average temperature for Kuwait City for the last 50 years is 19.3°C, with a maximum mean temperature of 25.6°C and a mean minimum temperature of 13.2°C. Beyond the limits of Kuwait City, the suburban area consists of flat undulating desert with more than 50% eolian sand surface deposits. During the spring rain season the area supports the growth of ephemeral vegetation. However, desert areas are heavily overgrazed by camels, sheep, and goats, leaving the soil nearly bare most of the time. Local landscaping, greening, and beautification projects were initiated in the 1960s as a source of national pride and to enhance economic productivity.

New York City, in contrast, is characterized by a temperate climate modulated by the thermal inertia of the Atlantic Ocean. Annual temperatures generally range from −20°C to 40°C with seasons dictated by temperature rather than precipitation cycles. Contrary to popular belief, the New York metro area is characterized by abundant urban and suburban vegetation in the form of mature deciduous street trees and numerous parks and public green spaces.

Vegetation fraction images were generated from a three-component mixing model based on high albedo substrate, vegetation and dark surface that is physically consistent with the spectral characteristics that might be expected for an urban environment (Small 2001). Fractional abundance images resulted from a unit sum constrained least squares inversion of the linear mixing model using the spectra of endmembers. The surface brightness temperatures were derived from the Landsat ETM+ thermal band 6 as explained in Section 14.4.

Figure 14.1 shows vegetation fraction and surface temperature images for Kuwait City and New York City derived from Landsat 7 imagery. The Kuwait City image was acquired under desert springtime conditions on March 6, 2001, while the New York image was acquired under late summer drought conditions on August 6, 2001. The datasets have been subjected to the same processing and enhancement techniques and therefore the dark and light tones in the vegetation fraction and surface temperature images are comparable. The only exception is the Kuwait City vegetation fraction image in which the brightness to contrast ratio has been increased by 30% so that features are observable. From Fig. 14.1, the distribution of vegetation fractions observed in Kuwait City is generally lower than that in New York City which has abundant vegetation in the form of large deciduous trees, wetlands and closed canopy forest in parks and cemeteries. In the New York City image, parks and public greenspace are easily distinguished. In addition, interurban differences in street and courtyard vegetation can be consistently detected. Vegetation in Kuwait City consists of palm and shade trees, shrubs, groundcovers, and grass, which are irrigated year round (Kwarteng 2002a, b). In spite of the low vegetation cover observed in Kuwait City, the relative amounts of the vegetation fractions in different

Fig. 14.1 Vegetation fraction and surface temperature images derived from Landsat 7 images for Kuwait City and New York City recorded on March 6, 2001 and August 14, 2002, respectively. Each scene is 30 × 30 km. Grayscale ranges from 0% to 100% for vegetation fractions (*left*) and from 285 to 315 K (12–42°C) for surface temperature (*right*). Water bodies in the two images, namely the Arabian Gulf and the Upper New York Harbor, the Hudson River and its tributaries have zero vegetation fractions and the minimum temperatures and appear in dark tones in the images. The brightness to contrast ratio in the Kuwait City vegetation image has been increased by 30% so that features are visible

residential areas can be detected in the images. Even though New York City has abundant vegetation, the surface temperatures are higher than Kuwait City, primarily due to the different seasons. Notwithstanding the lower vegetation fractions, the cooling effect is apparent in the clearly defined residential areas in the Kuwait City image.

Each of the scatterplots of surface temperature and vegetation fractions for New York City and Kuwait City show a cloud of a triangular distribution of pixels

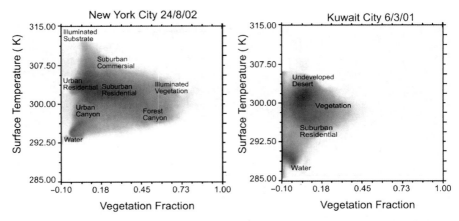

Fig. 14.2 Scatterplots of Surface Temperature and Vegetation Fraction for New York City and Kuwait City. Darker areas correspond to greater number of pixels. Low temperature, unvegetated areas generally correspond to water bodies. Note the Arabian Gulf is cooler in March than in New York Harbor and the surrounding rivers in August. New York also has higher temperatures, in spite of abundant vegetation, because the image was acquired during drought conditions when evapotranspiration was low

(Fig. 14.2). Pixels near the base of the triangle, which is parallel to the surface temperature axis, attain a wider range of temperatures compared to the pixels in the vertex of the triangles. Dark areas in Fig. 14.2 represent higher pixel density than lighter tones.

According to Carlson et al. (1995), the warm edge is a sharply defined boundary representing the locus of highest temperatures and different vegetation fractions. On the other hand, the cold edge represents the locus of lowest temperatures for the vegetation fractions. The triangular shape distribution in the scatterplots defines the physical limits imposed by the vegetation cover, soil water content and different combinations of surface materials. In rural areas, the shape of the Temperature/Vegetation (TV) distribution results from spatial variations in vegetation cover and soil moisture availability (Gillies et al. 1997; Crombie et al. 1999). In urban areas, there is generally little exposed soil so the shape of the distribution is dictated by variations in vegetation cover, albedo and shadow. The TV distributions for New York and Kuwait City do not show the familiar triangular distributions usually seen in rural areas. This is primarily a result of urban land cover heterogeneity and the negligible fraction of exposed moist soil. The warm edge shows a less pronounced cooling effect in New York as the image was acquired during drought conditions with low moisture availability for many areas. The absence of a well-defined cold edge is a consequence of extensive area of the partial shadow at all vegetation fractions. The pervasive presence of shadow in the urban mosaic is a consequence of the characteristic 20–30 m spacing of trees, streets and buildings (Small 2003) coinciding with the spatial scale of the Landsat IFOV. The increase in minimum temperature is more pronounced in the Kuwait City because of the relatively low water temperatures when the image was recorded in the morning of March 6, 2001. Slightly negative vegetation fractions in the Kuwait City image correspond to bare soil that are not accurately represented in the three endmember mixture model. The vegetation fractions in New York City are generally lower than 0.75 as most of the vegetation takes the form of deciduous trees which have as much as 25% internal shadow from canopy structure. In contrast, vegetation fractions in Kuwait City are less that 0.45 because most of the vegetation patches are much smaller than the 30 m IFOV of the ETM+ sensor.

More land cover categories, identified by locating the geographic position of the pixels in each part of the mixing space, are observed in the New York City than the Kuwait City image primarily due to the morphology and location of the two cities as mentioned previously. Apart from a few tall buildings in the central business district area in the upper-central part along Kuwait Bay, the majority of houses in Kuwait City consists of one or two storey buildings which could be residential, commercial or both. The study area in New York City on the other hand consists of several residential areas with 2–20 story buildings and commercial districts containing tall buildings creating urban canyons.

Chapter Summary

The world's population growth in urban areas will continue to escalate and have a profound effect on environmental conditions and processes. Urban sprawl presents several challenges in the area of land use planning, housing, pollution and development, and changes in microclimate. The most successful applications of remote sensing to the urban environment generally involve measurement of physical quantities related to environmental conditions such as vegetation abundance and surface temperature. Urban data recorded by operational satellites such as Landsat and SPOT consist of spectrally mixed pixels because of the relatively coarse satellite resolution than individual features in urban mosaics. Spectral mixture analysis provides a physically based system to quantify spectrally heterogeneous urban reflectance. Vegetation fraction images are generated from a three-component mixing model based on high albedo substrate, vegetation and dark surface that is physically consistent with the spectral characteristics that might be expected for an urban environment. Surface temperatures can be derived from satellites' thermal infrared measurements through Planck's law. Surface temperature distribution and vegetation fraction analysis using Landsat ETM+ data reveal the different energy flux and surface properties for New York City and Kuwait City, located in temperate and desert environments, respectively. Scatterplots of surface temperature and vegetation fractions define the physical limits imposed by the vegetation cover, soil water content, and different combinations of surface materials in each city.

LEARNING ACTIVITIES

Study Questions

- How does urban vegetation influence comfort level, especially in warm climates?
- What are the advantages and disadvantage of satellite-derived thermal measurements and *in situ* air temperatures observations for an urban area?
- How do you expect the urban heat island effect in a typical temperate city like New York City to vary compared to an arid city like Kuwait City as both cities expand over a period of time?
- Why is spectral mixing analysis (SMA) more appropriate and useful for urban vegetation mapping than the normalized difference vegetation index (NDVI)?
- From a hypothetical triangular scatterplot generated from surface temperature and vegetation fraction, sketch and discuss the trend of urbanization?
- Discuss the pros and cons of spectral mixture techniques and non-spectral mixture techniques for urban vegetation mapping?
- How does the physical environment in urban area alter the flux of mass and energy compared to the surrounding non-urban areas?
- Discuss the major public issues associated with urban vegetation and urban thermal environments?

References

Adams JB, Smith MO, Johnson PE (1986) Spectral mixture modeling: a new analysis of rock and soil types at the Viking Lander 1 site. J Geophys Res 91:8089–8122

Adams JB, Smith MO, Gillespie AR (1989) Simple models for complex natural surfaces: A strategy for hyperspectral era of remote sensing, vol 1. In: Proceedings of IGARSS 1989, Vancouver, Canada, pp 16–21

Adams JB, Sabol DE, Kapos V, Filho RA, Roberts DA, Smith MO, Gillespie AR (1995) Classification of multispectral images based on fractions of endmembers: application to land cover change in the Brazilian Amazon. Remote Sens Environ 52:137–154

Akbari H (2002) Shade tees reduce building energy use and CO emissions from power plants. Environ Pollut 116:S119–S126

Akbari H, Rosenfeld A, Taha H, Gartland L (1996) Mitigation of summer urban heat islands to save electricity and smog. In: Proceedings of the 76th annual American meteorological society meeting, Atlanta, GA, 28 Jan–2 Feb 1996. Report No. LBL-37787, Lawrence Berkeley National Laboratory, Berkeley, CA

Akbari H, Pomerantz M, Taha H (2001) Cool surfaces and shade trees to reduce energy use and improve air quality in urban areas. Solar Energy 70(3):295–310

Berry BL (1990) Urbanization. In: Turner BL, Clark WC, Kates RW, Richards JF, Matthews JT, Meyer WB (eds) The Earth as transformed by human action. Cambridge University Press, Cambridge, pp 103–119

Bornstein R, LeRoy M (1990) Urban barrier effects on convective and frontal thunderstorms. Preprint volume, Fourth AMS Conference on Mesoscale Processes, Boulder, CO, 25–29 June

Bornstein R, Lin QL (2000) Urban heat islands and summertime convective thunderstorms in Atlanta: three case studies. Atmos Environ 34(3):507–516

Buettner KJK, Kern CD (1965) The determination of infrared emissivities of terrestrial surfaces. J Geophys Res 70:1329–1337

Carlson TN, Boland FE (1978) Analysis of urban–rural canopy using a surface heat flux/temperature model. J Appl Meteorol 17:998–1013

Carlson TN, Augustine JA, Boland FE (1977) Potential application of satellite temperature measurements in the analysis of land use over urban areas. Bull Am Meteorol Soc 58(12):1301–1303

Carlson TN, Gillies RR, Perry EM (1994) A method to make use of thermal infrared temperature and NDVI measurements. Remote Sens Rev 9:161–173

Carlson TN, Gillies RR, Schmugge TJ (1995) An interpretation of methodologies for indirect measurement of soil content water. Agric Forest Meteorol 77:191–205

Chameides WL, Kasibhatla PS, Yienger J, Levy II J (1994) Growth of continental-scale metro-agro-plexes, regional ozone pollution, and world food production. Science 264:74–77

Crombie MK, Gillies RR, Arvidson RE, Brookmeyer P, Well GJ, Sultan M, Harb M (1999) An application of remotely derived climatological field for risk assessment of vector-borne disease: a spatial study of filariasis prevalence in the Nile Delta, Egypt. Photogramm Eng Remote Sens 65(12):1401–1409

Curriero FC, Heiner KS, Samet JM, Zeger SL, Strug L, Patz JA (2002) Temperature and mortality in 11 cities of the eastern United States. Am J Epidemiol 155(1):80–87

Dash P, Gottsche FM, Olesen FS, Fischer H (2002) Land surface temperature and emissivity estimation from passive sensor data: theory and practice – current trends. Int J Remote Sens 23(13):2563–2594

Elmore AJ, Mustard JF, Manning SJ, Lobell DP (2000) Quantifying vegetation change in semiarid environments: precision and accuracy of spectral mixture analysis and the normalized difference vegetation index. Remote Sens Environ 73:87–102

Flynn LP, Harris AJL, Wright R (2001) Improved identification of volcanic features using Landsat ETM+. Remote Sens Environ 78:180–193

Forster B (1983) Some urban measurements from Landsat data. Photogramm Eng Remote Sens 49:1693–1707

Fowler D (1985) Deposition of SO_2 onto plant canopies. In: Winner WE, Mooney HA, Goldstein RA (eds) Sulfur dioxide and vegetation. Stanford University Press, Stanford, CA, pp 389–402

Gallo KP, McNab AL, Karl TR, Brown JF, Hood JJ, Tarpley JD (1993) The use of NOAA AVHRR data for assessment of urban heat island effect. J Appl Meteorol 39:899–908

Gallo KP, Tarpley JD, McNab AL, Karl TR (1995) Assessment of urban heat islands: a satellite perspective. Atmos Res 37:37–43

Gillespie AR, Smith MO, Adams JB, Willis SC, Fischer AF, Sabol DE (1990) Interpretation of residual images: spectral mixture analysis of AVIRIS images, Owens Valley, California. In: Proceedings of second airborne visible/infrared imaging spectrometer (AVIRIS) workshop, Pasadena, CA, pp 243–270

Gillies RR, Carlson TN, Cui J, Kustas WP, Humes KS (1997) Verification of the 'triangle' method for obtaining surface soil water content and energy fluxes from remote measurements of the Normalized Difference Vegetation Index (NDVI) and surface radiant temperature. Int J Remote Sens 18(15):3145–3166

Gover M (1938) Mortality during periods of excessive temperature. Public Health Rep 53:1122–1143

Goward SN, Cruickshanks GD, Hope AS (1985) Observed relation between thermal emission and spectral radiance of a vegetated landscape. Remote Sens Environ 18:137–146

Howard L (1833) The climate of London reduced from meteorological observations made in the metropolis and various places around it, 2nd edn. A. Arch, Cornhill, Longman & Co., London

Jauregui E (1997) Heat island development in Mexico City. Atmos Environ 31(22):3821–3831

Johnson PE, Smith MO, Taylor-George S, Adams JB (1983) A semiempirical method for analysis of the reflectance spectra for binary mineral mixtures. J Geophy Res 88:3557–3561

Kim HH (1992) Urban heat island. Int J Remote Sens 13:2319–2336

Kinney PL (1999) The pulmonary effects of outdoor ozone and particle air pollution. Semin Respir Crit Care Med 20:601–607

Kressler F, Steinnocher K (2001) Monitoring urban development using satellite images. In: Proceedings of the second international symposium on remote sensing of urban areas, Regensburg, Germany

Kwarteng AY (2002a) Remote sensing monitoring of greenery development in Kuwait City. In: Proceedings of the 3rd international symposium on remote sensing of urban areas, Istanbul, Turkey, 11–13 June, pp 337–345

Kwarteng AY (2002b) The use of remote sensing imagery to monitor greenery development in Kuwait City. In: Al-Awadi NM, Taha FK (eds) New technologies for soil reclamation and desert greenery. Amherst Scientific, Amherst, MA, pp 157–177

Landsat Project Science Office (2004) Landsat 7 science data user's handbook. Goddard Space Flight Center, NASA, Washington, DC. http://ltpwww.gsfc.nasa.gov/IAS/handbook/handbook_toc.html

Landsberg HE (1981) The urban climate. Academic, New York

Li B, Avissar R (1994) The impact of spatial variability of land-surface characteristics on landsurface heat fluxes. J Climatol 7:527–537

Lo CP, Quattrochi DA, Luvall JC (1997) Application of high-resolution thermal infrared remote sensing and GIS to assess the urban heat island effect. Int J Remote Sens 18:287–304

Lu D, Weng Q (2004) Spectral mixture analysis of the urban landscape in Indianapolis with Landsat ETM+ imagery. Photogramm Eng Remote Sens 70:1053–1062

Miller RB, Small C (2003) Cities from space: potential applications of remote sensing in urban environmental research and policy. Environ Sci Policy 6:129–137

Nichol JE (1994) A GIS based approach to microclimate monitoring in Singapore's high-rise housing estates. Photogramm Eng Remote Sens 60:1225–1232

Nichol JE (1996) High-resolution surface temperature patterns related to urban morphology in a tropical city: a satellite-based study. J Appl Meteorol 35:135–146

Nichol JE (1998) Visualisation of urban surface temperatures derived from satellite images. Int J Remote Sens 19:1639–1649

Nichol JE (2003) GIS and remote sensing in urban heat island in the Third World. In: Mesev V (ed) Remotely sensed cities. Taylor & Francis, New York, pp 243–264

Norman JM, Divakarla M, Goel NS (1995) Algorithms for extracting information from remote thermal-IR observations of the Earth's surface. Remote Sens Environ 51:157–168

Nowak DJ (1994) Air pollution removal by Chicago's urban forest. In: McPherson EG, Nowak DJ, Rowntree RA (eds) Chicago's urban forest ecosystem: results of the Chicago Urban Forest Climate Project (NE-186). Forest Service, US Department of Agriculture, Department of Agriculture, Radnor, PA, pp 63–81

Nowak DJ, Civerolo KL, Rao ST, Sistla G, Luley CJ, Crane DE (2000) A modeling study of the impact of urban trees on ozone. Atmos Environ 34(10):1601–1613

O'Neill BC, Balk D, Brickman M, Ezra M (2001) A guide to global population predictions. Demogr Res 4(8) www.demographic-research.org

Oke TR (1987) Boundary layer climates, 2nd edn. Methuen, London/New York, pp 262–302

Owen TW, Carlson TN, Gilles RR (1998) An assessment of satellite remotely sensed land cover parameters in quantitatively describing the climatic effect of urbanization. Int J Remote Sens 19:1663–1681

Parker JH (1981) Use of landscaping for energy conservation. Department of Physical Sciences, Florida International University, Miami, FL

Pech RP, Davies AW, Lamacraft RR, Graetz RD (1986) Calibration of Landsat data for sparsely vegetated semi-arid rangelands. Int J Remote Sens 7:1729–1750

Pielke RA Sr, Marland G, Betts RA, Chase TN, Eastman JL, Niles JO, Niyogi DS, Running SW (2002) The influence of land-use change and landscape dynamics on the climate system: relevance to climate-change policy beyond the radiative effect of greenhouse gases. Philos Trans: Math Phys Eng Sci 360(1797):1705–1719

Price JC (1979) Assessment of the urban heat island effect through the use of satellite data. Mon Weather Rev 107:1554–1557

Price JC (1987) Calibration of satellite radiometers and the comparison of vegetation indices. Remote Sens Environ 21:15–27

Price JC (1990) Using spatial context in satellite data to infer regional scale evapotranspiration. IEEE Trans Geosci Remote Sens 28(5):940–948

Rao PK (1972) Remote sensing of urban heat islands from an environmental satellite. Bull Am Meteorol Soc 53:647–648

Rashed T, Weeks JR, Stow D, Fugate D (2002) Measuring temporal compositions of urban morphology through spectral mixture analysis: toward a soft approach to change analysis in crowded cities. In: Proceedings of the 3rd international symposium on remote sensing of urban areas, Istanbul, Turkey 11–13 June

Roberts DA, Batista G, Pereira J, Waller E, Nelson B (1998) Change identification using multitemporal spectral mixture analysis: applications in Eastern Amazonia. In: Elvidge C, Lunetta R (eds) Remote sensing change detection: environmental monitoring applications and methods. Ann Arbor Press, Ann Arbor, MI, pp 137–161

Rosenfeld AH, Akbari H, Bretz S, Fishman BL, Kurn DM, Sailor D, Taha H (1995) Mitigation of urban heat islands: materials, utility programs, updates. Energy Build 22:255–265

Roth M, Oke TR, Emery WJ (1989) Satellite-derived urban heat island from three coastal cities and the utilization of such data in urban climatology. Int J Remote Sens 10:1699–1720

Roy S, Avissar R (2000) Scales of response of the convective boundary layer to land-surface heterogeneity. Geophys Res Lett 27:533–536

Singer RB (1981) Near-infrared spectral reflectance of mineral mixtures: systematic combinations of pyroxenes, olivine and iron oxides. J Geophys Res 86:7967–7982

Singer RB, McCord TB (1979) Mars: large scale mixing of bright and dark surface materials and implications for analysis of spectral reflectance. In: 10th lunar and planetary science conference American geophysical union, Houston, TX, pp 1835–18480

Small C (2001) Estimation of urban vegetation abundance by spectral mixture analysis. Int J Remote Sens 22(7):1305–1334

Small C (2003) High spatial resolution spectral mixture analysis of urban reflectance. Remote Sens Environ 88:170–186

Small C (2004) The Landsat ETM+ global mixing space. Remote Sens Environ 93(1–2):1–17

Small C, Pozzi F, Elvidge CD (2005) Spatial Analysis of Global Urban Extent from DMSP OLS Night Lights. Remote Sens Environ 96:277-291

Smith WH (1984) Pollutant uptake by plants. In: Treshow M (ed) Air pollution and plant life. Wiley, New York

Smith MO, Ustin SL, Adams JB, Gillespie AR (1990) Vegetation in deserts: I. A regional measure of abundance from multispectral images. Remote Sens Environ 31:1–26

Taha HS, Douglas S, Haney J (1997) Mesoscale meteorological and air quality impacts of increased urban albedo and vegetation. Energy Build 25(2):169–177

Taha HS, Chang C, Akbari H (2000) Meteorological and air quality impacts of heat island mitigation measure in three US cities (Report No. LBL-44222). Lawrence Berkeley National Laboratory, Berkeley, CA

Tucker CJ, Sellers PJ (1986) Satellite remote sensing primary production. Int J Remote Sens 7:1395–1416

United Nations (2001) The state of the world cities 2001. United Nations Centre for Human Settlements, Nairobi, Kenya

Voogt JA, Oke TR (2003) Thermal remote sensing of urban climates. Remote Sens Environ 86:370–384

Vukovich FM (1983) An analysis of the ground temperature and reflectivity pattern about St. Louis, Missouri, using HCMM satellite data. J Appl Meteorol 22:560–571

Wagrowski DM, Hites RA (1997) Polycyclic aromatic hydrocarbon accumulation in urban, suburban and rural vegetation. Environ Sci Tech 31(1):279–282

Welch R (1982) Spatial resolution requirements for urban studies. Int J Remote Sens 3(2):139–146

Wu CS, Murray AT (2003) Estimating impervious surface distribution by spectral mixture analysis. Remote Sens Environ 84(4):493–505

Chapter 15
Remote Sensing of Urban Land Use Change in Developing Countries: An Example from Büyükçekmece, Istanbul, Turkey

Derya Maktav and Filiz Sunar

This application chapter discusses rural–urban land use changes in developing countries using Büyükçekmece, a suburb of Istanbul, Turkey, as an example. Specifically, this chapter demonstrates the utility of remote sensing techniques, multitemporal satellite data, a 1984–1997 population database, and ground data to illustrate the impact of urban growth on land cover and land use changes in general and on agricultural land in particular.

Learning Objectives

Upon completion of this chapter, you should be able to:

❶ Explain the differences between land use and land cover in rural and urban areas, and describe the context of urbanization and informal settlements in developing countries

❷ Articulate ways for using remote sensing in assessing urban land use change in developing countries

❸ Speculate on change detection techniques for monitoring rural–urban changes

D. Maktav (✉) and F. Sunar
Faculty of Civil Engineering, Istanbul Technical University (ITU), Department of Geomatics, 34469, Maslak, Istanbul, Turkey
e-mails: maktavd@itu.edu.tr; fsunar@itu.edu.tr

T. Rashed and C. Jürgens (eds.), *Remote Sensing of Urban and Suburban Areas*,
Remote Sensing and Digital Image Processing 10,
DOI 10.1007/978-1-4020-4385-7_15, © Springer Science+Business Media B.V. 2010

15.1   Introduction

Land use changes are potentially the most significant driver of environmental changes, causing both the transformation and modification of the Earth's surface (Turner et al. 1994). Considering that human population pressure, hence resource demands, are continually increasing due to demographic and economic changes in human societies, land use is likely to further change, intensify, and ultimately threaten the sustainable development and management of our natural resources (Mander and Jongman 2000). To facilitate sustainable development and management, analyses of land use changes and their effects on the environment are indispensable. Remote sensing is an important tool that is now frequently employed to assist in such analyses. In this context, it should be noted that the remote sensing data and techniques used to assess land use changes in both developed vs. developing countries are identical, but that the causes and effects of land use changes in developed vs. developing countries are frequently different. This chapter focuses on developing countries, where urban-rural land use changes are primarily the result of rapid natural population growth, migration from rural areas to mega cities, and fast industrialization. An application of remote sensing is demonstrated to show the impact of urban growth on land use changes, especially on the agricultural land based on a case study in the Büyükçekmece in a suburban of Istanbul in Turkey. The study is based on the 1984–1997 population database, ground data, multi-temporal satellite data and remote sensing methods.

15.2   Rural–Urban Land Use Changes

A differentiation between the concepts "land cover" and "land use" is imperative if planning and management activities are to be successful. The term land cover describes the type of biophysical cover observed on the surface of the earth, and includes features such as water, vegetation, or bare soil. The term land use refers to the manner in which certain land cover types are employed or managed by humans, and includes uses such as rangeland, agriculture, or transportation. Land cover and land use are intrinsically linked, for example, an area with forest cover may be used for recreational or logging purposes. Urban land use and land cover information is required for a great variety of applications including residential, industrial, and commercial site selection; population estimation; tax assessment; or the development of zoning regulations (Jensen 2000; Lillesand et al. 2004; Green et al. 1994; Cullingworth 1997).

The United Nations (UN) simply defines "urban areas" as "localities with 20,000 or more inhabitants." However, other factors such as the level and type of economic activity may also be taken into consideration in defining urban areas (also see Chapter 3). For example, an urban area may be defined in terms of the built-up area or in terms of the functional area. The functional area includes areas for which services and facilities are provided, and may thus embrace not only the built-up area but also free-standing settlements outside the urban area and tracts of surrounding countryside if the population in these surrounding areas depends on the urban center for services and employment. An urban area may also be defined using population or buildings density as an indicator of urbanization (Barba and Rabuco 1997; Gross and Monteiro 1989).

Areas that are not classified as "urban" are typically represented as "rural areas." However, there is no standard definition of rural areas that is generally accepted in policy, research, and planning, which is at least partially due to the increasing integration of rural and urban areas through commuting patterns and urban and suburban expansion. Different definitions are in use that are each based on different criteria, different levels of analysis, and different methodologies. Most definitions, however, classify rural areas based on population density, level of urbanization, adjacency and relationship to an urbanized area. Some definitions also take into account the principal economic activity in an area (RPRI 2004; Cromartie and Swanson 1996).

the term *land cover* describes the type of biophysical cover observed on the Earth's surface, while land use refers to the manner in which certain land cover types are employed or managed by humans.

For the purposes of this chapter, "rural areas" are defined according to the International Institute for Environment and Development (IIED 2004), which define rural–urban interactions as "linkages across space (such as flows of people, goods, money, information and wastes) and linkages between sectors (for example, between agriculture and services and manufacturing). In broad terms, they also include 'rural' activities taking place in urban areas (such as urban agriculture) and activities often classified as 'urban' (such as manufacturing and services) taking place in rural settlements." As a result of continued population growth, departure from agricultural systems, and industrialization, urban areas have been sprawling at the expense of rural areas (e.g., forest land), ultimately causing the degradation of the physical environment. In order to ensure that urban areas do not encroach on valuable agricultural land, that agriculture occurs in suitable locations, and that urban development does not cause the degradation of adjacent agricultural lands, it is crucial to analyze rural–urban land use changes (CCRS 2003).

15.3 Urbanization in Developing Countries

A developing country (or a least developed country) is a country that has not reached the stage of economic development characterized by the growth of industrialization. In developing countries, national income is less than the amount of

a *developing country* is a country that has not reached the stage of economic development characterized by the growth of industrialization

money needed to pay for domestic savings, population growth is usually faster than in developed countries, and most people have a lower standard of living with access to fewer goods and services than most people in high-income countries. In developing countries, urbanization is a key process. Three main causes have been identified for urban population growth (Gross and Monteiro 1989; Barba and Rabuco 1997):

- Rapid overall natural population growth
- Rural-to-urban migration
- Reclassification of rural areas as urban areas

Natural population growth in cities, in addition to transformation of rural to urban areas, accounts for an average of 61% of the *urban population* growth in developing countries. Rural-to-urban migration accounts for 39%. However, differences in urban population growth exist within and between countries and regions in the world. For example, in Latin America, where the urbanization level is already high, natural population growth is likely the most important contributor to urban population growth (Rossi-Espagnot 1984). In contrast, sub-Saharan Africa and parts of Asia are primarily characterized by high levels of rural-to-urban migration and urban growth.

Developing countries contain a rapidly increasing proportion of the world's largest metropolitan areas. In 1975, 10 of the largest metropolitan areas were in developing countries. In the 1980s, 22 of the 35 largest metropolitan areas, containing about 45% of the world's metropolitan population, were in developing countries. By 2000, 25 of the largest urban populations were in developing countries. By the year 2010, it is projected that over 50% of the world's population will inhabit urban areas, whereby the majority of the urban population growth is expected to be concentrated in developing countries.

The growth rates of urban populations vary across regions. In Africa, the world's most rapidly urbanizing region, the annual urban population growth rate reached as high as 5.5% during the period 1985–1990. By 2025, it will still be around 3%. Alternatively, in Latin America, the average urban population growth rate declined from 3.9% (1970–1975) to 2.9% (1985–1990) and likely to approximately 1.45% (projected for 2025). In 1990, Latin America was the most urbanized region in the developing world, with 72% of its people living in urban areas. In Asia, the annual rate of urbanization was 3.1% during the period 1985–1990. This rate is expected to decline to 1.1% during the period 2020–2025 (Cepede 1984; Gross 1990; Gross and Monteiro 1989).

According to the World Bank's report, "World Development Indicators 2000," the share of urban residents in the world's total population (both developed and developing countries) rose from 40% in 1980 to 46% in 1999. China, the world's most crowded country, accommodates the highest urban population: in 1999, 400 million Chinese were urban residents. Second and third to China are India (279 million urban residents) and the US (210 million urban residents), respectively. Turkey, the country in which this chapter's study area is located, ranks is the 14th most urbanized country, whereby the proportion of people living in urban areas climbed from 44% (19.6 million in 1980) to 74% (47.7 million in 1999). The random and rapid urbanization in Turkey during the past two decades has caused pressing problems such as higher death tolls in earthquakes and inadequate education and health services (WBG 2000).

Formal settlements "refer to land zoned residential in city master plans or occupied by formal housing" (UN-HSP 2002). In contrast, informal settlements are defined as: "(i) residential areas where a group of housing units has been constructed on land to which the occupants have no legal claim, or which they occupy illegally, and (ii) unplanned areas where housing is not in compliance with current planning and building regulations" (UN-HSP 2002). Informal settlements in developing countries are typically located on the fringes of cities and characterized by a "dense proliferation of small, make-shift shelters built from diverse materials, degradation of the local ecosystem and by severe social problems" (Mazur and Qangule 1995).

informal settlements represent illegal or unplanned residential areas not in compliance with regulations, and are typically characterized by a complex mix of social and environmental problems

According to a report of the UN Human Settlements Programme on human settlements (UN-HSP 2002), 30–60% of urban residents in developing countries live in informal settlements. However, accurate population estimates and maps of these areas are scarce or nonexistent, making it difficult for authorities to enhance the situations of these areas. Aerial photography, satellite data or land use maps can be used to evaluate the area occupied by informal settlements. This is typically complemented by mapping applications incorporated into Geographical Information Systems (GIS) and global positioning systems (GPS), which are increasingly used to provide ground truth data.

Assessing Urban Land Use Change in Developing Countries

Over the years, aerial photography has been successfully utilized for mapping, monitoring, planning and development of urban sprawl, urban land use and urban environment. For reasons of their widespread availability, stereo and revisit capability, frequency of update and cost, reliable and accurate data, however, the focus of urban remote sensing research has shifted more towards the use of digital satellite images such as IKONOS, QUICKBIRD, EROS, LANDSAT, SPOT etc. (Donnay et al. 2001). As discussed in earlier chapters, some characteristics of the satellite sensors, for example spatial and spectral resolutions, influence their applications for mapping change in urban areas.

For example, crop and harvest forecasts are especially useful in developing countries. LANDSAT data can be used very successfully in identifying crop types and predicting harvest times with 90–95% accuracy. This facilitates the accurate forecasting of possible famines and helps put emergency measures into place well in advance. Another example is related to the informal settlement areas in the cities of the developing countries where remote sensing can play an important role by virtue of its repetitive and synoptic coverage that helps create a base map for many governmental organizations in a very rapidly and haphazardly growing urban area. One can monitor urban growth, locate slums and identify the physical characteristics of the slum areas in developing countries by means of interpretation of high resolution satellite data (Sur et al. 2003; Seto and Duong 2002; Ehlers et al. 2002). The range of image processing techniques generally used in land use change analysis encompasses various operations, including geometric, radiometric and atmospheric corrections, image compression and enhancement, spatial filtering and many of the image processing techniques discussed in earlier chapters. Change detection procedure, where two or more images are compared to determine differences, involves the use of multispectral data sets to discriminate areas of land use change between dates of imaging. The reliability of the change detection process may strongly be influenced by a number of environmental factors that might change between image dates. Two of the main methods are: image differencing, where data from one date are simply subtracted from those of the other (the difference in areas of no change should be zero). On the other hand, image ratioing involves computing the ratio of the data from two dates of imaging. Here, ratios for areas of no change should have a value of 1 (Lillesand et al. 2004; Jensen 2000; Sunar 1998; Jurgens 2000; Treitz and Rogan 2004; El-Raey et al. 1995; Gibson and Power 2000; Green et al. 1994; Mass 1999).

Vegetation indices are another common family of techniques used to monitor change in environmental conditions within urban settings. Vegetation indices are defined as dimensionless, radiometric measures that function as indicators of the relative abundance and activity of green vegetation, often including leaf-area index, percentage green cover, chlorophyll content, green biomass, and absorbed photo synthetically active radiation. There are more than 20 vegetation indices in use in the literature. Many are functionally equivalent in information content, while some provide unique biophysical information. One of the most commonly used vegetation indices is the Normalized Difference Vegetation Index (NDVI). It is formulated as (NIR − R)/(NIR + R), where NIR, and R represent data from infrared and red bands. The NDVI is preferred to other vegetation indices for global vegetation monitoring because it helps compensate for changing illumination conditions, surface slop, aspect, and other extraneous factors (Lillesand et al. 2004; Jensen 2000; Harrison and Jupp 1990).

15.4   Example: Analysis of Urban Growth in Istanbul, Turkey, Using Multitemporal Satellite Data

Istanbul, the third-largest city in Europe, extends over both banks of Bosphorus, the strait that separates the European and Asian continents. The city is home to about one-fifth of Turkey's population (12 million inhabitants) and it contributes with a higher share in Turkey's economy (Gür et al. 2003). During the last five decades, unplanned migration and industrialization have metamorphosed the city to such an extent that it seems to have almost forgotten its 2,500 years of historic heritage (Baytın 2000; ESA 2001).The following sections demonstrate how multitemporal satellite imagery, digital image processing techniques, a 1984–1998 population database, and ground data have been used to characterize the effects of urban growth on land use and land cover changes in Istanbul in general and on agricultural land in the district of Büyükçekmece, a suburb of the mega-city Istanbul, in particular.

15.4.1   Study Area

Büyükçekmece (centered at 41° 03′ N, 28° 45′ E), one of the 32 administrative districts of Istanbul, has been subject to rapid urbanization over the last few decades, primarily due to increased migration from the Black Sea regions of Turkey (DIE 2000). Büyükçekmece encompasses an area of 225 km², including the Büyükçekmece Lake, and is located along the north shore of the Marmara Sea. According to the 1997 census, the district's population is approximately 300,000. Population densities are highest in the coastal regions. Büyükçekmece had eight administrative subdistricts (Kavaklı, Yakuplu, Kıraç, Gürpınar, Esenyurt, Mimarsinan, Kumburgaz, and Tepecik) and contained five villages (Hoşdere, Türkoba, Çakmaklı, Karaağaç, and Ahmediye) (Fig. 15.1). However, the administrative boundaries of Büyükçekmece has been restructured over the years. For example, the village Güzelce is part of the Kumburgaz sub-district, and village Hoşdere was renamed Bahçeşehir after growing to become a separate sub-district in 1999.

15.4.2   Data and Methods

This case study utilized a variety of datasets. Population data for 14 administrative units in Büyükçekmece and covering the time period between 1970 and 1997 were derived from the Governmental Statistical Institute reports (DIE 2000), and are shown in Table 15.1 below. Population changes over the 1970–1985 and 1985–1997 time spans are listed in Table 15.1 and shown graphically in Fig. 15.2.

GIS data layers (vector data) of the Büyükçekmece district, its villages, subdistricts, and the coastline were obtained from the Büyükçekmece Municipality in AUTOCAD DXF format. Multi-temporal satellite data (raster data), including

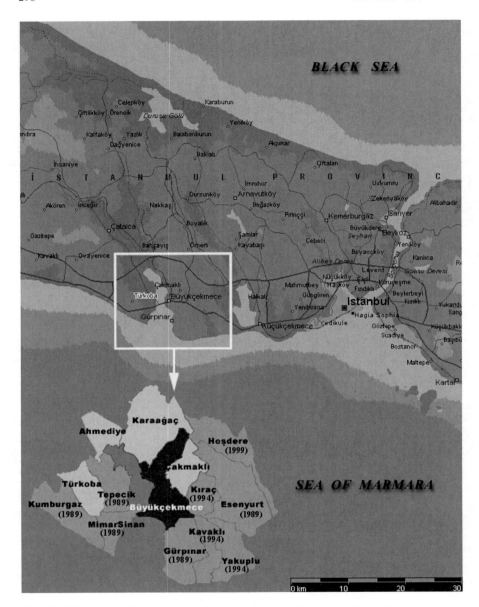

Fig. 15.1 Map of the Büyükçekmece district (numbers and brackets indicate the administrative transformation dates of villages into sub-districts)

LANDSAT Thematic Mapper (TM), SPOT P and IKONOS XS & P image data were also obtained and are described in more detail in Table 15.2.

The multi-temporal LANDSAT TM data (1984 and 1998) were rectified based on 1:25,000 scale topographic maps using an automated registration process

Table 15.1 Census data distribution in Büyükçekmece (1970–1997)

Admin. unit	1970	1975	1980	1985	1990	1997	Changes (%)	
							1970–1985	1985–1997
BÇekmece	3,913	5,204	8,121	11,310	22,394	41,644	189.04	268.21
Kavaklı	501	628	866	1,021	2,170	24,475	103.79	2,297.16
Yakuplu	974	1,045	1,252	1,664	2,841	23,878	70.84	1,334.98
Hoşdere	–	802	864	924	1,538	12,915	–	1,297.73
Kıraç	371	435	544	826	2,239	10,353	122.64	1,153.39
Gürpınar	1,305	1,578	2,812	3,584	10,191	20,702	174.64	477.62
Esenyurt	923	1,631	6,636	21,290	70,280	100,565	220.66	372.36
MimarSinan	2,296	2,232	3,138	4,083	7,690	15,204	77.83	272.37
Kumburgaz	928	1,270	2,750	2,569	7,118	8,329	176.83	224.21
Güzelce	722	999	2,111	1,366	–	–	89.2	–
Tepecik	1,607	3,134	4,805	7,382	12,240	14,588	359.36	97.62
Türkoba	339	505	364	436	712	2,392	28.61	448.62
Çakmaklı	344	801	525	709	1,633	3,675	106.11	418.34
Karaağaç	–	325	451	399	681	868	–	117.54
Ahmediye	435	473	664	802	1,183	1,300	84.37	62.09
Total	14,658	21,062	35,903	58,365	120,516	239,244	25.1	309.9

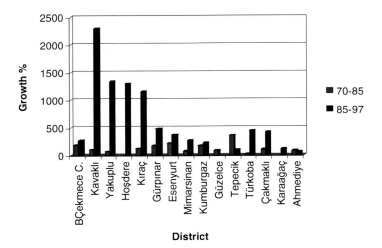

Fig. 15.2 Changes in population growth over the periods 1970–1985 and 1985–1997

Table 15.2 Characteristics of the satellite data used in the case study

Satellite	Sensor	Spatial resolution (m)	Date
LANDSAT	TM	30 (except TIR band)	June 12,1984
			Apr 16, 1998
SPOT	HRV	10	Apr 16,1989
			July 22, 1998
IKONOS	XS	4	Feb 13, 2002
	P	1	Feb 13, 2002

(see detail of the registration process in Maktav et al. 2000; Sunar et al. 2000;
Taberner et al. 1999). In the registration algorithm, matching between scenes is carried out using local correlations in the frequency domain. The result is a correlation
map and the location of the elements with maximum correlation provides the necessary x and y shift to give the best fit. With this automated procedure, over 1,600 points
in an almost complete matrix distribution described by a polynomial with a fit to
within ±0.5 pixel RMSE were produced. Because of the incompatibility of the automatic process due to different resolutions of the two different sensors, a first-degree
polynomial equation was used for the geometric registration process of the LANDSAT
TM and IKONOS data standard techniques with 15 ground control points. As a
re-sampling process, cubic convolution was used with ±0.5 and 3 pixel registration
accuracy for LANDSAT TM and IKONOS XS images, respectively. Because of
being same sensor and of seasonal compatibility (April and June) no atmospheric and
radiometric corrections were applied for the LANDSAT TM images.

LANDSAT TM data, excluding the thermal band, were classified separately
using a supervised classification technique. For both dates the following classes
were considered: settlement, fields, lake and sea (Büyükçekmece Lake and some
of the Marmara Sea coast), forest, stone quarries, and industrial areas (Fig. 15.3).

Fig. 15.3 Sample pictures for some of land use classes utilized in the case study (fields, stone quarries, settlement, and industrial areas)

The percentage of their distribution over the 14 administrative units of Istanbul was calculated from land use classification results for both dates. Classification accuracy analysis for each classification was performed to calculate the confusion matrix and Kappa coefficient using 50 ground truth sample points, which were selected independent of training areas.

Using the multi-temporal LANDSAT TM images, change detection methods were applied to reveal changes in land use between 1984 and 1998. In this study, both the image differencing and image ratioing methods were utilized. Band 3 (0.63–0.69 μm) of the first LANDSAT TM data (1984) was subtracted/divided from band 3 of the second LANDSAT TM data (1998). Band 3 was selected because it is believed to be the best band for cultural/urban feature identification. The resulting difference image was then classified to reveal changes in areas of different land uses between the years 1984 and 1998. In addition, the NDVI was used to evaluate urban change in a focused area, Mimarsinan – one of Istanbul's sub-districts known of main land use changes, between 1998 and 2002.

15.5   Results and Discussion

The population growth between 1970 and 1985 in some sub-districts of Büyükçekmece exploded after 1985 to reach a growth rate of over 1000% during the period 1985–1997 (Table 15.1 and Fig. 15.2). Attractive coastal sub-districts such as in Kavaklı, Yakuplu, and Kıraç, for example, witnessed a population growth rate of 70–125% during the period 1970–1985 and increased to 1,150–2,300% during the period 1985–1998. Even in Hoşdere, which was only a village in 1997, the increase in population was approximately 1,300% between 1985 and 1997. Likewise, the increase in Esenyurt was 221% between 1970 and 1985 then reached 372% in the period 1985–1997. The population in this sub-district has been continuously increasing from 1985 (21,000) to 1997 (100,000) and reached a greater population than the coastal sub-districts and city centre. The main reason for the increase in population is related to the huge amount of migration from other parts of Turkey rather than natural population growth. It is obvious that such a population explosion would cause great land use changes in the area.

The results obtained from the classified LANDSAT TM image of 1984 (Fig. 15.4a, b, Table 15.3) showed that 93.4% (21,013.8 ha) of the Büyükçekmece was covered with agricultural fields. For the fields located in Büyükçekmece, its sub-districts and villages the proportion was approximately 86–98% (except in Ahmediye). Before 1984 the whole area was covered with watermelon, muskmelon, grain, and sunflower fields, with farming and agriculture being the main land use activities. On the other hand, in Büyükçekmece there were only 3.6% settlement areas, having the densest building in Mimarsinan with 10.2% and a minimum proportion of settlement in Ahmediye with only 0.6%. The district had virtually no forested areas (only 2.4%). In this district, there were some stone quarries located in a 40.2 ha area and its percentage of coverage area within the total district was only 0.2%. There were no industrialized areas established prior to 1984.

According to the results obtained from classified LANDSAT TM data dated 1998 (Fig. 15.5a, b, Table 15.4) the percentage area of the fields averaged 67%. Fields in Kumburgaz and Türkoba located on the west of the Büyükçekmece Lake, and Karaağaç on the northeast of the lake covered 83–86% of the areas, but only 50–78% in other parts of the district (with the exception of Ahmediye, where they are only 29.3% of the area).

In Büyükçekmece, settlements covered 5,242 ha, which is about 1/4 of the total district. Industrialized areas comprising 3–5% of the study area were mostly located at the eastern part of the Büyükçekmece Lake in Kavaklı, Yakuplu, and Kıraç, and in Mimarsinan at the southwestern part of the lake. All of these sub-districts have shores or coastlines, except Kıraç. Forest areas covered approximately 6% of the Karaağaç, but they were less than 3.2% in all other areas. In the whole district, percentage of the forest areas average 1.7%, with half of it in the Karaağaç. Percentage of land used for stone quarries in the district was 1.5% (340 ha).

In this example, the overall performance of classification is a compound of the accuracies of the individual classifications which, in turn, depend largely on the consistency, homogeneity and separability of the original training classes and how representative they are (Coppin and Bauer 1996). Hence, classification accuracy analysis was performed after each classification process with error matrix and Kappa analysis using 50 randomly selected test points which are independent of training areas from the existing field maps (Table 15.5).

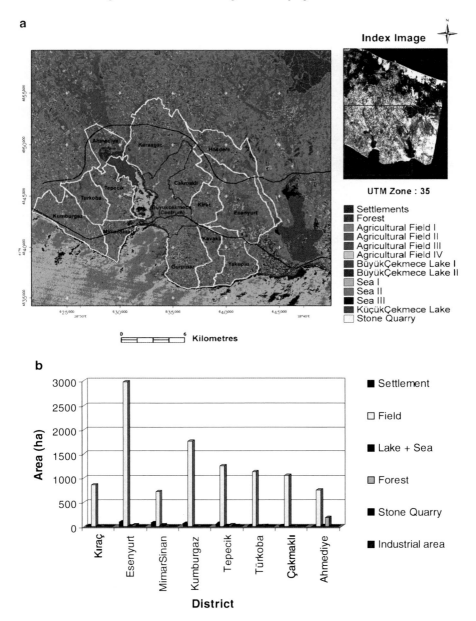

Fig. 15.4   Analysis results for the Landsat TM image of 1984. (**a**) Classified 1984 LANDSAT TM imagery, (**b**) calculated areal extents of the land use classes

In Büyükçekmece, land use for settlements over these years increased by almost 20%, from 3.6% to 23.3%. Table 15.6 shows a comparative analysis of the classification results and displays the extreme increase in the built area within the

Table 15.3 Land use results obtained from classified 1984 LANDSAT TM data

1984	Settlement		Fields		Lake-sea		Forest		Stone quarry		Ind.		Total	
	ha	%	ha	%	ha	%	ha	%	ha	%	ha	%	ha	%
BÇekmece	88.1	4.1	2,034.5	93.5	19.0	0.9	30.9	1.4	2.8	0.1	0	0	2,175.3	9.7
Kavaklı	48.4	4.9	919.3	93.1	0.6	0	19.0	1.9	0.6	0.1	0	0	987.4	4.4
Yakuplu	85.6	5.9	1,348.6	93.3	3.0	0.2	3.0	0.2	4.9	0.3	0	0	1,445.1	6.4
Hoşdere	40.9	2.2	1,735.1	92.7	06	0	94.	5.1	1.4	0.1	0	0	1,872.7	8.3
Kıraç	24.0	2.7	858.7	97	0	0	2.1	0.2	0	0	0	0	884.8	3.9
Gürpınar	120.0	6.6	1,662.8	91.7	13.9	0.8	11.9	0.7	4.8	0.3	0	0	1,813.6	8.1
Esenyurt	94.4	3	2,976.4	96	0	0	30.6	1	0.6	0	0	0	3,102.0	13.8
MimarSinan	85.3	10.2	719.6	86.2	26.2	3.1	2.5	0.3	1.3	0.1	0	0	834.9	3.7
Kumburgaz	69.8	3.8	1,765.5	95.2	6.3	0.3	9.9	0.5	2.9	0.2	0	0	1,854.4	8.2
Tepecik	58.6	4.2	1,250.8	90.5	15.8	1.1	36.4	2.6	20.4	1.5	0	0	1,381.9	6.1
Türkoba	11.6	1	1,132.9	97.1	0	0	21.9	1.9	0	0	0	0	1,166.4	5.2
Çakmaklı	14.8	1.4	1,061.9	98.3	0	0	3.9	0.4	0	0	0	0	1,080.6	4.8
Karaağaç	62.3	2.1	2,801.8	95	3.7	0.1	82.6	2.8	0	0	0	0	2,950.3	13.1
Ahmediye	5.5	0.6	745.9	78.2	11.3	1.2	190.9	20	06	0.1	0	0	954.3	4.2
Total	809.4	3.6	21,013.8	93.4	99.9	0.4	540.3	2.4	40.2	0.2	0	0	22,503.6	99.9

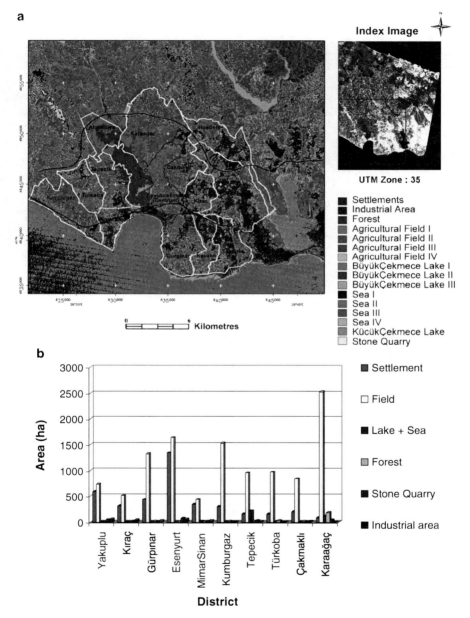

Fig. 15.5 Analysis result for the Landsat TM image of 1998. (**a**) Classified 1998 LANDSAT TM imagery, (**b**) calculated areal extents of the land use classes

Table 15.4 Land use results obtained from classified 1998 LANDSAT TM data

1998	Settlement		Fields		Lake-sea		Forest		Stone quarry		Ind.		Total	
	ha	%	ha	%	ha	%	ha	%	ha	%	ha	%	ha	%
BÇekmece	4,56.1	21	1,488.3	68.4	1,38.3	6.4	40.9	1.9	34.9	1.6	16.0	0.8	2,175.3	9.7
Kavaklı	3,55.1	36	591.2	59.9	0.2	0	1.4	0.1	7.3	0.7	32.3	3.3	987.4	4.4
Yakuplu	5,93.4	41.1	736.1	50.9	1.1	0.1	6.6	0.5	49.8	3.4	58.2	4	1,445.1	6.4
Hoşdere	4,78.8	25.6	1,243.9	66.4	1.1	0.1	59.4	3.2	45.3	2.4	44.2	2.4	1,872.7	8.3
Kıraç	317.8	35.9	509.7	57.6	0	0	4.4	0.5	7.8	0.9	45.4	5.1	884.8	3.9
Gürpınar	4,36.8	24.1	1,330.7	73.4	1.0	0.1	10.4	0.6	7.4	0.4	27.3	1.5	1,813.6	8.1
Esenyurt	1,346.8	43.4	1,636.0	52.7	0.1	0	1.1	0	66.2	2.1	51.9	1.7	3,102.0	13.8
MimarSinan	344.8	41.3	437.1	52.4	17.7	2.1	0.2	0	9.0	1.1	26.1	3.1	834.9	3.7
Kumburgaz	299.6	16.2	1,531.3	82.6	3.0	0.2	11.9	0.6	4.9	0.3	3.8	0.2	1,854.4	8.2
Tepecik	158.9	11.5	958.4	69.4	2,13.1	15.4	14.3	1	28.6	2.1	8.7	0.6	1,381.9	6.1
Türkoba	155.9	13.4	974.6	83.6	0.1	0	27.1	2.3	4.8	0.4	3.8	0.3	1,166.4	5.2
Çakmaklı	199.9	18.5	841.1	77.8	0.2	0	12.7	1.2	6.2	0.6	20.6	1.9	1,080.6	4.8
Karaağaç	83.1	2.8	2,521.7	85.5	1,11.8	3.8	1,84.3	6.2	44.6	1.5	4.9	0.2	2,950.3	13.1
Ahmediye	15.2	1.6	279.4	29.3	630.5	66.1	7.4	0.8	21.8	2.3	0	0	954.3	4.2
Total	5241.9	23.3	15079.4	67	1,118.1	5	382.1	1.7	338.2	1.5	343.9	1.5	22503.6	99.9

Table 15.5 LANDSAT TM image classification accuracy analysis

1984[a]

| Class | Ground truth data (%) | | | | | |
	Forest	Settlement	Stone quarry	Field	Lake + sea	Total
Forest	100.0	0.00	0.00	31.33	0.00	12.28
Settlement	0.00	87.74	0.00	0.00	0.00	6.60
Stone quarry	0.00	12.26	100.0	13.67	1.55	12.35
Field	0.00	0.00	0.00	55.00	0.00	11.45
Lake + sea	0.00	0.00	0.00	0.00	98.45	57.32
Total	100.0	100.0	100.0	100.0	100.0	100.0

1998[b]

| Class | Ground truth data (%) | | | | | | |
	Forest	Settlement	Stone quarry	Industry	Field	Lake + sea	Total
Forest	76.00	0.00	0.00	0.00	0.00	0.00	1.67
Settlement	0.00	97.17	13.85	43.90	0.00	0.00	32.43
Stone quarry	0.00	0.00	21.54	0.00	0.00	0.00	1.23
Industry	0.00	2.83	64.62	56.10	0.00	0.00	6.57
Field	24.00	0.00	0.00	0.00	100.00	0.00	19.89
Lake + sea	0.00	0.00	0.00	0.00	0.00	100.00	38.21
Total	100.0	100.0	100.0	100.00	100.0	100.00	100.00

[a]Kappa coefficient: 0.8201; overall accuracy: 88.8%
[b]Kappa coefficient: 0.8953; overall accuracy: 92.5%

sub-districts Esenyurt (1,252.3 ha, 40.4%), Yakuplu (507.8 ha, 35.2%), Mimarsinan (259.4 ha, 31.1%), Kıraç (293.7 ha, 33.2%) and Kavaklı (306.6 ha, 31.1%), most of which are located on the eastern side of Bütyükçekmece Lake. An opposing trend could be observed in the settlements of Ahmediye and Karaağaç, where the built area in each district increased only by about 1%, which corresponds to a total area of less than 100 ha. Analysis of the land classified as "field" in Büyükçekmece revealed a loss of 26.4% of agricultural fields over the period of 14 years. The areas with maximum loss of fields were again located on the eastern side of Büyükçekmece Lake: Esenyurt (1,340.4 ha, 43.3%), Yakuplu (612.5 ha, 42.4%), Kıraç (349 ha, 39.4%), and Kavaklı (328.1 ha, 33.2%). Apparently, there is a correlation between the loss of fields and the increase in settlement areas (Fig. 15.6).

As for fame lands, it is evident from the analysis that villas and new apartment houses in Büyükçekmece rapidly depleted agricultural lands, an observation supported by ground truth data. Using the same analytical methods, it was found that Ahmediye experienced a 49% loss of fields but only a 1% increase of settlements. Interpreting the two different years of LANDSAT TM images, one can easily detect that a significant portion of land in Ahmediye had been submerged by the water of the Büyükçekmece Lake over the study period. The reason for this interesting event is the enlargement of the Büyükçekmece Lake from a lagoon to a lake following

Table 15.6 Comparative analysis of the tables obtained from the classified LANDSAT TM images

	Population 1985–1997	Settlement 1984–1998		Field 1984–1998		Lake + sea 1984–1998		Forest 1984–1998		Stone quarry 1984–1998		Industry 1984–1998	
	%	ha	%	ha	%	ha	%	ha	%	ha	%	ha	%
BÇekmece	268.2	367.9	16.9	−546.2	−25.1	119.3	5.5	10	0.5	32.1	1.5	16.8	0.8
Kavaklı	2,297.2	306.6	31.1	−328.1	−33.2	0.1	0	−17.6	−1.8	6.6	0.6	32.3	3.3
Yakuplu	1,335.0	507.8	35.2	−612.5	−42.4	−1.9	−0.1	3.6	0.3	44.9	3.1	58.2	4
Hoşdere	1,297.7	437.9	23.4	−491.2	−26.3	0.4	0.1	−35.3	−1.9	43.9	2.3	44.2	2.4
Kıraç	1,153.4	293.7	33.2	−349	−39.4	0	0	2.4	0.3	7.6	0.9	45.4	5.1
Gürpınar	577.6	316.8	17.5	−332.1	−18.3	−12.9	−0.7	−1.5	−0.1	2.6	0.1	27.3	1.5
Esenyurt	472.4	1252.3	40.4	−1,340.4	−43.3	0.1	0	−29.5	−1	65.6	2.1	51.9	1.7
Mimarsinan	272.4	259.4	31.1	−282.5	−33.8	−8.5	−1	−2.3	−0.3	7.8	1	26.1	3.1
Kumburgaz	224.2	229.8	12.4	−234.3	−12.6	−3.3	−0.1	1.9	0.1	2.1	0.1	3.8	0.2
Tepecik	97.6	100.3	7.3	−292.4	−21.1	197.4	14.3	−22.1	−1.6	8.2	0.6	8.7	0.6
Türkoba	448.6	144.4	12.4	−158.3	−13.5	0.1	0	5.2	0.4	4.8	0.4	3.8	0.3
Çakmaklı	418.3	185.1	17.1	−220.9	−20.5	0.2	0	8.8	0.8	6.2	0.6	20.6	1.9
Karaağaç	117.5	20.8	0.7	−280.1	−9.5	108.1	3.7	101.8	3.4	44.6	1.5	4.9	0.2
Ahmediye	62.1	9.7	1	−466.6	−48.9	619.2	64.9	−183.5	−19	21.2	2.2	0	0
Total	309.9	4,432.5	19.7	−5934.4	−26.4	1,018	4.6	−158.3	−0.7	298	1.3	343.9	1.5

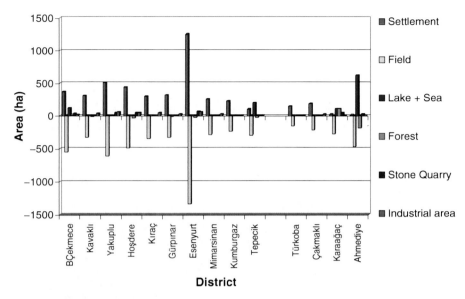

Fig. 15.6   Land use changes in Büyükçekmece and its sub-districts between 1984 and 1998

the construction of a dike to provide water for Istanbul in 1985. In addition to Ahmediye, the newly created lake also inundated parts of Tepecik, Karaağaç and Büyükçekmece. Overall, approximately 1,000 ha area was flooded: 619.2 ha (64.9%) in Ahmediye, 197.4 ha (14.3%) in Tepecik, and 119.3 ha (5.5%) in Büyükçekmece. Since the other parts of the study area are not adjacent to the lake, other changes of less than 1% correspond to slight sea level changes of the Marmara Sea. Only one 1% change in the Mimarsinan corresponds to a joint change caused by the lake creation plus sea level rises.

After 1990, industry in Büyükçekmece had started to grow. In 1998, in Kıraç, Kavaklı and Mimarsinan, 3–5% of industrialized areas were seen, which meant a total of 103.8 ha. This brought the total area of land used for industry in Büyükçekmece to 344 ha. The main reasons of the growth of the industrialized areas here are the planned structure of the region and the very convenient air, highway (Trans European Motorway (TEM)), and sea connection possibilities. Other land use is accounted for by stone quarries and open pits, which cover 298 ha of the study area.

Results of the change detection analysis obtained using image clearly showed that over the period 1984–1998, there has been a great increase in settlements mainly at the eastern part of the Büyükçekmece Lake, that is in Esenyurt, Kavaklı, Yakuplu and Kıraç and at the southwestern part of the lake in Mimarsinan (in yellow). Areal values of the changed areas (settlement plus industry) for the period 1984–1998 were calculated as 4,790 ha based on the classified image obtained using the change detection method (Table 15.6).

The IKONOS data analysis focused on the evaluation of urban changes in Mimarsinan (834.9 ha), one of the sub-districts with main land use changes, between 1998 and 2002 using NDVI. In the analysis, the NDVI was computed for the LANDSAT TM (1998) and IKONOS XS (2002) scenes of Mimarsinan using red and near-infrared band. A threshold was then applied to both NDVI images to indicate areas lacking vegetation. A region of interest (RO) containing all pixels with values that meet the NDVI threshold was established and converted to vector format (DXF) for comparison (Fig. 15.7a, b). The areal extent of

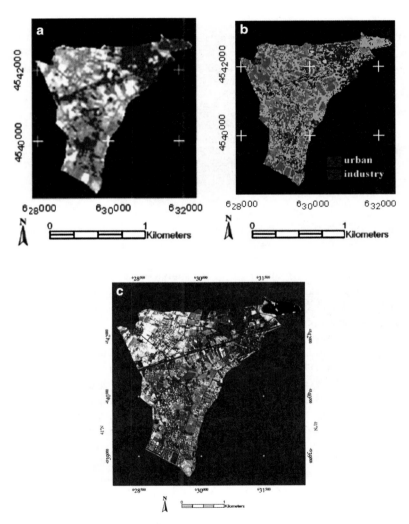

Fig. 15.7 Urban structures extracted from NDVI images of (**a**) LANDSAT TM (1998), (**b**) IKONOS XS (2002), and (**c**) Maximum likelihood classification result of LANDSAT TM (1998)

urban structures was calculated as 33.7 ha in difference between 1998 and 2002. In this calculation the urban pixels surrounding marshland (in red, upper right corner) were not included.

To evaluate the compatibility of the urban structures extracted from the IKONOS NDVI image, the urban classes in Mimarsinan (urban & industry) obtained using maximum likelihood classification of LANDSAT TM (1998) image (Fig. 15.7c) were overlaid on the IKONOS NDVI (2002) image. Thus, the areal extent of the urban growth in 2002 was compared with the LANDSAT TM classification result (1998) revealing a difference of 28.3 ha in greenness between these two dates and which can be attributed to a subtle trend towards an increase in urbanization especially in southern and eastern coastline of Mimarsinan.

15.6   Conclusions

As the earth's population increases, cities will continue to grow and spread. However, this continual growth comes with a range of environmental drawbacks which are typically magnified in developing countries. Rapid and haphazard urban sprawl and increasing population pressure in megacities often infringes upon agricultural and/or forest land. Analyzing agricultural and urban land use is important for ensuring that development does not encroach on productive agricultural land, and to likewise ensure that agriculture is occurring on the most appropriate land and will not degrade due to improper adjacent development or infrastructure. The multi-temporal multi-spectral analysis of satellite remote sensing data facilitates this process of monitoring land use in both agricultural and urban settings. In this study, in Büyükçekmece, in a sub-district of Istanbul in Turkey located 50 km far from city centrum, parallel to the continuous population growth (>40% in some regions) due to (i) intensive migration from the other parts of Turkey, (ii) good transport network such as Trans-European Motorway (TEM) which crosses Büyükçekmece, and (iii) natural population growth, rapid urbanization over a period of 15 years has been detected. Also, a decrease or total loss of agricultural fields in various sections of the region due to the rapid increase of settlement areas (mainly mass housing) has been determined. Besides, industrialized areas, which did not exist in 1984 have been extended due to good urban planning, convenient air, highway (TEM) and sea connections in the district.

Today, for Büyükçekmece, all these would not be considered as a negative condition due to the great areal extent (22,500 ha) of this sub-district and due to the still existing uninhabited areas, which can still be used for construction of new buildings. Thus, construction of the multi storey mass houses still continue rapidly. Nowadays, there are some suggestions regarding to applying visa for the city Istanbul which will also include Büyükçekmece to prevent this intensive imigration.

Chapter Summary

As the earth's population increases, cities will continue to grow and spread. However, this continual growth comes with a range of environmental drawbacks which are typically magnified in developing countries. Examples of these drawbacks include rapid and haphazard urban sprawl and increasing population pressure in megacities often infringes upon agricultural and/or forest land. Analyzing agricultural and urban land use is important for ensuring that development does not encroach on productive agricultural land, and to likewise ensure that agriculture is occurring on the most appropriate land and will not degrade due to improper adjacent development or infrastructure.

The multi-temporal multi-spectral analysis of satellite remote sensing data facilitates the process of monitoring land use in both agricultural and urban settings but the application varies between developed and developing countries. In alignment with other chapters in the book discussing the context of developing countries, this chapter discussed the impact of rapid urban growth on land use changes, especially on the agricultural land in Turkey and the way in which remote sensing is used to assess and monitor these changes using a case study from Büyükçekmece in Istanbul. Büyükçekmece has been experiencing expansion of informal settlements and continuous population growth (40% in some regions) due to (i) intensive migration from the other parts of Turkey, (ii) the building of new transport networks such as Trans-European Motorway (TEM) which crosses Büyükçekmece, and (iii) natural population growth and rapid urbanization over a period of 15 years.

The chapter showed how land use changes associated with this growth could be detected using simple change analysis techniques such as image differencing and vegetation indices. Comparing with many analytical techniques introduced in earlier chapters in the book, the techniques introduced in this chapter are considered much less sophisticated. Nevertheless, and as previously discussed in Chapter 13, the results of the analysis showed how it is possible with relatively simple techniques to utilize urban remote sensing in generating reliable measures and new layers of information that otherwise not readily available in developing countries.

LEARNING ACTIVITIES

Study Questions

- The case study discussed in this chapter utilized NDVI – a ratio-based vegetation index – to understand change in agriculture lands between 1998 and 2002 in a sub district within Istanbul.

- Speculate on the range of other ratio and vector based vegetation indices that could have also been used and compare them to the NDVI discussing how these indices are calculated and the way in which they are used to measure vegetation abundance.
- Discuss the implication of plant behavior on vegetation ratio-based indices on any time series/change analysis study, where the sun-angle may change significantly throughout time.

- The above case study utilized the red band of LANDSAT TM image to calculate urban change through image differencing under the assumption that this band helps distinguish a range of urban features associated with informal settlements in Istanbul. However, another band or subset of bands could be used in other urban application or city settings to map change. Describe a strategy you might develop for quantitatively choosing the optimal subset of bands in relation to another city of your choice or other urban application areas (urban flood modeling, urban drought, urban sustainable development, etc.).

References

Barba C, Rabuco L (1997) Overview of ageing, urbanization and nutrition in developing countries and the development of the reconnaissance project. Food Nutr Bull 18(3). http://www.unu.edu/unupress/food/v183e/ch03.htm. Accessed 20 June 2003

Baytın Ç (2000) Dwelling in Turkey's 50 years of modern urbanization period: sample Istanbul. In: Proceedings of ENHR 2000 conference in Gavle, 26–30 June

CCRS (Canada Center for Remote Sensing) (2003) Tutorial on fundamentals of remote sensing, http://www.ccrs.nrcan.gc.ca-/ccrs/learn/tutorials/fundam/fundam_e.html. Accessed 15 May 2003

Cepede M (1984) Nutrition and urbanization. Food Nutr Bull 10:43–51

Coppin BR, Bauer ME (1996) Digital change detection in forest ecosystems with remote sensing imagery. Remote Sens Rev 13:207–234

Cromartie J, Swanson L (1996) Census tracts more precisely define rural populations and areas. Rural Dev Perspect 11(3):31–39

Cullingworth B (1997) Planning in the USA: policies, issues and processes. Routledge, London

DIE (2000) Reports of DIE. Ankara, Turkey

Donnay JP, Barnsley MJ, Longley PA (eds) (2001) Remote sensing and urban analysis. Taylor & Francis, London

Ehlers M, Schiewe J, Tufte L (2002) Urban remote sensing: new developments and challenges. In: Proceedings of the 3rd international symposium on remote sensing of urban areas, Istanbul, Turkey, pp 130–137

El-Raey M, Nasr SM, El-Hattab MM, Frihy OE (1995) Change detection of Rosetta Promontory over the last forty years. Int J Remote Sens 16:825–834

ESA (2001) Megacities geospace. Verlag, Lothar, Beckel (Eds). Second English Edition, Austria

Gibson PJ, Power CH (2000) Introductory remote sensing (digital image processing and applications). Taylor & Francis, London

Green K, Kempka D, Lackey L (1994) Using remote sensing to detect and monitor land cover and land use change. Photogramm Eng Remote Sens 60(3):331–337

Gross R (1990) Urbanization and nutrition in Brazil. In: Pongpaew P, Sastroamidjojo S, Prayurahong B, Migasena P, Rasad A (eds) Human nutrition: better nutrition in nation building.

Proceedings of the 2nd SEAMEO-TROPMED seminar on nutrition. Bangkok, SEAMEO-TROPMED, pp 31–36

Gross R, Monteiro CA (1989) Urban nutrition in developing countries: some lessons to learn. Food Nutr Bull 11:14–20

Gür M, Çağdaş V, Demir H (2003) Urban–rural interrelationship and issues in turkey. In: Proceedings of the 2nd Fédération Internationale des Géomètres (FIG) regional conference, Marrakech, Morocco. http://www.fig.net/pub/morocco/proceedings/TS1/TS1_6_gur_et_al.pdf. Accessed 10 May 2004

Harrison BA, Jupp DLB (1990) Introduction to remotely sensed data. CSIRO, Australia

IIED (International Institute for Environment and Development) (2004) http://www.iied.org. Accessed 15 July 2004

Jensen JR (2000) Remote sensing of the environment: an earth resource perspective. Prentice Hall, NJ

Jurgens C (2000) Change detection: Erfahrungen bei der vergleichenden multitemporalen Satellitbildauswertung in Mitteleuropa, PFG, Jahrgang 2000. Heft 1:5–18

Lillesand TM, Kiefer RW, Chipman JW (2004) Remote sensing and image interpretation, 5th ed. New York: John Wiley & Sons, Inc.

Maktav D, Sunar F, Taberner M, Akgün H (2000) Monitoring urban expansion in the Büyükçekmece District of Istanbul using satellite data. In: Proceedings of the ISPRS, Amsterdam, Holland, pp 1484–1492

Mander Ü, Jongman RHG (2000) Advances in ecological sciences, consequences of land use change. WIT, Southampton

Mass JF (1999) Monitoring land cover changes: a comparison of change detection techniques. Int J Remote Sens 20(1):139–152

Mazur RE, Qangule VN (1995) African migration and appropriate housing responses in Metropolitan Cape Town. Western Cape Community-Based Housing Trust, Draft Report

RPRI (Rural Policy Research Institute) (2004) http://www.rupri.org/resources/context/rural.html. Accessed 10 July 2004

Rossi-Espagnot A (1984) Primary health care in urban areas: reaching the urban poor in developing countries. A state of the art report. UNICEF/WHO Rep No. SHS844, World Health Organization, Geneva

Seto KC, Duong ND (2002) Using a multisensor approach to monitoring urban growth in Greater Hanoi, 1975–2001. In: Proceedings of the 3rd international symposium on remote sensing of urban areas, Istanbul, Turkey, pp 561–565

Slater PN (1980) Remote sensing: optics and optical systems. Addition-Wesley, Reading, MA

Sunar F (1998) Analysis of changes in multidate data set: a case study in İkitelli area, Istanbul, Turkey. Int J Remote Sens 19(2):225–235

Sunar F, Maktav D, Taberner M, Kapdasll S (2000) Monitoring the changes at the Büyükçekmece Lake Istanbul using multitempotral satellite data. In: Proceedings of the 2nd ICGESA international conference on GIS for earth science applications, Menemen, Izmir, Turkey (in CD)

Sur U, Jain S, Sokhi BS (2003) Identification/Mapping of slum environment using IKONOS satellite data: a case study of Dehradun, India, http://www.gisdevelopment.net/application/environment/pp/mi04011.htm. Accessed 20 Sept 2003

Taberner M, Sunar F, Maktav D (1999) Monitoring vegetation biomass of a coastal ecosystem using multidate optical satellite data. In: Proceedings of the EUROPTO series, remote sensing for earth science, ocean and sea, vol 3868, 20–24 September, Florence, Italy, pp 68–79

Treitz PM, Rogan J (2004) Remote sensing for mapping and monitoring land-cover and land-use change – an introduction. Prog Plann 61(3):269–279

Turner BL, Meyer WB, Skole DL (1994) Global land-use/land-cover change: towards an integrated study. Ambio 23(1):91–95

UN-HSP (UN Human Settlements Programme) (2002) http://www.unchs.org/. Accessed 12 Dec 2002

WBG (The World Bank Group) (2000) Entering the 21st century, world development report 1999/2000. http://www.worldbank.org-/wdr/2000/pdfs/tableA.2.pdf. Accessed 13 Mar 2003

Chapter 16
Using Satellite Images in Policing Urban Environments

Meshgan Mohammad Al-Awar and Farouk El-Baz

This chapter discusses the role of remote sensing technology in the monitoring and management of security in cities and in assuring the timely policing of urban environments. The chapter presents application examples from the Dubai's Police in the United Arab Emirates to show how the utilization of geo-referenced satellite images on top of GIS platforms can allow the immediate location of the needed response.

Learning Objectives

Upon completion of this chapter, you should be able to:

❶ Speculate on general role of remote sensing technology in policing urban environments
❷ List and discuss the range of geospatial techniques used in the proper analysis of crime data
❸ Identify basic skills and knowledge required for the efficient use of remote sensing in community policing

---

16.1 Introduction

Crime rates in most cities are on the rise, fueled by a large and growing population with differences in density, diversity in racial and ethnic backgrounds, and decreasing economic health. Law enforcement agencies are required to respond to these increasing crime rates. In doing so, they face numerous challenges that frequently

M.M. Al-Awar (✉) and F. El-Baz
Research and Studies Center, Dubai Police Academy, 53900 Dubai, United Arab Emirates
e-mail: meshkan@dubaipolice.gov.ae; drmeshkan@yahoo.com

F. El-Baz
Center for Remote Sensing, Boston University, 725 Commonwealth Avenue, Boston, MA, 02215-1401, USA

T. Rashed and C. Jürgens (eds.), *Remote Sensing of Urban and Suburban Areas*,
Remote Sensing and Digital Image Processing 10,
DOI 10.1007/978-1-4020-4385-7_16, © Springer Science+Business Media B.V. 2010

stretch their resources to the limit. To help optimize the use of these resources, new technologies and powerful systems are now utilized to assist in the management of urban environments. Geospatial technologies in particular have proved to be very valuable in facilitating work in areas related to traffic analysis, city management, firefighting, policing, and crime analysis (El-Baz 1998).

The value of remote sensing and GIS stems from the way in which these technologies help law enforcement agencies approaching community problem-oriented policing philosophies and practices from a geographic perspective. For example, techniques for "crime mapping" have been brought to the center of crime prevention practice and policy through both technological advances in computer mapping and information systems, and theoretical innovations in crime prevention. Crime mapping is a vital step in preventing and suppressing crime; identifying and distributing crime-related information to urban communities, police executives, and patrol officers; reducing the fear of crime; enhancing the quality of life in urban communities, and improving community problem-orienting policing (Weisburd and McEwen 1997).

> **geospatial technologies help law enforcement agencies approaching community problem-oriented policing philosophies and practices from a geographic perspective**

Community problem-oriented policing requires the consideration of two major elements: community engagement, and problem-solving (Banas and Trojanowicz 1985; Bayley 1989). Community engagement involves a continuous dialog between the police and the public. Problem-solving is the police primary service to the public and occurs in four steps: (1) identifying problems in the neighborhood; (2) detecting the conditions that create those problems; (3) developing and implementing solutions; and (4) assessing the impact of these solutions. As community engagement raises police accountability, community engagement and problem-solving become inseparable. The need for effective collaborative problem-solving efforts is very obvious in this context (Brann and Whalley 1992).

16.2   Spatial Analysis of Crime

16.2.1   Data Versus Information in Community Policing

Information is one of the most valuable tools available for effective crime fighting (Harries 1999). Data are simply a collection of observations that, by themselves, have little meaning to humans. Information is filtered or processed data that are meaningful to humans. For example, a raw satellite image contains a collection of pixels (data), which needs to be processed in order to obtain a map that shows a selected number of classes or themes and their relationships across space (i.e., information). The ability to collect reliable observation data

> **Information is filtered or processed data that are meaningful to humans in a given context**

and the way in which these observations are translated effectively to information are the foundation upon which proactive community oriented policing can be developed and implemented. To this end, computerized information systems (e.g., robotics, remote sensors, supercomputers, digital images, satellite communications and GIS) become important means for both data collection and information generation (Fatah and Higgins 1999; Olligschlaeger 1997).

Digital technology (digital images) for crime applications has been developed to provide highly accurate records for crime scenes. This includes the position and morphology of evidence discovered at the crime scene, post blast and fire scenes, enhanced field testing applications, and narrowing down the area of investigation. These protocols unquestionably lower the overall cost of crime scene processing while providing high quality information. Remote sensors, whether portable, airborne or space-borne, could be used in the detection of narcotics, explosives and other trace evidence and are very promising in protecting crime scene personnel.

Similarly, GIS permits information layering to produce detailed descriptions and visual representations of events, and analyses of relationships among variables. As such, GIS provides powerful technological tools to obtain spatial information for urban police departments and other law enforcement agencies (Harries 1999). High speed, high capacity computer systems and the relatively inexpensive access to high-resolution satellite imagery are all factors that are encouraging the speed and routine of incorporating GIS applications in the analysis of crime. The focus is on methodological issues in spatial statistical analyses of crime data rather than a comprehensive treatment of the existing theoretical and empirical research.

16.2.2   Crime Mapping

Sorting out the relationship between place and crime requires analytical methods that can isolate the impacts of both. There are two dominant theoretical perspectives on the distribution of crime, both of which consider time and space. The first theory is the Rational Choice Theory which was developed by economists and introduced into criminology in the 1970s (Cornish and Clarke 1986). This theory explains how desired goals are obtained for the lowest cost. Therefore, it emphasizes specific crime events as well as the offender's behavior or decisions. The second theory is the Routine Activities Theory (Cohen and Felson 1979). This theory is an intuitive theory that focuses on situations of crimes. It takes a macro-level view of crime, i.e. it emphasizes on space and time besides the victim's decisions and behavior. This theory was later refined by Felson (1986, 1994) and extended to Crime Pattern Theory in Brantingham and Brantingham (1993). A broad and detailed discussion of these two theories is beyond the scope of this

there are two dominant theoretical perspectives on the distribution of crime: the routine activities theory, and the rational choice theory

chapter (refer to the substantial literature on crime mapping and the aforementioned references for guidance).

The earliest applications of computerized crime mapping appeared in the mid-1960s (Pauly et al. 1967). Among the most remarkable research emphasizing crime mapping was Frisbie el al.'s Crime in Minneapolis: Proposals for Prevention (1977), which attempted to bridge the gap between the academic crime mapping and analysis/applications specifically aimed at crime prevention. In this regard, Frisbie et al.'s publication represented the first work that recognized computerized visualization of crime data as a management tool.

GIS applications in policing took off in the late 1980s and early 1990s as desktop computing and applications software became more accessible. Foresman (1998) recognized five stages of GIS development, including: (1) the Pioneer Age, which lasted from the mid-1950s to the early 1970s and was characterized by some primitive hardware and software, (2) the Research and Development Age, which lasted into the 1980s and overlapped (3) the Implementation and Vendor Age, which lasted into the 1990s, when (4) the Client Applications Age began, which was followed by (5) the Local and Global Network Age.

Major advances in crime mapping have occurred during the last 15 years (Fisher and Winograd 1999). Although a Crime Mapping Research Center (CMRC), currently known as Mapping and Analysis for Public Safety (MAPS), was established at the U.S. National Institute of Justice (NIJ) in 1997, only 13% of the U.S. police departments were using computerized crime mapping in 1997–1998 (Mamalian and La Vigne 1999). Obviously, these were the large urban departments since higher population density means higher potential for crime in a given area. However, the number of police agencies with crime mapping capabilities has grown rapidly, most notably in countries like the U.S., Canada, U.K., Australia and South Africa where the technological infrastructures and developments of GIS are more advanced. Nevertheless, geospatial technologies are becoming more and more integrative tools in crime mapping applications in many developing countries. The e-government initiative in Dubai and the whole United Arab Emirates (UAE) is an obvious example. As these initiatives and technologies gather momentum, GIS applications are used in the identification of geographically dependent citizen services. Additionally, the growing knowledge-base economy has accelerated the implementation of GIS programs, which offer managers, politicians and the general public views of dense fields of data that could previously be presented only on paper.

Contemporary trends in crime mapping call for an increased integration of data and technologies in order to extract as much value as possible. One possibility to do this would be the combined and integrated use of GIS, global positioning systems (GPS), and management information systems (MIS). Such integration would allow close to real-time crime mapping, because the geo-coding step would be automated (Sorensen 1997). Moreover, GPS offers the possibility of accurately reporting places using real-world coordinates, making the application of crime mapping less dependent on the collection and verification of street addresses. GPS, in its more advanced modes, can provide accuracy to one centimeter and thus pinpoint locations such as the precise spot of an auto theft in a parking lot (Harries 1999).

However, all of the above mentioned technologies are only tools and their benefits to the society and law enforcement depend on the human agents who have innovated them.

Crime Hotspots

As the use of GIS evolves in law enforcement and includes an increasing number of "spatial data analysis" tools, new and innovative applications are emerging. In law enforcement, GIS is primarily used to visualize crime occurrences and determine if they are of regular or varied numerous patterns. The concentration of crime in certain places, also known as criminal spatial behavior or crime hot spots, was first identified in Brantingham and Brantingham (1982). These crime hot spots represented the potential value of places in the analysis of crime patterns, particularly in exploring the variables that affect crime patterns (Sherman et al. 1989; Sherman 1992; Weisburd and Green 1994). In the context of law enforcement, the hot spot concept is typically applied to street crime rather than white-collar crime, organized crime, or terrorist crime. For example, even though white-collar crimes may sometimes overwhelm street crime, their economic impact tends to be ignored, possibly because they do not cause the same kind of community fear and anxiety as street crimes.

Crime hotspots can be described both geographically and temporally. Several factors need to be explored to determine the spatial extent as well as the temporal persistence of a hot spot (that is, how long does a hotspot remain "hot"?). For example, If a city experiences several terrorist bombings or school shootings within 1 year, the city as a whole could be considered a hotspot – an idea that challenges the typical hot spot definition. Other examples of these factors include incident accrual rates within a spot, how these rates are measured in relation to all confirmed crimes within the city, and all calls for service, specific crimes or other conditions that are reported from a particular spot.

An investigation sponsored by the Crime Map Research Center (currently MAPS) in 1998 found that most hotspot analysis methods fall into one of the following five categories (Jefferis 1999): visual interpretation, chropleth mapping, grid cell analysis, cluster analysis, and spatial autocorrelation. Some of the aforementioned methods involve user-defined search criteria, whereby variations in the criteria affect outcomes.

16.3   Application Examples

In the U.S., the NIJ (Office of Justice Programs, U.S. Department of Justice), in cooperation with the Environmental Systems Research Institute Inc. (ESRI) and many U.S. police departments has supported many GIS crime applications through special awards. One of these applications is the Crime Analysis Extension Application. This extension covers three broad categories of activities that constitute community-oriented policing (crime analysis, patrol and administration). Accordingly it provides tools regarding crime data and thorough geographical analysis. Another application which the NIJ has supported is the development of Community Policing Beat Book Application, which serves community policing and has a variety of features that provide information about the community and its management to make fast and informed decisions regarding the crime field (Fig. 16.1). Another application is the SARA 4 Step Mapping Process for police problem solving as indicated by Read and Tilly (2000) and Goldstein (1977). In this application, GIS mapping is broken down into four stages for watching the problematic situations through scanning, analysis, response, and assessment (Fig. 16.2).

One of the most recent examples of GIS creativity in crime analysis is its utilization in triangulating gunfire or what is generally called Shot Spotter (Fig. 16.3). Shot Spotter is a product that was specifically developed for law enforcement as an "early warning system" for detecting gunfire locations across large urban areas using a small number of inexpensive and easy use sensors. In the U.S., California was the first state using this technology as in the Police Department of Redwood City, California (Fields 2000).

The potential uses of spatial data in crime analysis have been significantly developed in recent years. These uses can be extended to monitoring parolees and people who

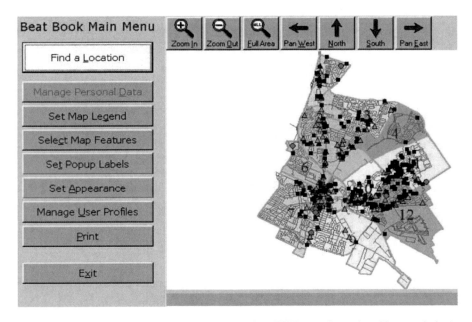

Fig. 16.1 Community policing beat book application (ESRI.com. Reproduced by permission)

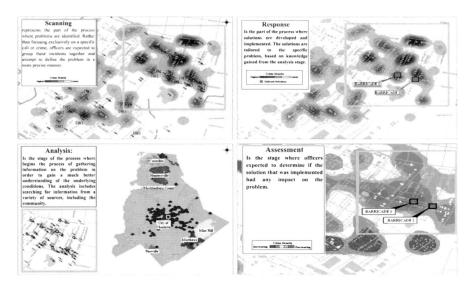

Fig. 16.2 The SARA 4 step crime mapping process (www.cicp.org. Reproduced by permission)

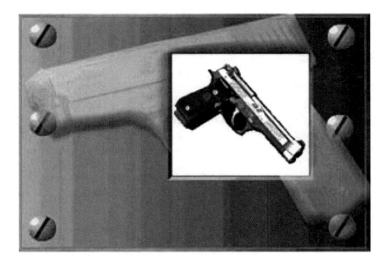

Fig. 16.3 Shot Spotter

are on probation if the terms of their release require limitations on their mobility (Harries 1999). Environmental criminology (developed in 1980s) is also important for many aspects of crime mapping. In the U.S., NASA and the FBI have used spatial analysis to investigate environmental pollution and drug cases by comparing historical data through change detection techniques (Berger 1998). Mobile GIS is also a growing technology for field spatial data collection such as LBS (Location Based Services), which provides GIS and spatial data information via mobile and field units.

The growing use of GIS in crime mapping has also increased public access to crime data, especially through the most accessible and popular medium, the World Wide Web. Moreover, complex GIS applications can lead to scenario planning to develop hazard plans and public safety for floodwater management, earthquake or other disasters. GIS can also be relied upon as an effective rescue effort to assist fire and police departments. It will also be particularly useful for code enforcement officials who wish to locate property owners.

16.4 Dubai Police Practice[1]

With the launching of e-government initiative in Dubai/UAE, technologies such as remote sensing and GIS can clearly contribute to all phases of the city's initiative to improve the command level decision-making in the Dubai Government. This section presents some of the efforts being taken in this regard by Dubai Police.

16.4.1 Image Processing of IKONOS Data of Dubai

Figure 16.4 shows a snapshot of one of the IKONOS images acquired for a portion of Dubai City covering the area from Dubai's airport to Dubai Creek. Panchromatic and multi-spectral IKNONS data were fused together based on the Hue Saturation Value (HSV). In this image sharpening technique, the higher resolution panchromatic data sharpens the detail of the multi-spectral data by transforming the RGB colors to an HSV color space. During this transformation process, band values in the low-resolution multi-spectral are replaced by values from the high-resolution panchromatic images based on the hue and saturation properties of the multi-spectral data. The result is a multi-spectral image with a higher resolution than that of the original raw data.

Figure 16.5 shows the different steps involved in the HSV transformation process. To get a better image with the pixel size of a high-resolution image but without signs of image "blockness," a bilinear re-sampling (BL) technique was applied to the HSV-sharpened image. The resolution of this image is one meter; however, at certain conditions of lighting and background color what is depicted by such images may surpass the theoretical limits of the ground resolution. For example, in Fig. 16.5c, striations less than one meter in width allow for the identification of the direction in which the grass of the golf course had been mowed.

---

[1]The figures shown in this chapter are for illustrative purposes only; they do not constitute legal or official documentation.

Fig. 16.4   A fused IKONOS image of Dubai, UAE

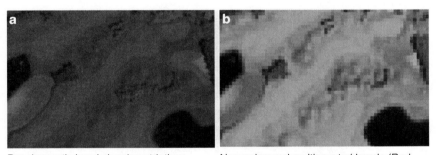

Panchromatic band showing striations   Non-enhanced multispectral bands (Red,
                                               Near IR and Green); striations not visible

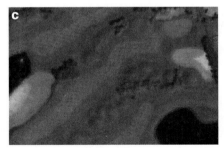

HSV-sharpened image (BL) showing striations

Fig. 16.5   Fusion of IKONOS data using HSV transformation

16.4.2   Crime Hotspot Analysis

This and the following sections discuss some examples for analyzing crime scenarios in Dubai using GIS technology. The hotspot clustering application was used to identify and mark the hotspots and displacement of different types of crime in Dubai City and ultimately allowed for the recognition of relationships between mapped crime patterns and socio-economic characteristics of the city's districts (Fig. 16.6).

Figures 16.7 and 16.8 show two examples of these applications. In the first example, a density map of different types of crimes in Dubai City was created, which showed clustering in certain districts of the city (Fig. 16.7). In these clusters, crime

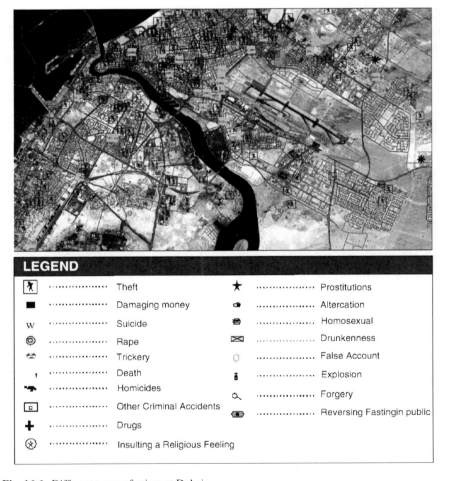

LEGEND

🏃	Theft	★	Prostitutions
■	Damaging money		Altercation
W	Suicide		Homosexual
◉	Rape	✕	Drunkenness
	Trickery	○	False Account
'	Death		Explosion
	Homicides	⚲	Forgery
▣	Other Criminal Accidents		Reversing Fastingin public
✚	Drugs		
⊛	Insulting a Religious Feeling		

Fig. 16.6 Different types of crime at Dubai

Fig. 16.7   Distribution of crimes in the eastern part of Dubai in 2001

Fig. 16.8   Burglary hot spots in Dubai

characteristics (e.g., differences in types, times, recovery) were then examined and reported to the law enforcement to take appropriate measures to reduce crimes (crime rate was reduced from 100 in 2001 to 66 in 2002 in the district shown in Fig. 16.7).

In the second example, a clearer picture of the above-mentioned analysis was created by (a) generating a more specified density map of only burglaries in another district in Dubai, and by (b) following up changes for 6 months. The mapping procedures and hot spot analysis once again allowed the Dubai Police to decrease the rate of burglaries (Fig. 16.8).

16.4.3 *Automated Vehicle Location (AVL)*

Because the Dubai Police is a community and problem solving-oriented agency, and starting with Intelligence Transportation System (McCormack and Legg 2000), it has recently developed another in-house GIS application called Automated Vehicle Location or AVL (Fig. 16.9). The AVL application is based on four databases: data server (network), map center, message center, and report generator (assessment). In this application, the city of Dubai is classified into different districts. Each has a special patrol code number, which is color-coded according to its function (on duty, in-service, at the scene, rescue, needs, etc., see Fig. 16.9).

The AVL system assisted in many aspects, especially in optimal vehicle dispatching, patrol administration, and traffic control (Fig. 16.10). In addition to traffic crimes and accidents, the AVL system is used in community outreach activities and the delivery of community-oriented services by improving the response time (Fig. 16.11). Although response time is a traditionally troublesome benchmark, it has been considered a relatively important measure of police efficiency and effectiveness since the 1930s. The relatively low-cost AVL system addresses the link between the GIS and safety in the urban environment, and can also benefit the Emergency Management System (EMS).

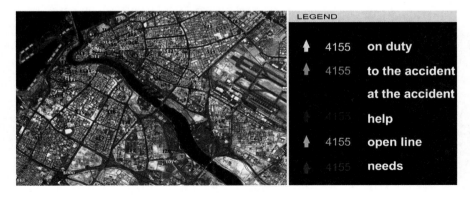

Fig. 16.9 (**a**) A snapshot for downtown Dubai showing the application of the Automated Vehicle Location system (AVL) and (**b**) Legend describing the patrol code numbers

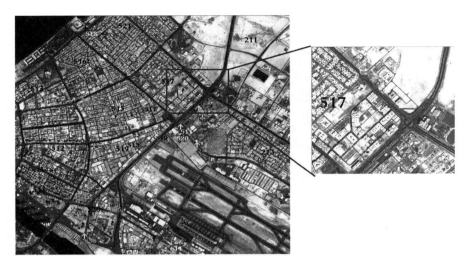

Fig. 16.10 The AVL system provides accurate information that helps dispatching the closest patrol and the response route to any incident

Fig. 16.11 Response time in 2001

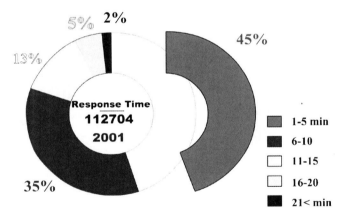

▨	1-5 min
■	6-10
□	11-15
□	16-20
■	21< min

Chapter Summary

The safety and economic well-being of urban communities require an efficient and effective police force. In most cities, law enforcement personnel face many challenges that stretch the resources to their limit. Space-age technologies promise to assist law enforcement officers and agencies in policing urban environments and to address security issues in any community. However, for the successful application of new technologies, it must be possible to identify what is currently available, or under development, that could be transferred to law enforcement activities in an efficient and timely manner. As we have shown in the examples of Dubai Police practices in Dubai City of the United Arab Emirates (UAE), remotely sensed data and geospatial technologies can be used for better command level decision-making and are most useful in the reconstruction and enhancement of crime scenes as well as the reduction of response time for law enforcement decision-making in a timely manner.

Policing, including crime mapping, is undergoing a tremendous change under the influence of the long-term impacts of technological changes. This is made clear by the increasing number of applications in crime mapping, which are becoming more sophisticated and integrated. GIS crime mapping is no longer an innovative tool to be discussed theoretically, and maps are not to be judged like works of art. Both have become valuable tools in the hands of experienced law enforcement personnel and practitioners, crime analysts and criminologists who try to identify and analyze crime through trials of what is possible to achieve the ultimate goal. In addition to technological and methodological innovations in crime mapping, a debate about the disclosure of crime data will continue into the foreseeable future (Harries 1999).

LEARNING ACTIVITIES

Internet resources for Crime Mapping and Analysis

- Tutorial by the National Archive of Criminal Justice Data http://www.icpsr. umich.edu/NACJD/cmtutorial.html
- CrimeStat http://www.icpsr.umich.edu/NACJD/crimestat.html/about.html
- Mapping and Analysis for Public Safety (MAPS) http://www.ojp.usdoj.gov/nij/ maps/
- Law enforcement decisions at Geodecisions http://www.geodecisions.com/ frameslaw.htm
- Cambridge Police Department http://www.ci.cambridge.ma.us/%7ECPD/analysis. html

Study Questions

- What is the role of a remote sensing expert in solving law enforcement (police) cases?

- Discuss the role of GIS in policing an urban environment.
- What future developments might make space-borne data more useful to law enforcement agencies in policing cities?

References

Banas D, Trojanowicz RC (1985) Uniform crime reporting and community policing – a historical perspective. Michigan, National Neighborhood Foot Patrol Center, Michigan State University, School of Criminal Justice, East Lansing

Bayley DH (1989) Model of community policing: the Singapore story. Department of Justice, Office of Justice Programs, National Institute of Justice, Washington, DC

Berger B (1998) NASA explores using crime-fighting satellites. Justnet News 33(44):12

Brann JE, Whalley S (1992) COPPS: the transformation of police organizations: community oriented policing and problem solving. California Attorney General's Crime Prevention Center, Sacramento, CA

Brantingham PL, Brantingham PJ (1982) Mobility, notoriety, and crime: a study of crime patterns in urban nodal points. J Environ Syst 11:89–99

Brantingham PL, Brantingham PJ (1993) Environment, routine, and situation: towards a pattern theory of crime. In: Clarke RV, Felson M (eds) Routine activity and rational choice: advances in criminological theory, vol 5. Transaction Publishers, New Brunswick, NJ, pp 259–294

Cohen LE, Felson M (1979) Social change and crime rate trends: a routine activities approach. Am Sociol Rev 44:588–605

Cornish LE, Clarke RV (eds) (1986) The reasoning criminal: rational choice perspectives on offending. Springer-Verlag, New York

El-Baz F (1998) The Arab world and space research: where do we stand. Emirates Center for Strategic Studies and Research, UAE

Fatah AA, Higgins KM (1999) Forensic sciences review of status and needs. Report by the National Institute of Justice. http://www.ncjrs.org/pdffiles1/173412.pdf

Felson M (1986) Linking criminal choices, routine activities, informal crime control, and criminal outcomes. In: Cornish D, Clarke RV (eds) Reasoning criminal: rational choice perspectives on offending. Springer-Verlag, New York, pp 119–129

Felson M (1994) Crime and everyday life: impact and implications for society. Pine Forge Press, Thousands Oaks, CA

Fields S (2000) ShotSpotter, Inc. and Dialogic Communications Corporation recognized for vision, leadership in using technology to benefit society. http://www.shotspotter.com

Fisher R, Winograd M (1999) U.S. Department of Justice, National Partnership for Reinventing Government. Report of the Task Force on Crime Mapping and Data-Driven Management "Mapping Out Crimes", USA. http://govinfo.library.unt.edu/npr/library/papers/bkgrd/crime-map-/content.html

Foresman TW (ed) (1998) The history of geographic information systems: perspectives from the pioneers. Prentice Hall, Upper Saddle River, NJ

Frisbie DW, Fishbine G, Hintz R, Joelson M, Nutter JB (1977) Crime in Minneapolis: proposals for Prevention. Community Crime Prevention Project, Governor's Commission on Crime Prevention and Control, St. Paul, MN

Goldstein H (1977) Policing a free society. Ballinger, Cambridge, MA. http://www.emeraldinsight.com/Insight/ViewContentServlet?Filename=/published/emeraldfulltextarticle/pdf/1810210202_ref.html

Harries K (1999) Mapping crime: principles and practice. National Institute of Justice. http://www.ncjrs.org/html/nij/mapping/pdf.html

Jefferis E (ed) (1999) A multi-method exploration of crime hot spot: a summary of findings. U.S. Department of Justice, National Institute of Justice, Crime Mapping Research Center, Washington, DC

Mamalian CA, La Vigne NG (1999) The use of computerized crime mapping by law enforcement: survey result. U.S. Department of Justice, National Institute of Justice, Washington, DC, FS 000237

McCormack E, Legg B (2000) Technology and safety on urban roadways: the role of ITS for WSDOT. Research Report, Washington State Transportation Center (TRAC). Washington State Transportation Commission, U.S. Department of Transportation. http://www.wsdot. wa.gov/ppsc/research/CompleteReports/WARD460_2its_urban.pdf

Olligschlaeger A (1997) Spatial analysis of crime using GIS-based data, Ph.D. Dissertation, H. John Heinz III School of Public Policy and Management, Carnegie Mellon University

Pauly GA, McEwen JT, Finch S (1967) Computer mapping – a new technique in crime analysis. In: Yefsky SA (ed) Law enforcement science and technology, vol 1. Thompson Book Company, New York, pp 739–748

Read T, Tilly N (2000) Not rocket science? Problem-solving and crime education. Home Office, London

Sherman LW (1992) Attacking crime: police and crime control. In: Tonry M, Morris N (eds) Crime and justice: a review of research, vol 15. University of Chicago Press, Chicago, IL, pp 159–230

Sherman LW, Gartin PR, Buerger ME (1989) Hot spots of predatory crime: routine activities and the criminology of place. Criminology 27:27–55

Sorensen SL (1997) SMART mapping for law enforcement settings: integrating GIS and GPS for dynamic, near-real time applications and analysis. In: Weisburd DL, McEwen JT (eds) Crime mapping and crime prevention. Criminal Justice Press, Monsey, NY, pp 349–378

Weisburd DL, Green L (1994) Defining the street level drug market. In: MacKenzie DL, Uchida CD (eds) Drugs and crime: evaluating public policy initiatives. Sage Publications, Beverly Hills, CA, pp 61–76

Weisburd DL, McEwen JT (eds) (1997) Crime mapping and crime prevention. Criminal Justice Press, Monsey, NY

Chapter 17
Using DMSP OLS Imagery to Characterize Urban Populations in Developed and Developing Countries

Paul C. Sutton, Matthew J. Taylor, and Christopher D. Elvidge

Nighttime Satellite imagery shows great potential for mapping and monitoring many human activities including: (1) population size, distribution, and growth, (2) urban extent and rates of urbanization, (3) impervious surface, (4) energy consumption, and (5) CO_2 emissions. Surprisingly the relatively coarse spectral, spatial, and temporal resolution of the imagery proves to be an advantage rather than a disadvantage for these applications.

Learning Objectives

Upon completion of this chapter, you should be able to:

❶ Describe the nighttime satellite image products and the factors which affect their quality
❷ Explain the myriad applications of the nighttime imagery for mapping numerous anthropogenic activities and processes
❸ Speculate on the way in which the relatively coarse spatial resolution of the DMSP OLS imagery can capture human activity related to population density, levels of economic development, and cultural practices

P.C. Sutton (✉)
Department of Geography, University of Denver, Denver, CO 80208, USA
e-mail: psutton@du.edu

M.J. Taylor
Department of Geography, University of Denver, Boettcher Center West,
2050 E. Iliff Ave, Denver, CO 80208-0183, USA
e-mail: mtaylor7@du.edu

C.D. Elvidge
Earth Observation Group, NOAA National Geophysical Data Center,
325 Broadway, Boulder, CO 80305, USA
e-mail: chris.elvidge@noaa.gov

T. Rashed and C. Jürgens (eds.), *Remote Sensing of Urban and Suburban Areas*,
Remote Sensing and Digital Image Processing 10,
DOI 10.1007/978-1-4020-4385-7_17, © Springer Science+Business Media B.V. 2010

17.1 Introduction

Nighttime satellite imagery provided by the Defense Meteorological Satellite Program's Operational Linescan System (DMSP OLS) provides a unique, fascinating, and powerful view of human activities at one km spatial resolution globally. The DMSP OLS data products are particularly valuable because of their time of observation, spatial resolution, global scope, and availability in time series. Most other satellite observations of the earth are obtained during daylight hours in which the signal received is reflected solar radiation, emitted thermal radiation, or 'bounced' radiation from active sensors. Nighttime satellite imagery captures radiation primarily from lightning, fires and most importantly: human sources such as city lights, lantern fishing, and gas flare burns (Fig. 17.1). All of these different sources of nocturnal emissions of radiation are distinguishable due to the way the data products are produced. Consequently these data products are used to map fire, lightning, lantern fishing, urban extent, population, population density, CO_2 emissions, anthropogenic environmental impact, impervious surface, and economic activity. This chapter will provide an overview of the DMSP OLS sensor and how the nighttime satellite image data products are produced, briefly describe several applications of these data products for mapping and estimating various human activities, present some of the problems and potential of this imagery for mapping exurbia in the conterminous United States, and, explore the use of this imagery for mapping population, rural electrification, and wealth in Guatemala.

nighttime satellite imagery captures radiation primarily from lightning, fires and human sources

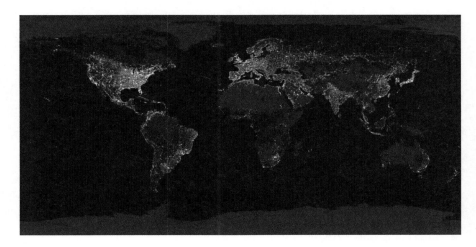

Fig. 17.1 Global composite DMSP OLS image of the Earth at night

17.2 Overview of the Nighttime Satellite Image Data Products

The DMSP platform was designed as a meteorological satellite for the United States Air Force. The DMSP system consists of two sun-synchronous polar orbiting satellites at an average elevation of 865 km above the earth. The swath width of a typical DMSP image is 3,000 km. One satellite observes the earth at dawn and dusk, the second observes the earth at approximately noon and midnight. The sensor on the system has two bands: a panchromatic visible–near infra-red (VNIR) band and a thermal infrared band. For a more detailed description of the DMSP platform see Dickinson's DMSP user's guide (Dickinson et al. 1974). Nighttime imagery provided by the DMSP OLS has been available since the early 1970s. The DMSP sensors are more than four orders of magnitude more sensitive to visible–near-infrared radiances than traditional satellite sensors optimized for daytime observation (Elvidge et al. 1995). However, the high sensitivity of the sensor was not implemented to see city lights. It was implemented to see reflected lunar radiation from the clouds of the earth at night. The variability of lunar intensity as a function of the lunar cycle is one of the reasons why the satellite system's sensors were designed with such a large sensitivity range.

DMSP OLS data is a good example of military technology being put to peaceful scientific and commercial applications

DMSP data was not available in digital format until 1992. Prior to that the images were produced on mylar film, interpreted by Air Force meteorologists, and archived or thrown away. This kind of data did not lend itself easily to analysis. Nonetheless, the imagery was so striking and potentially useful, several people conducted the tedious and difficult research necessary to quantify relationships between the DMSP OLS imagery and various other variables such as population and energy consumption (Welch 1980a, b; Foster 1983). The subsequent development of the digital DMSP archive has dramatically improved the access to, and utility of, the DMSP data.

The digital archive of the DMSP OLS data is housed at the National Geophysical Data Center in Boulder, Colorado (the NGDC is a subsidiary of NOAA). Algorithms developed by Elvidge et al. have produced a 1 km² resolution dataset of the city lights of the continental United States (Elvidge et al. 1995). Elvidge et al. (1995, 1997) developed algorithms to identify spatio-temporally stable VNIR emission sources utilizing images from hundreds of orbits of the DMSP OLS platform. The resulting hyper-temporal dataset is cloud-free because the infrared band of the system was used to screen out cloud impacted data. Later, a global version was prepared. Despite the cloud screening performed in these 'hyper-temporal' mosaics there is some difficulty separating the cold clouds from the cold snow-capped tops

DMSP OLS data products are not 'snapshot' satellite images, they are sophisticated compilations of *many* DMSP OLS 'snaphots' or orbits that meet many criteria such as being 'cloud-free' or non-ephemeral

of high elevation mountains. Temporally unstable lights are identified as lightning, fires, or lantern fishing on factors such as where they are (land or sea), how long they last, and spatial context (with clouds or not). A DMSP OLS mosaic of firs would consist of lights, on land, that were ephemeral. However, it is important to realize that all the fires you see in a DMSP OLS fire data product were not occurring at the same time. The dataset is in this sense more of a 'climatology' of fire.

These earlier datasets are referred to as the 'high-gain' data products. The primary drawback of these data was the issue of saturated pixels in urban areas. Both high-gain datasets were virtually binary with mostly black or dark pixels valued at 0 and brightly lit 'urban' pixels with a value of 63 (the DMSP sensor has a 6-bit quantization). These data lent themselves to aggregate estimation of urban 'cluster' populations but were not good at estimating population density variation within the urban areas. This drawback was identified and resulted in a special request to the Air Force regarding the use of the DMSP OLS platform.

low-Gain DMSP OLS Data shows a much greater diversity of light intensity and can be radiance calibrated to measure energy output

As mentioned before, the DMSP OLS platform was designed to observe reflected lunar radiation at night (primarily reflected from clouds). During the days just prior to and after a new moon there is very little lunar radiation striking the earth. Consequently, the sensor has its gain set to its maximum possible value. The NGDC requested that the Air Force turn down the gain on several orbits near the new moon. This request was honored by the Air Force and resulted in what is now referred to as the 'low-gain' data product. Turning down the gain produced dramatic results with respect to the saturation of the 'urban' pixels. The low-gain data show dramatic variation of light intensity within the urban areas, and it can be calibrated to at-sensor radiances. Hyper-temporal datasets similar to the previous data were made using the low-gain orbits (Elvidge et al. 1998). Another problem identified in the processing of these 'low-gain' data products is the problem of 'overglow' associated with snow cover on the land outside urban areas. The snow reflects light from urban areas and can increase the signal of light in areas outside urban centers that had snow on the ground during observation. This problem has been addressed by using observations in warmer months.

virtually all other remotely sensed data have challenges, obstacles, and issues pertaining to acquisition and processing that must be addressed prior to using the data

Presently global, low-gain DMSP OLS image products are available for 1992–1993, and 2000. Eventually products will be available for most of the years since 1992. The DMSP platform will eventually be retired and be replaced by the National Polar Orbiting Environmental Satellite (NPOES) which will obtain nighttime imagery at finer spatial and spectral resolution.

This perhaps overly detailed description of the processing of the DMSP OLS nighttime satellite image data products is provided to give the reader an appreciation of the challenges, obstacles, and issues associated with 'simply' preparing remotely sensed imagery for further analysis.

If you use a DMSP OLS data product be appreciative of the work of literally hundreds of people that were involved in the acquisition, storage, processing and analysis of the imagery that has been 'packaged' for you. It is important to appreciate these issues in order to make the most accurate and appropriate use of any remotely sensed imagery.

17.3   Summary of Several Applications of the Nighttime Imagery

There are many myriad applications of the DMSP OLS imagery. A brief description of several of them follows.

The city lights data product is clearly a measure of human activity on the planet that is a complex combination of population density, cultural variability, and levels of economic development. Fortuitously, the cultural variability and levels of economic development are less variable within national boundaries. This allows for interesting inter-national comparisons of the imagery. The nighttime imagery has been used to delineate urban extent nationally in the United Sates (Imhoff et al. 1997). Clearly, this delineation of urban extent can be applied globally by changing the light intensity threshold that defines 'urban'. This threshold changes from nation to nation primarily because of the varying levels of economic development and varying cultural practices associated with land use and energy consumption. Utilization of the imagery to delineate urban extent allows for the measurement of the areal extent of urban clusters and estimation of their population. This has been done in China (Lo 2001, 2002), the United States (Sutton et al. 1997), and globally (Sutton et al. 2001). The spatially explicit nature of the data has been used to compare the saturated areas of the high-gain data to theoretical models of urban population density (Sutton 1997). The DMSP OLS imagery has also been used for classifying urban areas in the creation of finer resolution land-cover datasets of the United States with Landsat data (Vogelmann et al. 2001).

city lights can be used as a proxy measure of a complex combination of population density, cultural variability, and levels of economic development

Because nighttime imagery is a good proxy measure of population it is also a good proxy measure of correlates of population such as economic activity and energy consumption. Consequently the DMSP OLS imagery has been used to measure CO_2 emissions nationally (Elvidge et al. 1996; Doll et al. 2000) and to map economic activity locally (Sutton and Costanza 2002). Variation in the light intensity within urban areas has also been used to map the impervious surface of the conterminous United States (Elvidge et al. 2004) and map intra-urban 'ambient' population density (Sutton et al. 2003). These applications of the DMSP OLS imagery take advantage of the time of observation, spatially explicit nature of the data, and global coverage of the DMSP OLS imagery. The following examples should enhance your appreciation of how the DMSP OLS data products can be used.

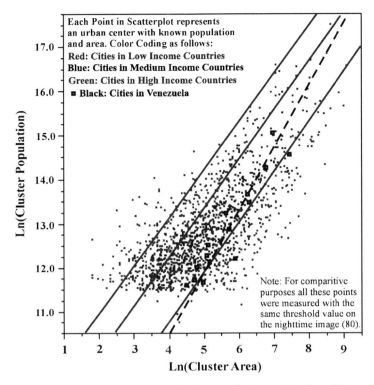

Fig. 17.2 Relationship between urban area and urban population as a function of Gross Domestic Product per capita

   Figure 17.2 below shows a scatter plot of the log of the area of urban clusters versus the log of the corresponding population of urban clusters around the world. This kind of information is easily extracted from an overlay of the nighttime imagery on a population density dataset such as LandScan (Dobson et al. 2000). Each point in the scatter plot is a 'blob' of light that is referred to as an urban cluster. The regression equations can be used to estimate the population of an urban cluster of unknown population or the estimated value can be compared to the actual value to provide a scale-adjusted measure of the extent to which that city suffers from urban sprawl (Sutton 2003). International comparisons of these kinds of regression equations show that countries with lower Gross Domestic Product per capita (GDP/Capita) tend to have the same log-log relationship between areal extent and population; however, the intercept is higher.
   This implies that a city of 800 km² in a poor country will have a higher aggregate population density than a city of 800 km² in a developed country. This makes sense when one considers how the availability and use of the automobile in developed countries has enabled urban populations to spread out to suburbs and beyond to exurbia. For a more detailed description of the use of these kinds of regression equations to estimate the populations of all the urban areas of the world see Sutton (1999).

Figure 17.3 below demonstrates how the varying levels of light intensity within an urban area (Los Angeles) can be used to approximate intra-urban 'ambient' population density. Census data typically records residential population density which is ironically a measure primarily of where people are at night when they are sleeping. To truly capture 'ambient' population density one would need spatially explicit human behavior data. The ground truth in Fig. 17.3 tries to capture this by averaging residential and employment population density data for Los Angeles. The correlation between the nighttime imagery is statistically significant with both employment and residential measures of population density. However the strongest correlation is with the average of the two which suggests that the nighttime imagery is probably a better measure of 'ambient' population density then either residence or employment based measures. To appreciate this, imagine an airport. The census will usually describe the area in and around an airport as having a very low population density because very few people live there. However, the nighttime imagery usually picks up airports as having a significant nighttime light signal. Airports have a high 'ambient' population density because of the many travelers that travel through them on a daily basis and the many people who work at airports.

the concept of 'ambient' population density is a representation of population density based on the mobility of people through time

Application of Nighttime Data for Urban Change Analysis

Another powerful aspect of the DMSP OLS image data products is their availability in time series. This allows for change detection. An obvious example would be attempts to measure increase in urban extent. The India–Pakistan border presents an interesting twist on this concept. Figure 17.4 shows changes in light emissions for the border between India and Pakistan. To create the image, a composite image derived from orbits in 1992–1993 was compared to a composite image derived from orbits in 2000. The interpretation of colors in the change-detection images is as follows: black represents bright lights (saturated) in both time periods, red represents lights much brighter in 2000, yellow indicates new lights in 2000, blue indicates lights missing or substantially dimmer in 2000, and grays indicate that no light was detected in both time periods. The bright linear feature on the left side of the image about halfway up isn't a river; it's the border. The blue area in the middle of the image seems to be confined to the state of Uttar Pradesh, raising interesting questions about how the demographic and economic changes that took place in Uttar Pradesh may have been different than those of other Indian states. The states of Haryana and the Punjab to the north of Delhi seemed to have dramatic increases in the spatial extent of lit areas (more red and yellow), whereas

Uttar Pradesh showed dramatic decreases. The abundant yellow areas through-out the image demonstrate the electrification of India during the 1993–2000 period. A cursory examination of the change detection image depicted in Fig. 17.4 showed many areas in the former Soviet Union with significantly reduced nocturnal emissions. Most areas of the world showed substantial increase in areas where new lights were seen. This kind of change detection was used to identify new exurban areas in the state of Colorado that were sub-ject to high fire risk (Cova et al. 2004). The potential of these time series data products for mapping, estimating, and measuring changes to land use from urbanization and perhaps even changing levels of economic development is just beginning to be exploited. Future nighttime satellite systems such as the National Polar Orbiting Environmental Satellite (NPOES) will have finer spec-tral and spatial resolution which will likely make these kinds of studies even more useful and exciting.

17.4   Case Studies

17.4.1   Case Study#1 (Developed Country): Mapping Exurbia in the Conterminous United States

It has been shown how the DMSP OLS imagery can be used to identify urban areas. However, there is much more to the low-gain data products than simply identifying urban areas. In developed countries there are vast expanses of land that emit a low light signal that would not be classified as urban by finer resolution imagery such as Landsat (Vogelmann et al. 2001) or be classified as urban by the standards of most census bureaus. In many cases these low-light areas could be called Exurban or areas beyond the suburban fringe of the urban areas that have lower population densities. This case study is an exploration of the exurban areas of the conterminous United States.

As an example inspect the southwest corner of the Denver, Colorado metropolitan area (Fig. 17.5). The areas of low-light in the DMSP OLS image that show up as vegetation in the 30 m resolution classification of Landsat imagery have significant populations and include three high schools and the second busiest 'Safeway' super-market in the state of Colorado. These low light areas capture significant popula-tions not identified as urban by the census bureau and identified as vegetation by 30 m resolution Landsat imagery. This area outside of Denver is populated by people who commute longer distances to Denver, tend to have private water and sewage systems and do not have or want street lighting. The inset of the 1 m resolution Ikonos imagery shows that these 'vegetated' areas according to the NLCD classifi-cation of Landsat imagery are in fact significantly impacted by human develop-ment. It is of interest to map these exurban areas because they often represent the

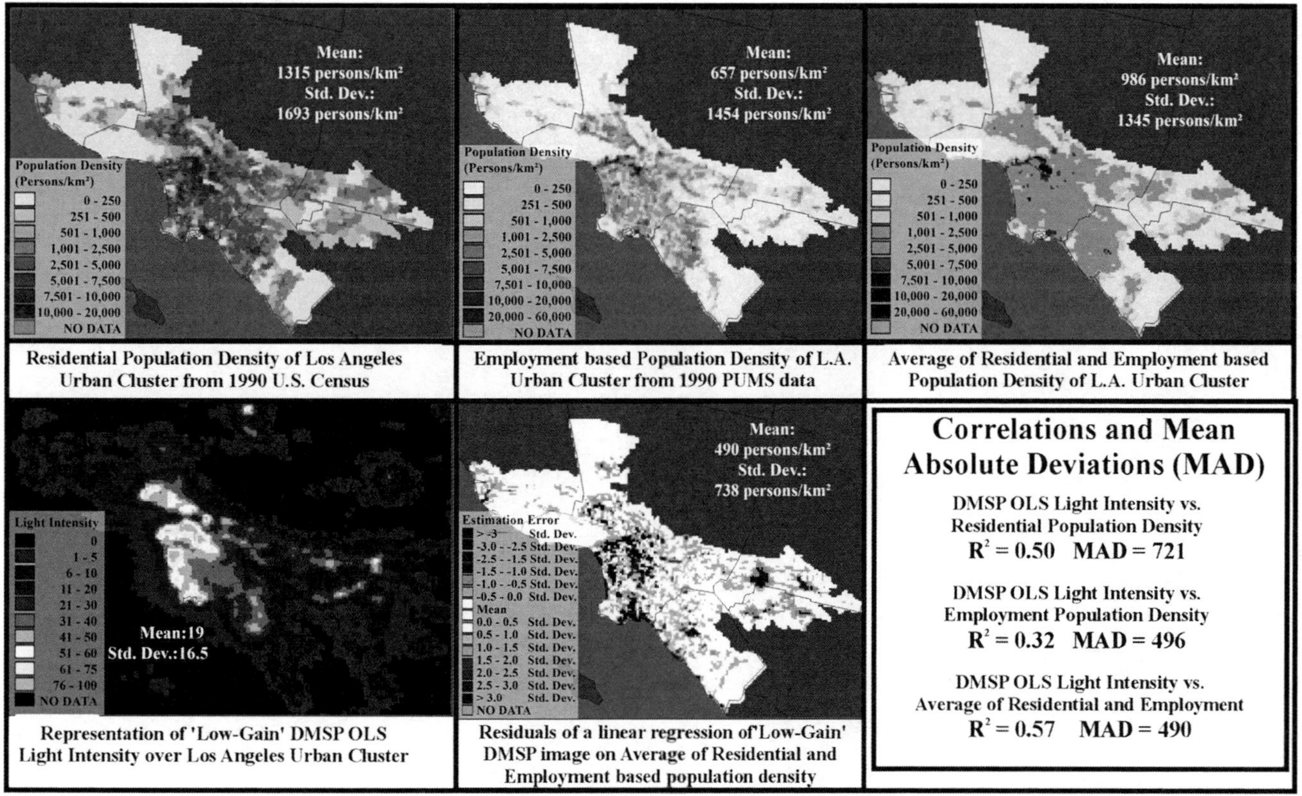

Fig. 17.3 Nighttime imagery and residence and employment based population density in the greater Los Angeles metropolitan area

Fig. 17.4 Changes to nighttime lights on the Asian Sub-Continent 1993–2000

urban–wildland interface of homes that are at threat to wildfires, present problems with human-wildlife interactions (mountain lion attacks, auto impacts with elk and deer, etc.). Many of the costs associated with living in exurbia are born by the exurbanites themselves; however, many are not (road maintenance, fire protection, etc.). There is some debate as to whether or not exurban development is a unique and unprecedented kind of development or if it is merely the beginning of suburbanization (Long and Nucci 1997; Nelson and Sanchez 1999).

In any case, it is an interesting question to ask how much of the United States would be classified as exurbia by the nighttime satellite imagery and how many people actually live in these areas. This has been done based on a simple classification of the low-gain DMSP OLS imagery (Fig. 17.6). The exurban areas in Fig. 17.6 contain 37% of the U.S. population on 14% of the land area. It has not been determined if they share the same attributes of the exurban areas in the southwest corner

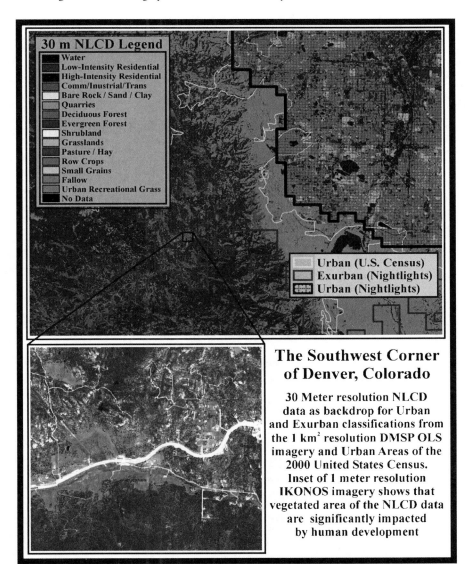

Fig. 17.5 The southwest corner of Denver, Colorado

of Denver. Nonetheless the image raises interesting questions regarding contemporary land use patterns in the United States. Traditionally urban was distinguished from rural. Since World War II this classification began to include 'suburban'. This exploration suggests that the new category of 'Exurban' might be appropriate. It is interesting to note that the population densities of urban (which includes suburban in this example), exurban, and rural roughly follow a descending power of

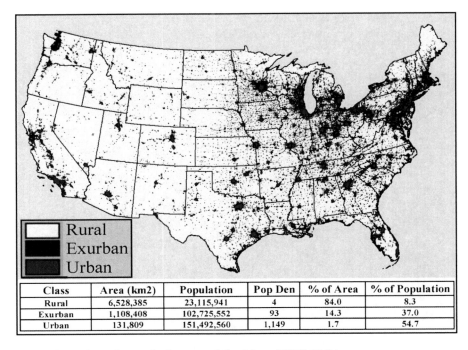

Class	Area (km2)	Population	Pop Den	% of Area	% of Population
Rural	6,528,385	23,115,941	4	84.0	8.3
Exurban	1,108,408	102,725,552	93	14.3	37.0
Urban	131,809	151,492,560	1,149	1.7	54.7

Fig. 17.6 Rural, exurban, and urban areas defined from DMSP OLS imagery

10. Perhaps the nighttime imagery may be used in the future as simple means of delineating 'exurban' areas.

17.4.2   Case Study #2 (Developing Country): Exploring the Use of Nighttime Imagery in Guatemala

Guatemala is one of the least developed countries in the western hemisphere. It is a nation slightly smaller in area than the state of Tennessee with a population of over 13 million people. The capital, Guatemala City has a population of over 3 million. Over 40% of the population is under 15 years old and approximately 60% of the population lives in poverty. Approximately 50% of the population works in the agricultural sector; however this only accounts for about 23% of GDP. The GDP per capita in terms of purchasing power parity is about $3,700 per year (CIA World Fact Book http://www.cia.gov/cia/publications/factbook/). These low levels of economic development have been exacerbated by a long running civil war that ended in 1996. Since then there have been many attempts at rural development including rural electrification programs. Using nighttime imagery for mapping population or economic activity is more problematic in less developed countries like Guatemala.

Infrastructure like street lights, roads, etc. are not nearly as developed as they are in wealthier nations; consequently nocturnal emissions captured by sensors like DMSP are greatly reduced. Moreover, household consumption of energy in Guatemala makes measurement of population density using DMSP difficult. For example, the government offers a social subsidy whereby users who consume less than 300 kilowatt hours (kWh) per month receive discounts on their energy bills paid by businesses and factories that consume larger amounts of energy. Also, despite a massive electrification program that reached large portions of Guatemala's rural residents, most of the rural residents cannot afford to pay for services and thus only use electricity to power a single light bulb for several hours each evening (Taylor 2005). The economic situation of Guatemala's majority prevents us, at this stage, from making useful population estimates using DMSP.

Figure 17.7 contrasts the nighttime image product derived from DMSP OLS imagery with the LandScan population dataset produced at Oak Ridge National Laboratory (Dobson et al. 2000). Urban clusters identified by a threshold in the

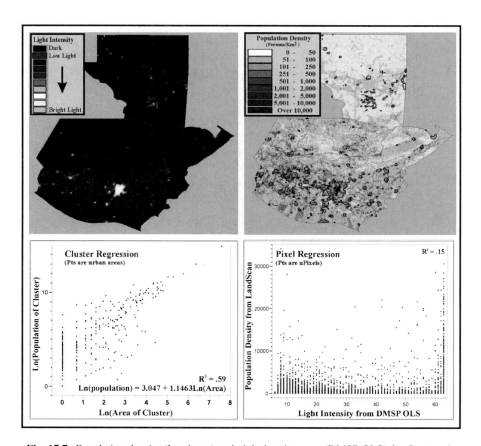

Fig. 17.7 Population density (Landscan) and nighttime imagery (DMSP OLS) for Guatemala

nighttime imagery are shown on the LandScan population density dataset. The regression on the left shows that there is a moderate log–log relationship between the population and area of the urban clusters (Stewart and Warntz 1958; Tobler 1969). The regression on the right contrasts with similar analyses shown in Fig. 17.3 in that there is only a very weak relationship between light intensity and population density on a pixel by pixel basis.

The clusters of light shown in Fig. 17.7 clearly illustrates; however, areas of high population density. The DMSP performs poorly as tool to measure rural population in Guatemala. The authors know of many rural areas (in a sense the Guatemalan equivalent of the North American exurbia) that do not appear using DMSP simply because rural folk in Guatemala use their light conservatively. Future research might combine night-time light data with rural fire data to get an idea of rural population. That is, use fire locations and areal extent detected by the satellite as an indicator of rural population density.

Figure 17.8 illustrates how different urban settings in Guatemala emit different light levels and how these light intensities are captured by the sensor. For example, a clear difference is noted in both the light levels and the ground shots between central Guatemala City and the northern reaches of the same city where recent population growth has taken new residents into unsuitable building areas that cannot support the same density of people and buildings as the traditional downtown area. Unlike many North American cities, Guatemala's downtown is a high-density residential and commercial zone. Overall, the imagery performs well in depicting population density in urban settings in Guatemala, regardless of the size of the community (see Fig. 17.8). Also, because Guatemala's population is still largely rural (60% rural, which is high for Latin America), the imagery also performs well in illustrating the distribution of Guatemala's rural population, who live in dispersed households or hamlets.

If we know where people live and we also have information about country-level GDP, we can begin to create new maps using DMSP to map GDP per capita at finer spatial resolutions. Basically, we can map areas of poverty and wealth in developing nations. Figure 17.9 is the result of an exploratory exercise combining nighttime satellite imagery and LandScan data in a new way. Basically the nighttime imagery is used to allocate the GDP of Guatemala to 1 km^2 spatial resolution. This is accomplished by spreading the roughly 50 billion of GDP around the country based on the lights. Agriculture is about 25% of Guatemla's GDP and this is uniformly allocated to all the 'dark' areas in the DMSP OLS image. The other 75% of GDP is linearly allocated to the lit areas of the DMSP OLS image based on the light intensity or DN value. This produces a map of GDP per km^2. The LandScan population density dataset is then divided into this map of GDP to produce a map of wealth and poverty or GDP per capita of Guatemala at 1 km^2 resolution (Fig. 17.9). This kind of analysis is merely exploratory; it has not been validated in any way and probably suffers from problems like the ecological fallacy and the modifiable areal unit problem (MAUP). The Ecological fallacy and MAUP are problems associated with interpreting the relationships between variables as their formal representations are

1) Central Guatemala City (DN= 500)

2) San Juan Ixcoy, Huehue. (DN=46)

3) Santa Eulalia, Huehue.
(DN=31)

4) Coban, Alta Verapaz
(DN=181)

5) Flores, Peten
(DN=301)

6) Nahuala, Solola
(DN=148)

7) Soloma, Huehue.
(DN =87)

8) Northern Guatemala City
(DN= 215)

Fig. 17.8 DMSP OLS image of Guatemala and corresponding ground and aerial photographs

aggregated or disaggregated. In this example both population and GDP have been disaggregated to 1 km^2 resolution in contestable ways (Aker 1969).

However, local knowledge of Guatemala does suggest that the map does capture some significant aspects of the spatial distribution of wealth and poverty in Guatemala. Urban areas and regions of known economic activity show up well on this map. Note the donut shape of high economic activity around Guatemala City. This pattern results because central Guatemala City is densely populated. Contrast

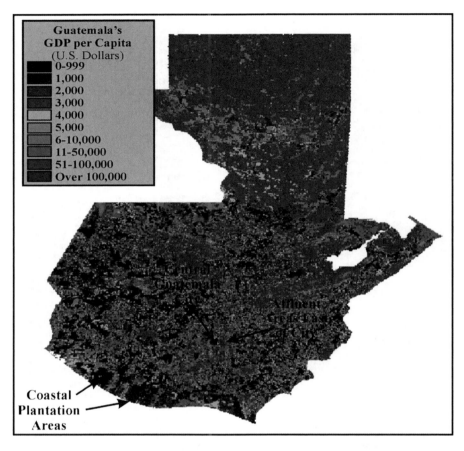

Fig. 17.9 Gross Domestic Product per capita in Guatemala at ~1 km² spatial resolution

this with the elite, low-density suburbs just east of the city. In many residential zones outside of Guatemala City's CBD, especially to the east, a quick drive through the neighborhoods reflects the patterns shown in Fig. 17.9. These eastern zones are dominated by the homes of Guatemala's elite, each house sitting on a large lot with commanding views of the densely populated city center below.

Mapping GDP in this fashion also reveals a well-recognized pattern of poverty in Guatemala. Basically, all Guatemalan government statistics tell us that the indigenous western, and northwestern regions are the poorest regions in Guatemala. This area is dominated by indigenous inhabitants and was also the region most impacted by the civil war in the 1980s and 1990s.

Another area of Guatemala is worthy of our focus. In this case the data depict areas of extreme poverty along Guatemala's pacific coast. On the map we see large blocks of GDP per capita below US$1,000 per person. This is an artifact of the mapping method and Guatemala's highly unequal distribution of land and wealth (Gini coefficient for land is 0.85, with 1 being perfect inequality) (Todaro 1994). The Pacific coast area is dominated by large cattle, coffee, sugar cane, and cotton estates. Population density on these estates is low most of the year. These areas do see, however, a large influx of seasonal workers to harvest crops, especially coffee. The specific dynamics of land use and ownership are not well represented by Fig. 17.9, but overall patterns are verifiable by an intimate knowledge of the country. We must temper this new method of mapping poverty and wealth in the developing world (where census data are often unavailable or unreliable) with the caution that it should only be conducted along with extensive ground truthing that can explain anomalies like the case illustrates above. Nonetheless, we believe that further research into the meaning and potential of this idea of mapping wealth and poverty with population and nighttime imagery should be explored.

Chapter Summary

Nighttime satellite imagery derived from the DMSP OLS has potential for monitoring and measuring many anthropogenic phenomena at a global scale. While nighttime satellite imagery is no substitute for an on the ground census of the population it can be used in innovative and interesting ways to supplement mapping human presence and activity on the earth. The case study exploring exurbia in the United States shows how the nighttime imagery may provide information that is not captured by finer resolution imagery such as Landsat. The case study exploring applications of DMSP OLS imagery in Guatemala demonstrates some of the potential and pitfalls of using this kind of imagery in developing countries. Many countries of the world lack the financial and/or institutional resources to conduct useful censuses. Models derived from readily available satellite imagery that have been validated in parts of the world where good ground based information is available could serve as reasonable proxy measures in countries that lack such information. The LandScan data product takes advantage of many of these ideas. In addition, if a country has some limited resources with which to conduct an incomplete census of its population, existing imagery for that country could be used to help design statistical sampling strategies for a limited census. These sampling strategies could be designed to maximize the effectiveness and accuracy of proxy measures derived from satellite imagery. Future nighttime image products from NPOES will have finer spatial and spectral resolution which are likely to improve the number and quality of applications to which this kind of imagery can be used.

LEARNING ACTIVITIES

Study Questions

- Look at Fig. 17.1 or find a similar image on the web (http://dmsp.ngdc.noaa.gov/ html/night_light_posters.html, or search the internet for DMSP, Nighttime lights of the world or something like that). Study the image and answer the following questions:

  - How are city lights, fires, lightning, lantern fishing, and gas flares separated from one another systematically?
  - What percentage of the earth's land surface appears to have city light emanating from it?
  - Do you think your 'eyeball' guess would correspond reasonably with an analytic inquiry using a GIS or remote sensing package?
  - Does the map of city lights correspond directly with population density?
  - How do varying levels of economic development around the world influence the amount of light seen in the image (compare U.S. to India, check out North Korea and Afghanistan)?

- Consider the concept of 'ambient' population density which the LandScan dataset attempts to represent. Ambient population density is a temporally averaged conception in which the population density of any particular area is a function of the mobility of the human population. Census data records population density on the basis of where people live. Typically census data records low population density for places like airports where many people work and travel through on a daily basis. Does the nighttime imagery provide a way for measuring ambient as opposed to residential population density? How? Why?

- In case study #1 the DMSP OLS imagery was used to map 'exurbia'. Surprisingly the coarse resolution of the DMSP OLS imagery (1 km) could 'see' human development that finer resolution Landsat imagery (30 m) could not.

  - How is the nighttime imagery able to 'leapfrog' this 'disadvantage' in spatial scale of measurement to capture human development and activity?
  - Do you think the nighttime imagery used in this way truly measures the areal extent of 'exurbia' (Explain Why or Why not)?
  - Do you think developing countries would have a similar phenomena of 'exurbia' (Explain Why or Why not)?

- In case study #2 the DMSP OLS imagery was used to explore various aspects of population and income distribution in Guatemala. Guatemala is one of the least developed countries in the western hemisphere.

- How do the low levels of economic development influence any attempts to use the nighttime imagery for population mapping?
- How does the percentage of the population involved in mixed or subsistence agriculture influence this kind of mapping?
- How would you incorporate this information into any model building process?
- The idea of using the nighttime imagery as a proxy for economic activity was also explored. Does the map of GDP/capita at 1 km resolution make sense? How could a relatively fine resolution (1 km) of GDP/capita map be used?

References

Aker HR (1969) A typology of ecological fallacies. In: Rokkam S, Dogan M (eds) Quantitative ecological analysis in the social sciences. MIT Press, Cambridge, MA, pp 69–86

Cova TJ, Sutton PC, Theobald DM (2004) Exurban change detection in fire-prone areas with nighttime satellite imagery. Photogramm Eng Remote Sens 70(11):1249–1257

Dickinson LC, Boselly SE, Burgmann WW (1974) Defense Meteorological Satellite Program User's Guide, Air Weather Service (MAC), U.S. Air Force

Dobson JE, Bright EA, Coleman PR, Durfee RC, Worley BA (2000) LandScan: a global population database for estimating populations at risk. Photogramm Eng Remote Sens 66(7):849–857

Doll C, Muller JP, Elvidge CD (2000) Night-time imagery as a tool for global mapping of socio-economic parameters and greenhouse gas emissions. Ambio 29(3):159–174

Elvidge CD, Baugh K, Kihn E, Kroehl H, Davis E, Davis C (1996) Relation between satellite observed visible–near infrared emissions, population, economic activity, and electric power consumption. Int J Remote Sens 18:1373–1379

Elvidge CD, Baugh KE, Kihn K, Kroehl H, Davis E (1997) Mapping city lights with nighttime data from the DMSP operational linescan system. Photogramm Eng Remote Sens 63(June):727–734

Elvidge CD, Baugh K, Dietz JB, Bland T, Sutton PC, Kroehl H (1998) Radiance Calibration of DMSP-OLS low-light imaging data of human settlements. Remote Sens Environ 68(1):77–88

Elvidge CD, Milesi C, Dietz JB, Tuttle BT, Sutton PC, Nemani R, Vogelmann JE (2004) U.S. constructed area approaches the size of Ohio. EOS Trans Am Geophys Union 85:2333

Foster JL (1983) Observations of the Earth using nighttime visible imagery. Int J Remote Sens 4:785–791

Imhoff ML, Lawrence WT, Stutzer DC, Elvidge CD (1997) A technique for using composite DMSP/OLS city lights satellite data to map urban area. Remote Sens Environ 61(3):361–370

Lo CP (2001) Modeling the population of China using DMSP OLS nighttime data. Photogramm Eng Remote Sens 67:1037–1047

Lo CP (2002) Urban indicators of China from radiance calibrated digital DMSP-OLS nighttime images. Ann Assoc Am Geogr 92(2):225–240

Long L, Nucci A (1997) The clean break revisited: is U.S. population again deconcentrating? Environ Plann A 29(8):1355–1366

Nelson AC, Sanchez TW (1999) Debunking the exurban myth: a comparison of suburban households. Hous Policy Debate 10(3):689–709

Stewart J, Warntz W (1958) Physics of population distribution. J Reg Sci 1:99–123

Sutton PC (1997) Modeling population density with nighttime satellite imagery and GIS. Comput Environ Urban Syst 21(3/4):227–244

Sutton PC (1999) Census from heaven: estimation of human population parameters using night-time satellite imagery. Int J Remote Sens 22:3061–3076

Sutton PC (2003) A scale-adjusted measure of Urban Sprawl using nighttime satellite imagery. Remote Sens Environ 86(3):353–363

Sutton PC, Costanza R (2002) Global estimates of market and non-market values derived from nighttime satellite imagery, land cover, and ecosystem service valuation. Ecol Econ 41:509–527

Sutton PC, Roberts D, Elvidge CD, Meij H (1997) A comparison of nighttime satellite imagery and population density for the continental United States. Photogramm Eng Remote Sens 63(11):1303–1313

Sutton PC, Elvidge CD, Obremski T (2003) Building and evaluating models to estimate ambient population density. Photgramm Eng Remote Sens 69(5):545–553

Taylor MJ (2005) Electrifying rural Guatemala: central policy and local reality. Environ Plann C 23(2):173–189

Tobler W (1969) Satellite confirmation of settlement size coefficients. Area 1:30–34

Todaro M (1994) Economic development, 5th edn. Longman, New York

Vogelmann JE, Limin Y, Larson CR, Wylie BK, Van DN (2001) Completion of the 1990s national land cover data set for the conterminous United States from Landsat thematic mapper data and ancillary data sources. Photogramm Eng Remote Sens 67(6):650–662

Welch R (1980a) Monitoring urban population and energy utilization patterns from satellite data. Remote Sens Environ 9(1):1–9

Welch R (1980b) Urbanized area energy utilization patterns from DMSP data. Photogramm Eng Remote Sens 46(2):201–207

Index

A

Absorption, 50–52, 56–58
Absorption features, 166, 173, 174
Accuracy assessment, 227–229
Across-track illumination correction, 167, 168
Advanced spaceborne thermal emission and reflection radiometer (ASTER), 219, 221–222, 224–231, 236–238
Airborne visible/infrared imaging spectrometer (AVIRIS), 51–54, 56–58, 60–62, 166
Albedo, 221–223, 234–236
Ambient population density, 333, 335, 346
Ancillary data, 141, 142, 160
Ancillary geospatial data, 238
Aqua satellite, 221–222
Arizona, 219, 221, 224
ARSIS concept, 204–209
Astronaut photography, 221
Atmospheric correction, 224

B

Bare soil, 40
Bayesian, 146, 148
Bilinear re-sampling (BL), 320
Biodiversity, 220, 223
Biomass, 221
Built environment, 34, 36–43

C

Census, 37–39, 42, 43, 130–134, 144, 148–151, 159
Central Arizona-Phoenix Long-Term Ecological Research Project (CAP LTER), 224
Change detection, 245, 250, 252–253, 263

Change detection analysis, 289, 294, 299, 307
100 Cites Project, 221
City size, 34, 35
Class area, 229, 232–233, 235
Classification, 181–184, 186–190, 219, 225–230
   hard, 141, 142, 144–147, 159, 160
   soft, 141, 142, 144–146, 159, 160
   spatial, 141–147, 158
   spectral, 141–146, 159–161
Clumpiness, 42
Community and problem solving policing, 314, 324
Compact development, 20
Contextual interpretation, 6
Contiguity index, 42
Conversion of agricultural lands, 2
Crime hot spots, 317, 322–324
Crime mapping, 314–317, 319, 320, 326
Crime pattern, 317, 322
Crime pattern theory, 315
Crime prevention, 314, 316

D

Data fusion, 7, 193–215
Data number (DN), 225
Data requirements, 79–81
Defense Meteorological Satellite Program Operational Linescan System (DMSP OLS), 221, 329–347
Demographic transition, 35
Density, 34–38, 42, 43
Desert cities, 245–264
Developing countries, 7–8, 245–264
Digital elevation model, 221
Dimensions of urbaneness, 5

E

Earth observation, 3, 5
Earth observing system (EOS), 220–222
eCognition, 181–182, 185–186, 189
Ecological analysis, 221, 224
Ecological functioning, 3
Economic activity, 330, 333, 340, 343, 347
Ecosystems, 85–88, 96, 107, 219–220, 223, 238
Edge density, 229, 232–233, 235
Encrustation, 195,198–203
Energy consumption, 329, 331, 333, 341
Enhanced thematic mapper plus (ETM+), 220
Environmental criminology, 319
Environmental Mapping and Analysis
        Program (EnMAP), 176
Environment for visualizing images (ENVI),
        148, 160, 161
Expert classification, 226–227
Expert systems, 128, 219, 226–227
Exurbia, 330, 334, 336–340, 342, 345, 346

F

Famine, 133
Form, 13–29
fPAR dataset, 222, 234–236
FRAGSTATS software, 229–231

G

Geographic information systems (GIS), 230,
        249–254, 256, 257, 260, 263
Geo-information, 5
Geometric correction, 169–170
Geospatial technologies, 5, 8
Grid, 219, 224, 229–237

H

Health, 127, 133, 134
Housing, 127, 133
Hue saturation value (HSV), 320, 321
Human sources, 330
Hyperspectral, 47–50, 52, 54, 56–58,
        165–178
Hyperspectral image analysis, 7

I

IKONOS, 48–49, 58–62, 145, 156, 158, 161,
        221, 234, 237
Image objects, 183, 185–188
Image segmentation, 181–185, 187, 189
Imaging spectrometry, 165, 170, 173

Impervious surface, 3, 6, 8, 40–42, 85–89,
        91–93, 95–97, 99, 102, 104–105,
        108, 110–112, 329, 330, 333
Interspersion/Juxtaposition index, 229,
        233, 235
Istanbul, 289–311

K

Kuwait city, 267, 269, 279–283

L

LAI dataset, 234, 236
Land cover, 47, 49, 51–62, 141–162, 219–221,
        223, 225–238
Land cover/land use change, 220–221, 234
Landsat, 148, 149
Landsat ETM+, 274, 280
Landscan, 334, 341, 342, 345, 346
Landscape
    dynamics, 2
    ecology, 223
    metrics, 39, 41, 43, 222–223, 237–238
    structure, 7
Land use, 15, 18–19, 21–23, 29, 63–64,
        220– 221, 227, 231, 234
    commercial, 153, 156, 158
    residential, 153, 156, 158
Leaf area index (LAI), 222, 234–236
Light detection and ranging (LIDAR),
        59–62, 221
Lightning, 330, 332, 346
Linear regression, 230, 238

M

Maximum likelihood, prior probabilities,
        148–151, 162
Mean patch size, 229, 231, 235, 237
Mesic, 225, 227, 229, 232, 234, 236
Migration, 35
Minimum distance to means (MDM),
        225–227
Minimum/Maximum noise fraction (MNF),
        172, 173, 175, 177
Moderate resolution imaging
        spectroradiometer (MODIS), 219,
        221–225, 227, 229–230,
        234–238
Modern technology, 35, 36
MOLAND, 221
Multiscale methods, 206, 207, 209
Multispectral scanner (MSS), 220

Multitemporal satellite data, 289, 295–299

N

Nearest neighbor, 142, 148, 152–158
Near-infrared (NIR) region, 49, 51, 58, 59, 220–222, 224–226, 234
Neighborhood, 153, 155, 160
New York city, 267, 279–283
Nighttime imagery, 221
Nighttime satellite imagery, 8, 131, 329–333, 336, 338, 342, 345
Nonagricultural activities, 34, 43
Normalized difference vegetation index (NDVI), 222, 225–227, 235–238

O

Oases, 237
Object-based image analysis, 7, 181–190
Object features, 186
Open spaces, 2–4, 7
Ordnance survey, 152

P

Patch dynamics, 223
Patch types, 223, 229
Peri-urban, 219–238
Peri-urban developments, 5
Phoenix, 219, 222, 224–225, 228–229, 231–232, 234, 236–237
Pixel-based image analysis, 181–183, 188
Planned and unplanned development, 67–69, 71, 72, 79
Population, 128, 130–133, 135, 136
Population size, 34, 35, 43
Post-classification recoding, 226
Pre-processing, 165, 167–170, 172, 176
Proxy measure, 333, 345, 347

Q

Quantify land cover, 86, 91
Quickbird, 221, 237

R

Radar, 221
Radiometric correction, 168–169
Radiometric resolution, 121–126, 132, 135, 136
Rational choice theory, 315

Reference dataset
Reflectance, 167–169, 171, 172, 174–176
Remotely-sensed data, 38–42
Remote sensing (RS), 1–8, 85–87, 94–98, 102, 108–109, 113, 245–264
Resilience, 3
Routine activities theory, 315
Rural, 33, 34, 36–40, 43
Rural-urban continuum, 37

S

Satellite imagery, 42
Scale, 142, 144–146, 148, 153, 155, 156, 159, 160
Sealed surface, 3
Security in cities, 8
Segmentation, 7
Self-sufficiency, 36
Semantic classes, 182, 187
Settlement density, 3
Short-wave-infrared (SWIR) region, 49, 51, 58, 220–225
Signal-to-noise ratio (SNR), 167, 174, 176
Social indicators, 5
Socioeconomic development, 2
Soft classification, 39, 40
Spatial and temporal change, 67–82
Spatial data analysis, 317–320
Spatial data infrastructures, 4
Spatial extensification, 36
Spatial heterogeneity, 223
Spatial information, 5, 7
Spatial metrics, 221, 225
Spatial mismatch, 38
Spatial resolution, 121–123, 125–127, 131, 132, 135, 136, 220, 222, 225, 237–238
Spatial structure, 2
Spatial unit of analysis, 39, 43
Spatial variance texture, 226
Spectra, 47–62
Spectral angle mapping, 174, 177
Spectral bands, 48–49, 56–57, 61
Spectral libraries, 170–171
Spectral mixture analysis (SMA), 40, 43, 129, 134, 172, 175, 177, 273–275, 279–280
Spectral properties, 6
Spectral resolution, 121, 122, 124–127, 135, 136, 220–221
Spectral separability, 54–57
Sprawl/urban sprawl, 13–22, 24–28
Suburban, 1–8, 13–21, 23, 25, 27–28
Supervised classification, 225
Surface temperature, 7–8, 222, 235–237

Surface temperature/emissivity, 221
Sustainability, 17, 21
Système Probatoire d' Observation de la Terre
    (SPOT), 221

T

Temporal resolution, 121, 122, 125–126, 136,
    220, 238
Ternary diagram, 85, 87–88, 94, 110, 111
Terra satellite, 221–222
Texture, 128–131
Thematic Mapper (TM), 220
Thermal-infrared region, 220–222, 224
Transport, 14–15, 18–19, 21–22, 25
Transportation, 36

U

United Kingdom, 149, 152, 155
Unsupervised ISODATA algorithm, 226
Urban, 1–8, 13–29, 33, 43, 85–113, 219–238
Urban agglomerations, 2
Urban areas, 141–162
Urban attributes, 67, 71–72
Urban climatology, 222
Urban cluster, 332–334, 341, 342
Urban core, 231, 236–237
Urbane, 33
Urban ecology, 3
Urban environment, 3, 6–8
Urban environmental conditions, 267–283
Urban Environmental Monitoring
    Project, 221
Urban extent, 8, 329, 330, 333, 335
Urban forms, 3
Urban fringe, 231–232
Urban growth, 2–3, 5, 8, 13–18, 29, 289, 290,
    292, 294–299, 309, 310
Urban heat island, 222, 237, 270, 272,
    276–279

Urbanization, 220, 223, 238, 245, 246,
    248, 253, 255, 256, 260,
    261, 263
Urbanizing landscape, 3
Urban material, 49, 51–54
Urban monitoring, 72, 80
Urbanness, 33, 37–39, 41–43
Urban objects, 67–82
Urban planning and management (UPM), 6,
    67–70, 80, 81
Urban remote sensing (URS), 3,
    5–8
Urban-rural dichotomy, 39
Urban-rural divide, 34
Urban-rural gradient, 39
Urban sprawl, 2
Urban surface temperature, 276
Urban transition, 34, 35
Urban vegetation, 7–8, 267,
    269–275, 280

V

Vegetation, 40, 41
Vegetation-impervious
    surface-soil (V-I-S) model,
    6, 40, 41, 85–113
Vietnam, 34
Visible (VIS) region, 49, 51, 58,
    220–226

W

Water/shade, 40, 41
Wavelet transform, 193, 196–198, 201, 202,
    205, 207
Wilderness areas, 40

X

Xeric, 225, 229